AF581888

Mechanical Properties in Progressive Mechanically Processed Metallic Materials

Mechanical Properties in Progressive Mechanically Processed Metallic Materials

Editors

Radim Kocich
Lenka Kunčická

MDPI • Basel • Beijing • Wuhan • Barcelona • Belgrade • Manchester • Tokyo • Cluj • Tianjin

Editors

Radim Kocich
VŠB—Technical University of Ostrava
Czech Republic

Lenka Kunčická
Czech Academy of Sciences
Czech Republic

Editorial Office
MDPI
St. Alban-Anlage 66
4052 Basel, Switzerland

This is a reprint of articles from the Special Issue published online in the open access journal *Materials* (ISSN 1996-1944) (available at: https://www.mdpi.com/journal/materials/special_issues/mechanical_metallic_materials).

For citation purposes, cite each article independently as indicated on the article page online and as indicated below:

LastName, A.A.; LastName, B.B.; LastName, C.C. Article Title. *Journal Name* **Year**, *Volume Number*, Page Range.

ISBN 978-3-0365-0076-8 (Hbk)
ISBN 978-3-0365-0077-5 (PDF)

Cover image courtesy of Lenka Kunčická.

Contents

About the Editors vii

Radim Kocich and Lenka Kunčická
Special Issue: Mechanical Properties in Progressive Mechanically Processed Metallic Materials
Reprinted from: *Materials* **2020**, *13*, 4668, doi:10.3390/ma13204668 1

Radim Kocich
Deformation Behavior of Al/Cu Clad Composite During Twist Channel Angular Pressing
Reprinted from: *Materials* **2020**, *13*, 4047, doi:10.3390/ma13184047 5

Lenka Kunčická and Zuzana Klečková
Structure Characteristics Affected by Material Plastic Flow in Twist Channel Angular Pressed Al/Cu Clad Composites
Reprinted from: *Materials* **2020**, *13*, 4161, doi:10.3390/ma13184161 19

Long Zhao, Xiangru Chen, Tieming Wu and Qijie Zhai
Study on Strengthening Mechanism of 9Cr-1.5Mo-1Co and 9Cr-3W-3Co Heat Resistant Steels
Reprinted from: *Materials* **2020**, *13*, 4340, doi:10.3390/ma13194340 35

Anna Olina, Miroslav Píška, Martin Petrenec, Charles Hervoches, Přemysl Beran, Jiří Pechoušek and Petr Král
Assessment of Retained Austenite in Fine Grained Inductive Heat Treated Spring Steel
Reprinted from: *Materials* **2019**, *12*, 4063, doi:10.3390/ma12244063 53

Jaromír Fumfera, Radim Halama, Radek Procházka, Petr Gál and Miroslav Španiel
Strain Range Dependent Cyclic Hardening of 08Ch18N10T Stainless Steel—Experiments and Simulations
Reprinted from: *Materials* **2019**, *12*, 4243, doi:10.3390/ma12244243 71

Adéla Macháčková
Decade of Twist Channel Angular Pressing: A Review
Reprinted from: *Materials* **2020**, *13*, 1725, doi:10.3390/ma13071725 99

Karel Dvořák, Adéla Macháčková, Simona Ravaszová and Dominik Gazdič
Effect of Imposed Shear Strain on Steel Ring Surfaces during Milling in High-Speed Disintegrator
Reprinted from: *Materials* **2020**, *13*, 2234, doi:10.3390/ma13102234 119

Jaroslav Málek, Jiří Zýka, František Lukáč, Monika Vilémová, Tomáš Vlasák, Jakub Čížek, Oksana Melikhova, Adéla Macháčková and Hyoung-Seop Kim
The Effect of Processing Route on Properties of HfNbTaTiZr High Entropy Alloy
Reprinted from: *Materials* **2019**, *12*, 4022, doi:10.3390/ma12234022 131

Libor M. Hlaváč, Adam Štefek, Martin Tyč and Daniel Krajcarz
Influence of Material Structure on Forces Measured during Abrasive Waterjet (AWJ) Machining
Reprinted from: *Materials* **2020**, *13*, 3878, doi:10.3390/ma13173878 147

Irena M. Hlaváčová, Marek Sadílek, Petra Váňová, Štefan Szumilo and Martin Tyč
Influence of Steel Structure on Machinability by Abrasive Water Jet
Reprinted from: *Materials* **2020**, *13*, 4424, doi:10.3390/ma13194424 163

Lenka Kunčická, Radim Kocich and Zuzana Klečková
Effects of Sintering Conditions on Structures and Properties of Sintered Tungsten Heavy Alloy
Reprinted from: *Materials* **2020**, *13*, 2338, doi:10.3390/ma13102338 . . . . **181**

Adéla Macháčková, Ludmila Krátká, Rudolf Petrmichl, Lenka Kunčická and Radim Kocich
Affecting Structure Characteristics of Rotary Swaged Tungsten Heavy Alloy Via Variable Deformation Temperature
Reprinted from: *Materials* **2019**, *12*, 4200, doi:10.3390/ma12244200 . . . . **193**

Pavel Strunz, Lenka Kunčická, Přemysl Beran, Radim Kocich and Charles Hervoches
Correlating Microstrain and Activated Slip Systems with Mechanical Properties within Rotary Swaged WNiCo Pseudoalloy
Reprinted from: *Materials* **2020**, *13*, 208, doi:10.3390/ma13010208 . . . . **207**

Petra Ohnistova, Miroslav Piska, Martin Petrenec, Jiri Dluhos, Jana Hornikova and Pavel Sandera
Fatigue Life of 7475-T7351 Aluminum After Local Severe Plastic Deformation Caused by Machining
Reprinted from: *Materials* **2019**, *12*, 3605, doi:10.3390/ma12213605 . . . . **223**

About the Editors

Radim Kocich Ph.D. Prof. Kocich has been dealing with plastic deformation of non-ferrous materials for about 25 years. His specialization is in methods of severe plastic deformation, one of which he invented (twist channel angular pressing, TCAP), and the design and optimization of thermomechanical processing. He also has extensive experience with numerical modeling of deformation processes via the finite element method, and experimental evaluation of the mechanical properties of materials. He has participated in numerous (inter)national scientific projects, (co)authored more than 260 papers, 2 books, 4 book chapters, acquired 3 patents, 1 utility design, and invented 7 software programs.

Lenka Kunčická, Ph.D. Dr. Kunčická has experience in the field of material forming, including thermomechanical processing and severe plastic deformation processes. Her focus is also on powder metallurgy and subsequent deformation of compacted specimens. In addition, she has experience with numerical modeling of plastic deformation tasks. She specializes in structure analyses via electron microscopy. She has authored or co-authored more than 60 publications, most of which are published in high-impact journals.

Editorial

Special Issue: Mechanical Properties in Progressive Mechanically Processed Metallic Materials

Radim Kocich [1,*] and Lenka Kunčická [2,*]

1 Faculty of Materials Science and Technology, VŠB-Technical University of Ostrava, 70833 Ostrava 8, Czech Republic
2 Institute of Physics of Materials, Czech Academy of Sciences, 61662 Brno, Czech Republic
* Correspondence: radim.kocich@vsb.cz (R.K.); kuncicka@ipm.cz (L.K.)

Received: 13 October 2020; Accepted: 15 October 2020; Published: 20 October 2020

Abstract: The research and development of modern metallic materials imparts not only the introduction of innovative alloys and compounds, but also the increasing lifetime of existing materials via optimized deformation processing. Among the essential features of progressive metallic materials used for modern applications are enhanced mechanical properties, but also other high-level functional characteristics, such as thermal–physical parameters, corrosion rate, and electric resistance. The properties of materials and alloys ensue from their structures, which can primarily be affected by the preparation/production process. The Special Issue "Mechanical Properties in Progressive Mechanically Processed Metallic Materials" was established to present recent developments and innovations particularly in the engineering field. The Special Issue comprises papers dealing with modern materials, such as metallic composites and pseudoalloys, as well as developments in various processing technologies.

Keywords: mechanical properties; functional properties; metallic systems; mechanical processing; structural phenomena

---

The demands on innovative materials given by the ever-increasing requirements of contemporary industry impart the usage of high-performance engineering materials, among the most innovative ones are multicomponent materials, such as gradient structures and composites, which are able to satisfy top-level individual requirements through the possibility to benefit from the advantages of all their components. The Special Issue contains two papers dealing with the preparation of modern aluminum/copper electro-conductive clad composites. One of the papers primarily focuses on numerical prediction and experimental evaluation of the deformation behavior of both the component metals during processing [1], whereas the other is a thorough study of the internal structure, texture in particular, of the reinforcing wires [2]. The composites are prepared via the innovative method of severe plastic deformation (SPD) of twist channel angular pressing (TCAP), the thorough review of which is performed in another paper [3].

Among the possible ways of how to effectively increase the utility properties of metallic materials is to decrease their grain size. In addition to using the SPD methods based on imposing severe shear strain resulting in grain refinement introducing enhancement of numerous properties, powder metallurgy can be applied to provide powders featuring fine grain sizes at the very beginning of the production process. The paper by K. Dvořák et al. [4] presents the possibility of using modern disintegrators to prepare fine powder particles. Several papers published within the Special Issue then deal with processing of metallic powders—assessing and optimizing the processing conditions and evaluating various phenomena of the final products. The paper by J. Málek et al. [5] deals with the effects of the selected processing route on the properties of a biocompatible HfNbTaTiZr high entropy alloy (HEA), for the specific advantageous properties of HEAs have been within the focus of researchers for the last few decades, whereas other researchers focused on a WNiCo tungsten heavy alloy, from evaluating and optimizing the sintering conditions for the initial powders [6], through numerical simulation of deformation behavior during plastic deformation processing [7], to thorough investigation of structural phenomena, such as microstrains and residual stress [8].

The introduction of thermomechanical treatment represented a breakthrough in grain refinement, consequently leading to significant improvement of the mechanical properties of metallic materials. Contrary to conventional production technologies, the main advantage of such treatment is the possibility to precisely control structural phenomena, including grain size, substructure development, texture, and volumes and types of grains boundaries, all of which affect the final mechanical and utility properties. The strengthening mechanisms in modern high-temperature resistant alloyed steels after heat treatment were studied by L. Zhao et al. [9], while A. Olina et al. [10] investigated the occurrence of retained austenite in a fine-grained spring steel processed via thermomechanical rolling. J. Fumfera et al. [11] then numerically and experimentally evaluated the cyclic hardening behavior in dependence with the strain range for a 08Ch18N10T austenitic stainless steel.

Last but not least, the semi-products fabricated from modern materials have to be finished to reach their final shapes and forms, which can be a challenge considering their enhanced properties and unique structures. Nevertheless, there are cutting and shaping methods that can be advantageously used for these purposes. The well-known machining process is suitable for the finishing of numerous products; however, the process introduces micro-deformations to the materials being processed and this factor should be considered as it can alter the lifetime and properties of the final product [12]. Additionally, for this reason, the abrasive water jet cutting technology can be considered as very favorable for cutting and shaping modern materials [13,14].

Conflicts of Interest: The authors declare no conflict of interest.

References

1. Kocich, R. Deformation Behavior of Al/Cu Clad Composite during Twist Channel Angular Pressing. *Materials* **2020**, *13*, 4047. [CrossRef] [PubMed]
2. Kunčická, L.; Klečková, Z. Structure Characteristics Affected by Material Plastic Flow in Twist Channel Angular Pressed Al/Cu Clad Composites. *Materials* **2020**, *13*, 4161. [CrossRef] [PubMed]
3. Macháčková, A. Decade of Twist Channel Angular Pressing: A Review. *Materials* **2020**, *13*, 1725. [CrossRef] [PubMed]
4. Dvořák, K.; Macháčková, A.; Ravaszová, S.; Gazdič, D. Effect of Imposed Shear Strain on Steel Ring Surfaces during Milling in High-Speed Disintegrator. *Materials* **2020**, *13*, 2234. [CrossRef] [PubMed]
5. Málek, J.; Zýka, J.; Lukáč, F.; Vilémová, M.; Vlasák, T.; Čížek, J.; Melikhova, O.; Macháčková, A.; Kim, H.-S. The Effect of Processing Route on Properties of HfNbTaTiZr High Entropy Alloy. *Materials* **2019**, *12*, 4022. [CrossRef]
6. Kunčická, L.; Kocich, R.; Klečková, Z. Effects of Sintering Conditions on Structures and Properties of Sintered Tungsten Heavy Alloy. *Materials* **2020**, *13*, 2338. [CrossRef]
7. Macháčková, A.; Krátká, L.; Petrmichl, R.; Kunčická, L.; Kocich, R. Affecting Structure Characteristics of Rotary Swaged Tungsten Heavy Alloy Via Variable Deformation Temperature. *Materials* **2019**, *12*, 4200. [CrossRef]
8. Strunz, P.; Kunčická, L.; Beran, P.; Kocich, R.; Hervoches, C. Correlating Microstrain and Activated Slip Systems with Mechanical Properties within Rotary Swaged WNiCo Pseudoalloy. *Materials* **2020**, *13*, 208. [CrossRef] [PubMed]
9. Zhao, L.; Chen, X.; Wu, T.; Zhai, Q. Study on Strengthening Mechanism of 9Cr-1.5Mo-1Co and 9Cr-3W-3Co Heat Resistant Steels. *Materials* **2020**, *13*, 4340. [CrossRef] [PubMed]
10. Olina, A.; Píška, M.; Petrenec, M.; Hervoches, C.; Beran, P.; Pechoušek, J.; Král, P. Assessment of Retained Austenite in Fine Grained Inductive Heat Treated Spring Steel. *Materials* **2019**, *12*, 4063. [CrossRef] [PubMed]
11. Fumfera, J.; Halama, R.; Procházka, R.; Gál, P.; Španiel, M. Strain Range Dependent Cyclic Hardening of 08Ch18N10T Stainless Steel—Experiments and Simulations. *Materials* **2019**, *12*, 4243. [CrossRef] [PubMed]

12. Ohnistova, P.; Piska, M.; Petrenec, M.; Dluhos, J.; Hornikova, J.; Sandera, P. Fatigue Life of 7475-T7351 Aluminum After Local Severe Plastic Deformation Caused by Machining. *Materials* **2019**, *12*, 3605. [CrossRef] [PubMed]
13. Hlaváč, L.M.; Štefek, A.; Tyč, M.; Krajcarz, D. Influence of Material Structure on Forces Measured during Abrasive Waterjet (AWJ) Machining. *Materials* **2020**, *13*, 3878. [CrossRef] [PubMed]
14. Hlaváčová, I.M.; Sadílek, M.; Váňová, P.; Szumilo, Š.; Tyč, M. Influence of Steel Structure on Machinability by Abrasive Water Jet. *Materials* **2020**, *13*, 4424. [CrossRef] [PubMed]

Publisher's Note: MDPI stays neutral with regard to jurisdictional claims in published maps and institutional affiliations.

Article

Deformation Behavior of Al/Cu Clad Composite During Twist Channel Angular Pressing

Radim Kocich

Faculty of Materials Science and Technology, VŠB–Technical University of Ostrava, 70800 Ostrava, Czech Republic; radim.kocich@vsb.cz; Tel.: +420-596-99-44-55

Received: 4 August 2020; Accepted: 10 September 2020; Published: 11 September 2020

Abstract: The research and development of modern metallic materials goes hand in hand with increasing their lifetime via optimized deformation processing. The presented work deals with preparation of an Al/Cu clad composite with implemented reinforcing Cu wires by the method of twist channel angular pressing (TCAP). Single and double pass extrusion of the clad composite was simulated numerically and carried out experimentally. This work is unique as no such study has been presented so far. Detailed monitoring of the deformation behavior during both the passes was enabled by superimposed grids and sensors. Both the sets of results revealed that already the single pass imparted significant effective strain (higher than e.g., conventional equal channel angular pressing (ECAP)), especially to the Al matrix, and resulted in notable deformation strengthening of both the Al and Cu composite components, which was confirmed by the increased punch load and decreased plastic flow velocity (second pass compared to first pass). Processing via the second pass also resulted in homogenization of the imposed strain and residual stress across the composite cross-section. However, the investigated parameters featured slight variations in dependence on the monitored location across the cross-section.

Keywords: clad composite; rotary swaging; finite element analysis; effective strain; residual stress

1. Introduction

As regards to intensive plastic deformation processing, the main focus has been on conventional Fe-based materials, although non-ferrous metals are given attention, too, primarily due to their wide applicability in various industrial branches. However, based on the increasing demands of the industry, the everlasting research of the forming processes increases their applicability also for innovative and newly developed materials. Nevertheless, the aims of the processing are similar regardless of the used material: the materials subjected to intensive plastic deformation are prepared to meet high demands, including the one on having ultra-fine-grained structures (UFG).

Among the preferred methods used to prepare UFG structures within (non)ferrous metals and alloys are severe plastic deformation (SPD) methods, such as equal channel angular pressing (ECAP) [1,2], twist channel (multi) angular pressing (TC(M)AP) [3,4], accumulative roll bonding (ARB) [5], high pressure torsion (HPT) [6], various combinations of torsion and extrusion [7,8], etc. The materials prepared via such methods typically exhibit very high strength at room temperature, but only limited plasticity. The decrease in plastic properties in very fine-grained materials is related to the combination of low deformation strengthening rate, and low value of strain rate sensitivity coefficient (m). In other words, the high deformation strengthening rate results in accumulation of dislocations inside the grains, which, in combination with high strain rate sensitivity coefficient m, significantly suppresses the development of material failure during deformation processing and thus increases plasticity of the material during forming. Nevertheless, ways how to achieve combinations of high strength and high plasticity exist; such material behavior is characterized as the strength and ductility paradox [9]. The methods leading to the achievement of the paradox are: increasing the imposed strain within UFG materials up to very high values, and performing very short annealing immediately after the intensive deformation processing. Short-time annealing contributes to ordering of defects within grain boundaries, which approaches them to the equilibrium state [9,10]. This phenomenon is crucial to prevent substantial grain growth. In addition, annealing at elevated temperatures imparts decreasing dislocation density (i.e., dislocations annihilation and rearrangement), which facilitates effective development and arrangement of new dislocations, contributing to further deformation strengthening and plasticity increase. Among the first materials, the applicability of the paradox for which was proven (i.e., which exhibited both high strength and superplastic behavior), was the Al4Cu0.5Zr alloy processed via SPD [11].

One of the alternative approaches for achieving favorable combinations of mechanic, electric, or magnetic properties is also preparation of composite materials, among the individual types of which are clad composites [12,13]. Various methods of preparation of the composites, as well as numerous bi-metallic and multi-metallic systems of clad composites, have been reported [14–17]. However, the majority of the available works focuses on the Cu/Al system. This particular combination of metals exhibits high thermal and electric conductivity (primarily provided by Cu), low density (primarily provided by Al), and the advantage of lower price when compared to single Cu [18–22]. The early researched methods of their preparation were based on bonding under elevated temperatures, e.g., explosive welding [23]. However, such technologies feature heat development during processing and are disadvantageous from the viewpoint of the possible introduction of local structure changes and formation of intermetallic phases negatively affecting not only the mechanic, but also utility properties of the produced clad composite. Moreover, the applicability of such production methods is limited since, if applied for thin sheaths and small billets, the heat-affected region could comprise the entire product.

The methods based on deformation processing, which have so far been applied to prepare clad composites, are e.g., rolling and related methods such as accumulative roll bonding [24] and asymmetrical rolling [25], drawing [26], forward extrusion [27], rotary swaging [28], HPT [29], ECAP [30], and their combinations [31]. The performed experiments revealed that the application of SPD methods leads to a favorable increase in the final utility properties of the produced clad composites. Among the examples is e.g., the Al-Cu system; an Al-Cu composite sheet was reported to be up to 60% lighter and 30–40% cheaper than an identical sheet fabricated from Cu while maintaining comparable thermal and electric conductivity [32]. Moreover, from the viewpoint of mechanical properties, an Al-Cu clad composite can exhibit far higher fracture toughness than the individual metals [33]. Nevertheless, only limited attention has been given to preparation of clad composites via ECAP and ECAP-based methods.

The presented report comprises a detailed study characterizing the deformation behavior of an Al-Cu clad composite during the TCAP process, which enables to impose higher values of effective strain during each pass when compared to conventional ECAP (previous studies dealing with processing of single-phase CP Al billets documented the effectivity of a single pass TCAP to be higher than even double pass ECAP [3,34]). The study features finite element numerical analysis supplemented with results acquired from real experiments consisting of single and double pass extrusion of the designed clad composite (Al sheath with reinforcing Cu wires) via experimental TCAP die (refer to Section 2 for more detailed information). The main focus is on the development of deformation parameters of both the component metals during repeated plastic deformation, as well as on the development of their mechanical properties. Besides, the study is supplemented with characterization of material plastic flow directly influencing the localization and distribution of the effective imposed strain and residual stress.

2. Materials and Methods

The aim of the work was to provide a detailed characterization of the effects of TCAP (twist channel angular pressing) processing on the Al/Cu clad composite (Figure 1).

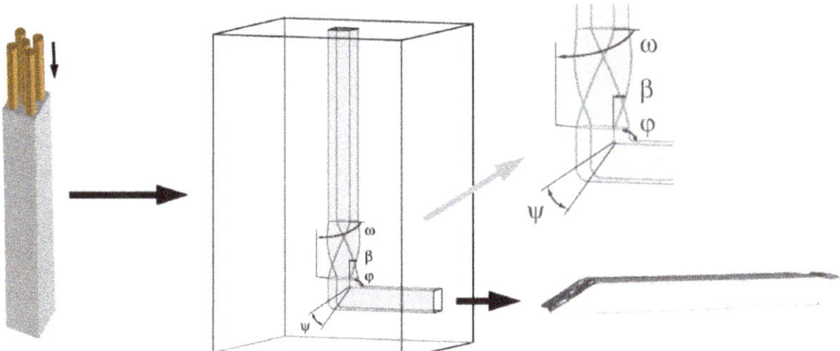

Figure 1. Schematic depiction of manufacturing Al/Cu composite by TCAP.

Since the work is unique, as this method has not been used before to process any composite material, the study does not only involve the experimental work, but also numerical analysis. The composite billet consisted of Al sheath and reinforcing Cu wires (Figure 1), and was processed via two subsequent TCAP passes with the selected deformation route *Bc*. The deformation route was based on the *Bc* deformation route known from conventional ECAP, i.e., the composite billet was subjected to +90° rotation before the second pass [35,36]. However, due to the implemented twist, the effects of the deformation route on the billet, moreover a composite one, are expected to be different than those of conventional ECAP. Among the selected investigated structure characteristics and development of mechanical properties, we also investigated selected processing parameters, such as the necessary punch load, values and distributions of the imposed strain, and residual stress.

The first part of the study deals with numerical analysis of extrusion of the clad composite via both the subsequent TCAP passes; the deformation behavior of the composite during processing was studied using Forge NxT commercial software. The simulations were assembled with a model and the geometrical dimensions and mechanical properties of the dies and billet in which were identical to the experimental parameters, which enabled subsequent direct comparison of the predicted results with the experimental ones. The extrusion was carried out on a hydraulic press with the use of MoS_2 as the lubricant, at the room temperature of approx. 25 °C and extrusion rate of 5 mm s^{-1}. The friction was in the simulation determined by the Coulomb friction with the values of $\mu = 0.02$. The values of the mentioned boundary conditions were selected based on the results previously acquired within the study mutual comparison of predicted and experimentally acquired values in which was successfully performed [35]. The geometry of the die used in the simulation was defined by the following angles: $\omega = 90°$ (twist rotation angle), $\varphi = 90°$ (channel bending angle), $\beta = 40°$ (twist slope angle), $\psi = 20°$ (angle of the arc of curvature in which the two channels intersect) (Figure 1).

The second part of the study deals with practical realization of the multiple TCAP process. The selected materials were commercially pure Cu (99.97%) with the chemical composition of 0.0074 Ni, 0.0058Sn, 0.0030Zn, 0.0031Fe, 0.0023Si, bal. Cu (in wt.%), and commercially pure Al (99.97%) with the chemical composition of 0.125Fe, 0.10Si, 0.020Cu, 0.020Zn, 0.015Mg, 0.015Mn, 0.015Ti, bal. Al (in wt.%). Before completing the composite, both the component metals were pre-annealed in a furnace at 500 °C for 30 min. The dimensions of the samples prepared for extrusion were matching the parameters given in the simulation, i.e., square 12 mm × 12 mm cross-section and 130 mm length

The structure analyses of the extruded clad composite samples were performed via the electron backscatter diffraction (EBSD) method (scanning electron microscopy (SEM)). Preparations of the samples for the analyses were carried out by grinding on SiC papers and final electrolytic polishing. SEM investigations were done using a Tescan Lyra 3 FIB/SEM microscope equipped with a NordlysNano EBSD camera (Oxford Instruments, Abingdon-on-Thames, Great Britain). The EBSD scans were acquired on samples tilted by 70° with the steps of 50 nm and the accelerating voltage of 20 kV.

To enable comparison with the predicted results, the presence of residual stress in the structures of the extruded billets was evaluated by analyzing the internal grains misorientations in the rainbow color scheme in the scale from 0° (negligible presence of residual stress), to 15° (occurrence of residual stress). Overall structure scan of the cross-section of the final extruded billet was performed on the polished sample using the OLYMPUS DSX1000 digital microscope (Shinjuku, Tokyo, Japan).

Last but not least, the mechanical properties of the composites were investigated via microhardness measurement performed on transversal cross-sectional cuts using a Zwick/Roell testing machine (Zwick/Roell, Ulm, Germany). The applied load was 200 g, and the loading time per indent was 10 s.

Numerical Simulation

The deformation behavior of both the component metals was predicted with the use of an elastic-plastic model defined via the Newton–Raphson convergent algorithm. The mesh of the clad composite billet consisted of 215,871 nodes. The billet was meshed with tetrahedral elements, whereas both the extruder and die were considered as rigid parts. Since severe shear deformations were expected to occur during the simulation, automatic re-meshing was activated. The stress-strain curves, depicted in Figure 2a, acquired for the experimentally used materials were determined on the basis of torsion tests performed using SETARAM, a servo-hydraulic torsion plastometer, at room temperature with the strain rates of 0.1 and 1 s^{-1}.

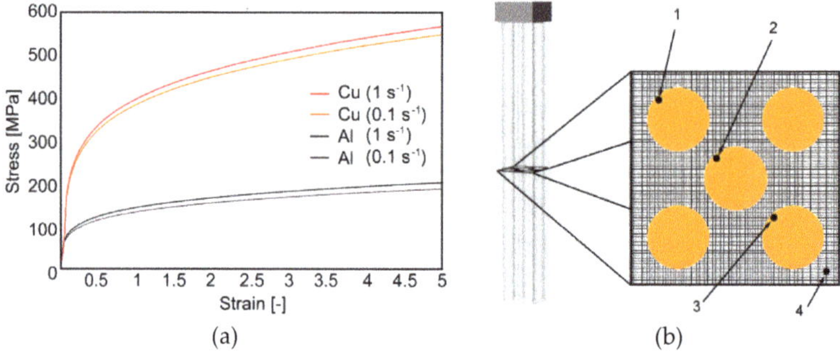

Figure 2. Stress-strain curves used for FEM (finite element method) computations (**a**); placement of analyzed sections within extruded composite (**b**).

The experimentally acquired stress-strain data were entered into the material flow stress database of the computational software. The Haensel–Spittel equation (Equation (1)) was then used to characterize material behavior during deformation processing:

$$\sigma_f = A e^{m_1 T} T^{m_8} \varepsilon^{m_2} e^{m_4/\varepsilon} (1+\varepsilon)^{m_5 T} e^{m_6 \varepsilon} \dot{\varepsilon}^{m_3} \dot{\varepsilon}^{m_7 T} \quad (1)$$

where ε is the equivalent von Mises strain, T is the temperature, $\dot{\varepsilon}$ is the equivalent von Mises strain rate, and A, m_1, m_2, m_3, m_4, m_5, m_6, m_7, m_8, and m_9 are regression coefficients. The values of the individual coefficients for Cu are, respectively, 411.19 MPa, −0.00121, 0.21554, 0.01472, −0.00935, and $m_5 \div m_8$ are 0. The values of the individual coefficients for Al are, respectively, 151.323 MPa, −0.00253, 0.21142, 0.03177, −0.00654, $m_5 \div m_8$ are 0.

The boundary conditions defined in the simulation were the temperature of 25 °C, and the values of parameters describing the temperature behavior of aluminum and copper, i.e., Young's modulus, Poisson's ratio, thermal expansion, thermal conductivity, heat transfer coefficient, specific heat, emissivity, and density. The parameters were defined for Al as the constants of 72 (GPa), 0.3, 2.4×10^{-5} (K^{-1}), 250 (W/(mK)), 1230 (J/kgK) 0.03 and 2800 (kg/m^3), and for Cu as 111 (GPa), 0.3, 1.7×10^{-5} (K^{-1}), 394 (W/(m K)), 100,000 (W/m^2 K), 398 (J/kg K), 0.7, and 8960 (kg/m^3).

In order to provide more specific characterization of the material plastic flow of both the composite components, a monitoring grid was superimposed through the billet being extruded, perpendicularly to the longitudinal axis of the composite (Figure 2b). The deformation behavior of the composite was monitored via four individual monitoring sensors localized within the superimposed grid (sensors 1–4, see Figure 2b). To enable detailed evaluation of the monitored phenomena, the superimposed grid was designed with very small square cells (0.1 × 0.1 mm). The purpose of this grid was to provide the possibility to characterize the influence of different plastic flows of the individual component metals on the development of temperature and its magnitude during processing, the development of the effective imposed strain and its (in)homogeneity across the billet cross-section, as well as the values and localization of residual stress. This research method is effective for characterization of the influences of the individual deformation zones within the die on the processed material.

3. Results

3.1. Temperature Development and Imposed Strain

As documented by the predicted results, the component metals exhibited differences in their behaviors during plastic deformation. The imposed strain is an important factor, the distribution of which non-negligibly influences also the distribution of temperature. However, for the Al/Cu clad composite, the temperature development was affected not only by the particular component metal, but also by the localization of the individual monitoring sensor in the deformation zones (Figure 3).

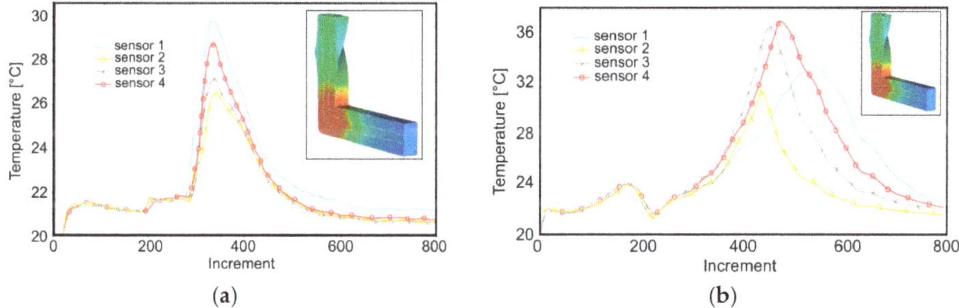

Figure 3. Development of temperature during: first pass (**a**); second pass (**b**).

The maximum value of temperature was detected in the main deformation zone (MDZ) for both the passes (see the detail in Figure 3). As documented by the temperature-time dependences for the monitored sensors, the maximum temperature value was approx. 50% higher than the average processing temperature during the first pass (Figure 3a). However, during the second pass, the maximum temperature increased up to ~37 °C (Figure 3b). The increase in the maximum temperature during the second pass can primarily be attributed to the effect of plastic deformation, i.e., deformation strengthening, of the Cu wires featuring higher flow stress than the Al sheath [37].

The occurring migration of the maximum temperature value is an intriguing phenomenon; whereas sensor 1 exhibited the maximum temperature during the first pass, sensor 4 exhibited the maximum temperature during the second pass (Figure 3a,b). This phenomenon was related to the applied deformation route *Bc*. In other words, the maximum temperatures occurred in the upper half of the cross-section of the extruded billet for both the passes, since the quicker plastic flow occurring in this region imparts the most intense shear. While sensors 1 and 2 were localized within the upper half of the cross-section during the first pass, during the second pass, as the result of the 180° rotation, sensor 4 was localized in the upper half of the billet cross-section.

By the effect of the inserted Cu wires, the effective strain was high also along the cross-sectional diagonals, although the wires themselves were not substantially affected by the imposed strain.

The development of distribution of the imposed strain was rather complicated, since it was influenced by multiple factors. During the first pass, the twist deformation zone (TDZ) of the die affected primarily the corner areas of the extruded composite (Figure 4a) (except at mutual Al/Cu interfaces).

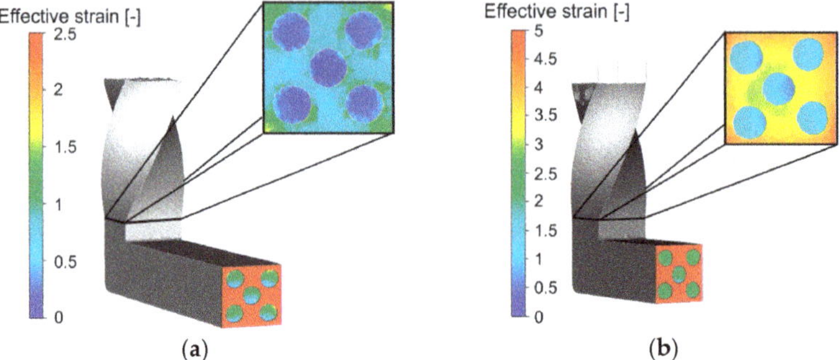

Figure 4. Effective imposed strain during: first pass (**a**); second pass (**b**).

The effect of the imposed strain on the Cu wires started to be evident after passing through the MDZ. As can be seen in Figure 4a, contrary to the Cu wires, the Al sheath exhibited more or less homogeneous distribution of high values (~2.5) of the effective strain after the first pass. The significantly lower effective strain values occurring within the Cu wires were caused by the minor influence of the TDZ on the wires; the strain gradient occurring between the Al sheath and Cu wires could not be completely eliminated during subsequent passing through the MDZ. On the other hand, the significant increase in the values of the imposed strain within the Cu wires occurring after passing through the MDZ points to the non-negligible deformation effect of this zone on both the composite components. The favorable combination of both the deformation zones within the die thus finally contributed to the more or less homogeneous distribution of the effective imposed strain throughout the composite cross-section after the second pass.

The imposed strain development was different during the second TCAP pass, the deformation effect of the TDZ during which was more notable than during the first pass. The predicted differences in the values of the imposed strain across the Al sheath cross-section were significantly lower, i.e., homogenization of the distribution of the effective strain occurred (Figure 4b). After passing through the MDZ, the Al sheath evidently exhibited increase in the values of the homogeneously distributed imposed strain (~5), whereas the Cu wires exhibited notable homogenization of the effective strain distribution (rather than significant increase in the imposed strain values).

As demonstrated by the predicted developments of the effective strain values in the individual monitored sensors within the composite, the developments of the imposed strain exhibited differences related to the individual component metal (similar to temperature behavior, see Figure 3a,b). During the first TCAP pass, the maximum values of the imposed strain were detected in sensor 4, i.e., in the Al sheath (Figure 5a).

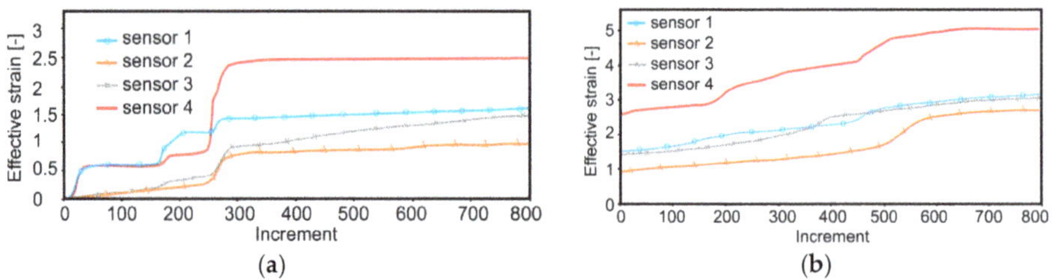

Figure 5. Imposed strain in individual sensors during: first pass (**a**); second pass (**b**).

The significant differences in the values of the imposed strain between the Al sheath and Cu wires originated primarily in the variations in the plastic flow intensity of both the component metals (Section 3.2).

Whereas during the first pass, the most significant effect on the imposed strain could be observed for the TDZ (jump increase in the imposed strain values were evident especially in sensors 1 and 4, see Figure 5a), during the second pass, the imposed strain in the monitored sensors exhibited gradual increasing (Figure 5b). This phenomenon was most probably related to the deformation strengthening of both the metals occurring during the first pass, by the effect of which the Al sheath became more effective transmitter of the imposed strain, which gradually diminished the differences in the values of the imposed strain within the individual wires. The predicted results showed that the second pass resulted in significant increase and homogenization in the imposed strain values, based on previous deformation strengthening. The Al sheath, being more susceptible to the imposed strain and thus exhibiting more intense plastic flow, featured higher effective strain values than the Cu wires. On the other hand, the differences in the values of the imposed strain between the individual Cu wires resulted from their localization. The most intense shear within the Cu was detected in sensor 1, i.e., in the upper half of the extruded composite.

The effective strain in sensors 2 and 3 exhibited relatively rapid increases in their developments during the second pass (Figure 5b). Whereas for sensor 3, this increase originated from its relocation caused by the selected deformation route, for sensor 2, this increase originated most probably from the occurring strengthening of the Al sheath in its vicinity. For the first pass, the moment in which the monitored sensors passed through the MDZ could clearly be identified (Figure 5a—increment 250 featured notable imposed strain jump increase). Despite the fact that minor jump increases in the imposed strain could be detected also during the second pass, their values were lower than during the first pass. On the contrary, most of the monitored locations (especially the Cu wires) exhibited gradual increases in the imposed strain.

3.2. Plastic Flow

Monitoring of the plastic flow behavior enabled detailed characterization of the influences of both the deformation zones on the extruded composite; their effects on the effective imposed strain were not identical, however, both the sections influenced non-negligibly not only the values of the imposed strain, but also its distribution. Already during the first TCAP pass, the TDZ evidently affected not only the (sub)surface, but also the axial region of the extruded composite (Figure 6a).

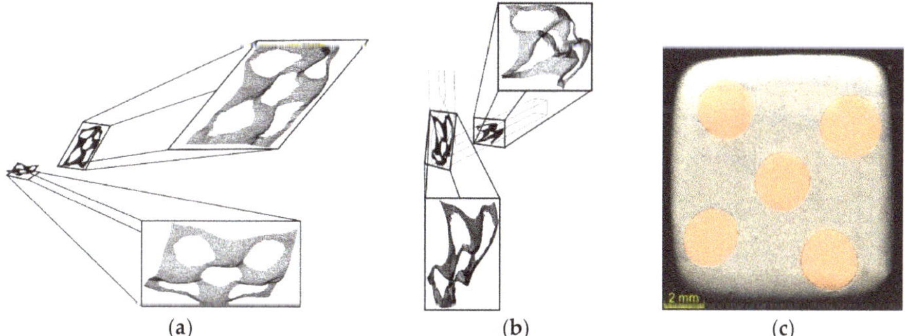

(a) (b) (c)

Figure 6. Material plastic flow of Al/Cu composite during: first pass (**a**), second pass (**b**); optical microscopy image of cross-sectional cut through billet extruded via two passes (**c**).

The effects of the TDZ were evident also on the Al/Cu interfaces, the superimposed grids at which exhibited evident serrations originating from rotations of the clad composite. Moreover, the original monitoring plane exhibited deformation in the extrusion direction. In other words, the deformations of the individual cells of the superimposed grid were not only caused by the tangential plastic flow, but also by its axial component which became dominant after passing through the MDZ (Figure 6a). The prediction thus revealed that significant plastic flow occurred within virtually all the regions of the Al sheath already during the first TCAP pass. On the other hand, the material flow in the upper half of the Al sheath was dominant. Together, with the significantly slower plastic flow of the axial Cu wire (decelerating effect), this phenomenon resulted in more intense slipping of both the metals in this region, and consequently in inhomogeneous distribution of the imposed strain, especially across the cross-sections of the Cu wires (as also confirmed by Figures 4a and 5a).

The cells of the superimposed grid, heavily deformed during the first pass, were subsequently deformed even more during the second pass. The effect of the previous deformation strengthening contributed to relatively uniform deformation of the entire superimposed grid; this was observed also for the Cu wires, the plastic flow of which was supported by the substantial plastic deformation introduced to the Al sheath. The second pass also resulted in full bonding of both the components, as can be seen in the optical microscopy scan of a cross-sectional cut through the extruded billet in Figure 6c; the metals did not exhibit slipping during processing.

The TDZ introduced rotations also during the second pass. However, during this pass, the rotations affected more intensively the axial region of the composite. Overall rotation of the cross-section of the billet, by 180°, was introduced by the selected deformation route *Bc*, which contributed to substantial reduction of the gradient of the imposed strain across the Cu wires cross-sections. This was also confirmed by the plastic behavior of the superimposed grid, the typical skewness of the cells of which reduced after passing through the MDZ (when compared to the first pass), as well as by the overall velocities of the plastic flow in the individual monitored sensors depicted in Figure 7a,b for the first and second TCAP pass, respectively.

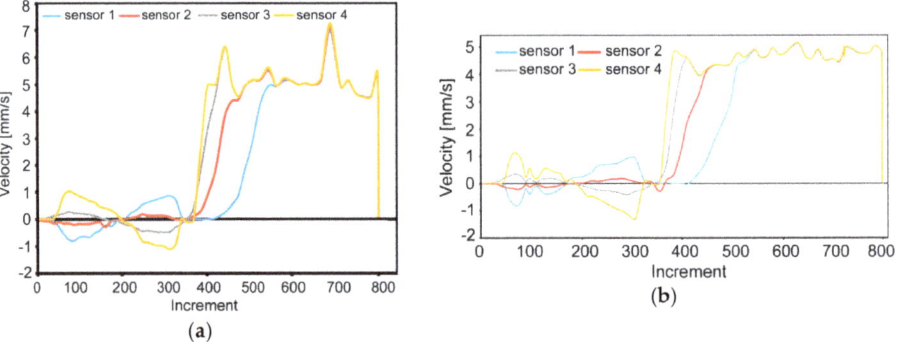

Figure 7. Development of plastic flow velocity during: first pass (**a**); second pass (**b**).

The graphical dependencies in Figure 7a,b depict the velocities of movement of the individual sensors in the monitored plane in relation to the extrusion direction (i.e., axis of the horizontal channel). The sensors enabled monitoring of the values and vectors of plastic flow velocities in the individual locations of the composite during the entire extrusion. As can be seen, passing of the composite through the TDZ caused reversion of the vector of plastic flow velocity for all the sensors during both, the first and second pass (Figure 7a,b). In other words, the negative values of velocity demonstrate the period of extrusion of the monitored volume of the material (i.e., the monitored sensor) in which reversed against the horizontal extrusion direction. Intriguing was also the notable decrease in the overall plastic flow velocities of the composite occurring between the first and second TCAP pass. This fact can be attributed to the deformation strengthening introduced during the first TCAP pass, which resulted not only in the decrease in the values of plastic flow velocity for all the sensors, but also in reduction of the observed oscillations.

3.3. Residual Stress

The effects of both the TCAP passes on the composite components were different from the viewpoint of residual stress. The composite exhibited prevailing tensile stress during the first pass though the TDZ. This phenomenon can be attributed to two main factors, the first one of which was the rotational movement of the Al sheath, the intensity of which decreased towards the axial region of the billet. The second factor was the dominant plastic flow occurring in the axial region of the composite (see Figure 6a). As the result of these factors, the maximum values of tensile stress were detected in the peripheral regions of the Cu wires (Figure 8a) after the first pass through the TDZ.

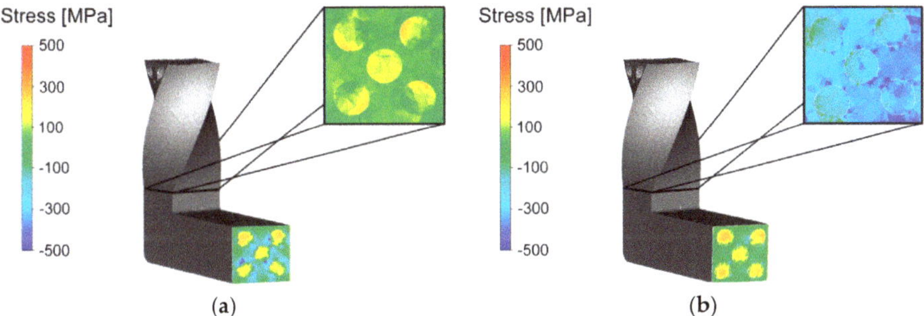

Figure 8. Distribution of residual stress during: first pass (**a**); second pass (**b**).

The stress state within the composite changed after passing through the channel bending, i.e., the MDZ. However, the stress character changed gradually from dominant tensile stress to dominant compressive stress during passing through the channel zone between the TDZ and MDZ due to the increasing resistance against the plastic flow caused by gradual filling of the MDZ (and the horizontal channel part) with the composite material. Behind the MDZ, the stress character changed again as axial differences in the plastic flows in different composite regions started to be notable. As evident from Figure 8a, the Cu wires in the extruded billet predominantly exhibited tensile stress, whereas the Al sheath in the extruded billet primarily featured compressive stress.

The development of residual stress was different during the second pass. The Cu wires featured local presence of tensile stress before entering the MDZ also during the second pass. On the other hand, the Al sheath exhibited predominant occurrence of compressive stress from the very beginning of extrusion. With continuing filling of the MDZ, the tensile stress within the Cu wires gradually transformed to compressive stress (Figure 8b); the gradual changing of the stress character was observed between the individual deformation zones, similarly to the first pass. After passing through the MDZ, homogenization of stress within both the Al sheath and Cu wires occurred. Nevertheless, the stress development was opposite for both the materials. Whereas the Cu wires featured gradual increase and, at the same time, homogenization of tensile stress, the Al sheath featured gradual homogenization of compressive stress. Contrary to the first TCAP pass, the occurring homogenization of stress decreased the overall stress gradient for both the component metals. Among the primary causes of this behavior was the occurrence of vortex-like flow within the Al sheath during extrusion [35]. In other words, the supplementary twist zone (compared to conventional ECAP) contributed to homogenization of the stress distribution throughout the composite cross-section. Indisputable influence had also the selected deformation route.

Experimental verification of the predicted data was performed via analyses of internal grains misorientations within both the components of the billet extruded via two consequent TCAP passes.

The results of the analyses for the peripheral Cu wire of the billets extruded via single and double TCAP pass are depicted in Figure 9a,b, respectively.

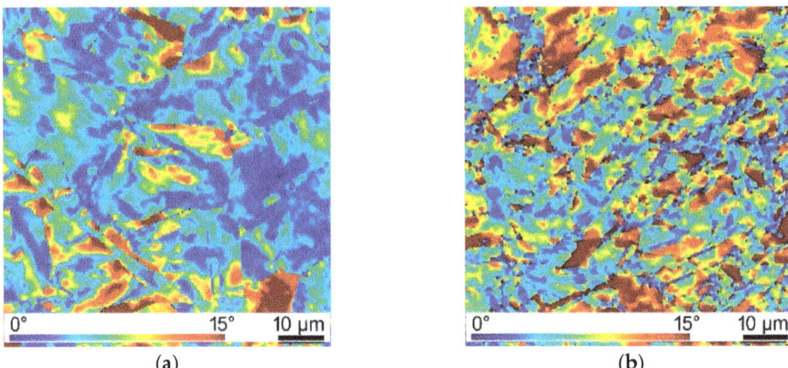

Figure 9. Maps of internal grains misorientations indicating residual stress within peripheral Cu wire of composite billet extruded via: first pass (**a**); second pass (**b**).

As can be seen, both the figures feature more or less homogeneous distribution of the areas featuring high misorientations (i.e., red color), pointing to the presence of residual stress. However, the occurrence of residual stress was more notable in the structure of the Cu wire from the composite extruded via two TCAP passes, which is in accordance to the predicted results (Figure 8a,b). In addition, the structure of the wire extruded via two TCAP passes exhibited finer grains and substructure development imparted by the severe imposed shear strain (see also Figure 4b).

3.4. Punch Load and Microhardness

The maximum loading force during the first pass reached to 40 kN (Figure 9). Whereas passing of the billet through the first deformation zone (TDZ) only resulted in a slight increase in the punch load, passing of the billet through the MDZ led to a significant increase in this parameter; the rapid increase in the punch load occurring approx. after 7 s of extrusion was followed with a mild linear increase to the maximum punch load value achieved in the time of extrusion of approx. 15 s. Such a steep increase in the loading force was most probably related to the increasing flow stress, primarily of the Cu wires. The subsequent decrease in the punch load after 15 s of extrusion was caused by gradual emptying of the twist zone of the channel and consequent decrease in friction between the Al sheath and channel walls.

The development of the punch load during the second TCAP pass featured even steeper increasing character from the very beginning of extrusion. Such behavior was imparted by two main influencing factors, both of which affected the development of the punch load during the entire pass. The first factor was the accumulation of deformation strengthening within both the component metals during the previous pass. In addition, the strain imposed during the first TCAP pass resulted in the changes of shapes of the ends of the inserted Cu wires (not shown here), which was the second influencing factor. In the moment in which the billet entered the twist deformation zone, the previous deformation strengthening accumulated in the Al sheath, but primarily in the Cu wires, resulted in the increase in the punch load by up to 50% (when compared to the first pass).

The already mentioned deformation strengthening, contributing also to the notable increase in the punch load during the second pass, was experimentally observed via microhardness measurements carried out along the diagonals across the cross-sectional cuts through both the composite billets (Figure 9b). The initial HV values of the original annealed component metals were 37.4 for Al, and 58.6 for Cu. The microhardness values evidently increased after both, the single and double TCAP pass.

However, the absolute increase in the HV values was more significant after the first pass. After the second pass, the increase in the HV values was less notable, however, the Cu wires within the billet extruded via route *Bc* exhibited the highest (maximum observed) HV values.

4. Discussion

The study featuring processing of the Al/Cu clad composite via the TCAP method revealed variations when compared to processing of single-component billets. Firstly, the deformation behavior of the investigated composite was influenced by the combination of the characteristic properties of both the component metals. For example, the comparison of the presented results with the results acquired during studies dealing with processing of single Al [34], or Cu [35], shows different temperature developments. Combining the Al and Cu within a single billet resulted in inhomogeneous temperature distribution, primarily due to the higher flow stress and thermo-physical parameters of the Cu wires (compared to the Al sheath). Moreover, the increase in temperature introduced by the intensive plastic flow of the Al sheath (especially in locations the vortex-like flow in which occurs [38–40]), generating heat originated from friction with the die and with the individual wires, should also be taken into account.

The deformation behavior, especially the effective imposed strain and plastic flow, of the composite was affected not only by the individual component metals, but also by the deformation zones within the die and the selected deformation route. Although the effects of the individual deformation zones on the values of the effective imposed strain are not equal, their mutual effect, which makes TCAP an advantageous method for processing of multicomponent materials, is non-negligible. Passing of the composite through the TDZ did not introduce as high strain as passing through the MDZ. The MDZ primarily influenced the axial region of the extruded billet, whereas in the TDZ, the highest effective strain values were detected in the (sub)surface regions of the Al sheath. The inter-regions and the axial region of the billet were more or less affected by the imposed strain in the TDZ, too, but the values of the imposed strain were significantly lower within the Cu wires. However, as the result of the variations in the velocities of the axial plastic flow of both the materials in the TDZ, the overall gradients in the observed parameters between the (sub)surface and axial billet regions were more or less reduced. In other words, among the main contributions of the TDZ is that it diminishes the differences between the axial and (sub)surface regions of the extruded composite. Therefore, the TDZ primarily contributes to homogenization of the imposed strain across the cross-section of the clad composite (especially the Cu wires).

During TCAP, the locations of the individual Cu wires with respect to the die channel vary, which is not possible for conventional ECAP [41]. By this reason, the plastic flow of the individual Cu wires is aggravated/supported during passing through and behind the MDZ. This behavior is closely related to the observed differences in the effective strain and residual stress values and distributions. Certain inhomogeneity of the imposed strain across the Cu wires cross-sections was also related to their orientations in the individual deformation zones during both the TCAP passes. Nevertheless, the final observed homogeneity of the imposed strain and residual stress was affected positively also by the applied deformation route (rotation of the billet with respect to the individual deformation zones resulted in alternation of the Cu wires and suppression of the differences occurring in the deformation history), as well as by the occurring deformation strengthening, which also affected the mentioned increase in temperature observed during the second pass.

The time dependence of the punch load featured notable oscillations for both the passes, however, the oscillations were more intense during the second pass (see Figure 10a).

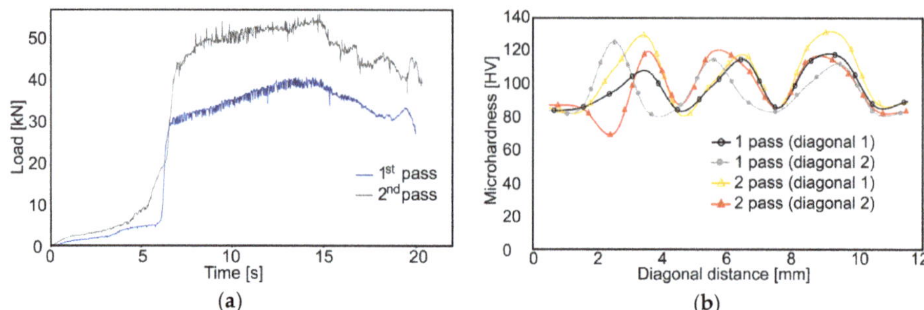

Figure 10. Material plastic flow of Al/Cu composite during both passes (**a**); microhardness measured experimentally across cross-sections of extruded billets (**b**).

Such serrated development is typically related to plastic flow instability [34,35], which was, in the presented study, most probably the direct effect of the different deformation strengthening (plastic flow) of the component metals. This supposition is supported by the fact that the oscillations were the most notable when the billet was passing through the MDZ, during which mutual slipping of both the component metals initiated by the differences in their plastic flows was the most intense. As discussed, the plastic flow of the Al sheath was more intense and the flows of the Cu wires exhibited localized tendencies to "delay"; this difference was caused by the Cu having higher flow stress than Al. The presupposition of deformation strengthening of both the component metals was confirmed by the data acquired from microhardness testing.

5. Conclusions

The presented study documented the results of numerical and experimental analysis of successful room-temperature preparation of an Al/Cu clad composite by single and double pass of twist channel angular pressing (TCAP). Both the TCAP passes introduced significant shear strain to the component metals, as the maximum effective imposed strain reached up to the value of 5 for the Al sheath of the billet extruded by two passes. The imposed strain contributed to deformation strengthening resulting not only in the increase in punch load by almost 50% and increase in microhardness up to 130 HV for the Cu wires, but also in temperature increase during processing. Nevertheless, the processing temperature was still only ~37 °C during the second pass, i.e., the processing conditions were safe from the viewpoint of introduction of possible structure changes. The results also showed that the processing via multiple passes increased the bonding quality of both the metals, as both the distributions of the imposed strain and residual stress homogenized after the second pass, which was ensured by the favorable combination of twist and bending deformation zones within the single unique die.

Author Contributions: The contributions of the individual authors were the following: methodology; numerical prediction and evaluation; experimental validation and microscopy investigations; writing—original draft preparation, review, and editing; project administration and funding acquisition—R.K. The author has read and agreed to the published version of the manuscript.

Funding: This research was supported by the 19-15479S Project of the Grant Agency of the Czech Republic.

Acknowledgments: The author would like to thank for and appreciate the kind cooperation of Lenka Kunčická during preparation of this manuscript.

Conflicts of Interest: The author declares no conflict of interest. The funders had no role in the design of the study; in the collection, analyses, or interpretation of data; in the writing of the manuscript, or in the decision to publish the results.

References

1. Kocich, R.; Kursa, M.; Szurman, I.; Dlouhý, A. The influence of imposed strain on the development of microstructure and transformation characteristics of Ni–Ti shape memory alloys. *J. Alloys Compd.* **2011**, *509*, 2716–2722. [CrossRef]
2. Kunčická, L.; Kocich, R.; Drápala, J.; Andreyachshenko, V.A. FEM simulations and comparison of the ecap and ECAP-PBP influence on Ti6Al4V alloy's deformation behavior. In Proceedings of the Metal 2013—22nd International Conference on Metallurgy and Materials, Brno, Czech Republic, 15–17 May 2013; pp. 391–396.
3. Kocich, R.; Kunčická, L.; Král, P.; Macháčková, A. Sub-structure and mechanical properties of twist channel angular pressed aluminium. *Mater. Charact.* **2016**, *119*, 75–83. [CrossRef]
4. Kocich, R.; Kunčická, L.; Macháčková, A. Twist Channel Multi-Angular Pressing (TCMAP) as a method for increasing the efficiency of SPD. *IOP Conf. Ser. Mater. Sci. Eng.* **2014**, *63*, 12006. [CrossRef]
5. Jamaati, R.; Naseri, M.; Toroghinejad, M.R. Wear behavior of nanostructured Al/Al2O3 composite fabricated via accumulative roll bonding (ARB) process. *Mater. Des.* **2014**, *59*, 540–549. [CrossRef]
6. Kawasaki, M.; Foissey, J.; Langdon, T. Development of hardness homogeneity and superplastic behavior in an aluminum–copper eutectic alloy processed by high-pressure torsion. *Mater. Sci. Eng. A* **2013**, *561*, 118–125. [CrossRef]
7. Kocich, R.; Szurman, I.; Kursa, M.; Fiala, J. Investigation of influence of preparation and heat treatment on deformation behaviour of the alloy NiTi after ECAE. *Mater. Sci. Eng. A* **2009**, *512*, 100–104. [CrossRef]
8. Khosravifard, A.; Jahedi, M.; Yaghtin, A.H. Three dimensional finite element study on torsion extrusion processing of 1050 aluminum alloy. *Trans. Nonferrous Met. Soc. China* **2012**, *22*, 2771–2776. [CrossRef]
9. Valiev, R.Z.; Alexandrov, I.V.; Zhu, Y.T.; Lowe, T.C. Paradox of Strength and Ductility in Metals Processed Bysevere Plastic Deformation. *J. Mater. Res.* **2002**, *17*, 5–8. [CrossRef]
10. Humphreys, F.; Hatherly, M. *Recrystallization and Related Annealing Phenomena*, 2nd ed.; Elsevier: Oxford, UK, 2004.
11. Langdon, T. Grain boundary sliding revisited: Developments in sliding over four decades. *J. Mater. Sci.* **2006**, *41*, 597–609. [CrossRef]
12. Cetin, A.; Krebs, J.; Durussel, A.; Rossoll, A.; Inoue, J.; Koseki, T.; Nambu, S.; Mortensen, A. Laminated Metal Composites by Infiltration. *Met. Mater. Trans. A* **2011**, *42*, 3509–3520. [CrossRef]
13. Polyanskii, S.N.; Kolnogorov, V.S. Cladded Steel for the Oil and Gas Industries. *Chem. Pet. Eng.* **2002**, *38*, 703–707. [CrossRef]
14. Motarjemi, A.K.; Kocak, M.; Ventzke, V. Mechanical and fracture characterization of a bi-material steel plate. *Int. J. Press. Vessel. Pip.* **2002**, *79*, 181–191. [CrossRef]
15. Jin, J.Y.; Hong, S.I. Effect of heat treatment on tensile deformation characteristics and properties of Al3003/STS439 clad composite. *Mater. Sci. Eng. A* **2014**, *596*, 1–8. [CrossRef]
16. Movahedi, M.; Kokabi, A.; Reihani, S.S. Investigation on the bond strength of Al-1100/St-12 roll bonded sheets, optimization and characterization. *Mater. Des.* **2011**, *32*, 3143–3149. [CrossRef]
17. Inoue, J.; Sadeghi, A.; Kyokuta, N.; Ohmori, T.; Koseki, T. Multilayer Mg: Stainless Steel Sheets, Microstructure, and Mechanical Properties. *Metall. Mater. Trans. A* **2017**, *48*, 2483–2495. [CrossRef]
18. Kocich, R.; Kunčická, L.; Davis, C.F.; Lowe, T.C.; Szurman, I.; Macháčková, A. Deformation behavior of multilayered Al–Cu clad composite during cold-swaging. *Mater. Des.* **2016**, *90*, 379–388. [CrossRef]
19. Kunčická, L.; Kocich, R.; Dvořák, K.; Macháčková, A. Rotary swaged laminated Cu-Al composites: Effect of structure on residual stress and mechanical and electric properties. *Mater. Sci. Eng. A* **2019**, *742*, 743–750. [CrossRef]
20. Kim, I.K.; Hong, S.I. Effect of heat treatment on the bending behavior of tri-layered Cu/Al/Cu composite plates. *Mater. Des.* **2013**, *47*, 590–598. [CrossRef]
21. Lee, T.H.; Lee, Y.J.; Park, K.T.; Jeong, H.G.; Lee, J.H. Mechanical and asymmetrical thermal properties of Al/Cu composite fabricated by repeated hydrostatic extrusion process. *Met. Mater. Int.* **2015**, *21*, 402–407. [CrossRef]
22. Li, F.S.; Xu, R.Z.; Wei, Z.C.; Sun, X.F.; Wang, P.F.; Li, X.F.; Li, Z. Investigation on the Electron Beam Welding of Al/Cu Composite Plates. *Trans. Indian Inst. Met.* **2020**, *73*, 353–363. [CrossRef]

23. Fronczek, D.; Wojewoda-Budka, J.; Chulist, R.; Sypien, A.; Korneva, A.; Szulc, Z.; Schell, N.; Zieba, P. Structural properties of Ti/Al clads manufactured by explosive welding and annealing. *Mater. Des.* **2016**, *91*, 80–89. [CrossRef]
24. Kim, I.K.; Hong, S.I. Roll-Bonded Tri-Layered Mg/Al/Stainless Steel Clad Composites and their Deformation and Fracture Behavior. *Metall. Mater. Trans. A* **2013**, *44*, 3890–3900. [CrossRef]
25. Li, X.; Zu, G.; Ding, M.; Mu, Y.; Wang, P. Interfacial microstructure and mechanical properties of Cu/Al clad sheet fabricated by asymmetrical roll bonding and annealing. *Mater. Sci. Eng. A* **2011**, *529*, 485–491. [CrossRef]
26. Jee, M.H.; Choi, J.U.; Park, S.H.; Jeong, Y.G.; Baik, D.H. Influences of tensile drawing on structures, mechanical, and electrical properties of wet-spun multi-walled carbon nanotube composite fiber. *Macromol. Res.* **2012**, *20*, 650–657. [CrossRef]
27. Kazanowski, P.; E Epler, M.; Misiolek, W.Z. Bi-metal rod extrusion—Process and product optimization. *Mater. Sci. Eng. A* **2004**, *369*, 170–180. [CrossRef]
28. Kocich, R.; Kunčická, L.; Macháčková, A.; Šofer, M. Improvement of mechanical and electrical properties of rotary swaged Al-Cu clad composites. *Mater. Des.* **2017**, *123*, 137–146. [CrossRef]
29. Ma, X.; Huang, C.; Xu, W.; Zhou, H.; Wu, X.L.; Zhu, Y. Strain hardening and ductility in a coarse-grain/nanostructure laminate material. *Scr. Mater.* **2015**, *103*, 57–60. [CrossRef]
30. Jafarlou, D.; Zalnezhad, E.; Ezazi, M.; Mardi, N.; Hassan, M. The application of equal channel angular pressing to join dissimilar metals, aluminium alloy and steel, using an Ag–Cu–Sn interlayer. *Mater. Des.* **2015**, *87*, 553–566. [CrossRef]
31. Sapanathan, T.; Khoddam, S.; Zahiri, S.H.; Zarei-Hanzaki, A. Strength changes and bonded interface investigations in a spiral extruded aluminum/copper composite. *Mater. Des.* **2014**, *57*, 306–314. [CrossRef]
32. Kwon, H.C.; Jung, T.K.; Lim, S.C.; Kim, M.S.; Kwon, H.C. Fabrication of Copper Clad Aluminum Wire (CCAW) by Indirect Extrusion and Drawing. *Mater. Sci. Forum* **2004**, *449–452*, 317–320. [CrossRef]
33. Kim, H.; Hong, S.I. Deformation and fracture of diffusion-bonded Cu–Ni–Zn/Cu–Cr layered composite. *Mater. Des.* **2015**, *67*, 42–49. [CrossRef]
34. Kunčická, L.; Kocich, R.; Král, P.; Pohludka, M.; Marek, M. Effect of strain path on severely deformed aluminium. *Mater. Lett.* **2016**, *180*, 280–283. [CrossRef]
35. Kocich, R.; Fiala, J.; Szurman, I.; Macháčková, A.; Mihola, M. Twist-channel angular pressing: Effect of the strain path on grain refinement and mechanical properties of copper. *J. Mater. Sci.* **2011**, *46*, 7865–7876. [CrossRef]
36. Macháčková, A. Decade of Twist Channel Angular Pressing: A Review. *Mater. Basel* **2020**, *13*, 1725. [CrossRef]
37. Russell, A.; Lee, K.L. *Structure-Property Relations in Nonferrous Metals*, 1st ed.; John Wiley & Sons: Hoboken, NJ, USA, 2005.
38. Orlov, D.; Beygelzimer, Y.; Synkov, S.; Varyukhin, V.; Tsuji, N.; Horita, Z. Microstructure Evolution in Pure Al Processed with Twist Extrusion. *Mater. Trans.* **2009**, *50*, 96–100. [CrossRef]
39. Kocich, R.; Greger, M.; Kursa, M.; Szurman, I.; Macháčková, A. Twist channel angular pressing (TCAP) as a method for increasing the efficiency of SPD. *Mater. Sci. Eng. A* **2010**, *527*, 6386–6392. [CrossRef]
40. Kocich, R.; Kunčická, L.; Král, P.; Strunz, P. Characterization of innovative rotary swaged Cu-Al clad composite wire conductors. *Mater. Des.* **2018**, *160*, 828–835. [CrossRef]
41. Djavanroodi, F.; Ebrahimi, M. Effect of die channel angle, friction and back pressure in the equal channel angular pressing using 3D finite element simulation. *Mater. Sci. Eng. A* **2010**, *527*, 1230–1235. [CrossRef]

Article

Structure Characteristics Affected by Material Plastic Flow in Twist Channel Angular Pressed Al/Cu Clad Composites

Lenka Kunčická [1,*] and Zuzana Klečková [2]

[1] Institute of Physics of Materials, ASCR, 61662 Brno, Czech Republic
[2] Department of Thermal Engineering, Faculty of Metallurgy and Materials Engineering, VŠB TU Ostrava 17. Listopadu 15, 70833 Ostrava-Poruba, Czech Republic; zuzana.kleckova@vsb.cz
* Correspondence: kuncicka@ipm.cz; Tel.: +420-532-290-371

Received: 23 August 2020; Accepted: 16 September 2020; Published: 18 September 2020

Abstract: The study focuses on structure analyses, texture analyses in particular, of an Al/Cu clad composite manufactured by single and double pass of the twist channel angular pressing (TCAP) method. Microscopic analyses were supplemented with numerical predictions focused on the effective imposed strain and material plastic flow, and microhardness measurements. Both the TCAP passes imparted characteristic texture orientations to the reinforcing Cu wires, however, the individual preferential grains' orientations throughout the composite differed and depended on the location of the particular wire within the Al sheath during extrusion, i.e., on the dominant acting strain path. The second TCAP pass resulted in texture homogenization; all the Cu wires finally exhibited dominant A fiber shear texture. This finding was in accordance with the homogenization of the imposed strain predicted after the second TCAP pass. The results also revealed that both the component metals exhibited significant deformation strengthening (which also caused bending of the ends of the Cu wires within the Al sheath after extrusion). The average microhardness of the Cu wires after the second pass reached up to 128 HV, while for the Al sheath the value was 86 HV.

Keywords: clad composite; rotary swaging; finite element analysis; effective strain; residual stress

---

1. Introduction

Composite materials can generally be characterized as materials consisting of two or more phases or components featuring different physical and chemical properties, separated by mutual interfaces [1,2]. The existence of the interfaces differentiates composites from alloys manufactured conventionally, e.g., by melting and subsequent casting. Combining different metals (featuring advantageous formability, strength, and thermal and electrical conductivities), possibly with other materials, such as ceramics (featuring favorable hardness, strength, tolerance to high temperatures, and low thermal expansion), brings about the possibility to produce composite materials with higher utility properties than those of numerous single-phase materials and alloys [3,4].

Quite a wide spectrum of composites has been presented so far and their characterization can be performed according to various criteria. Generally, composites consist of a matrix (i.e., first phase), and (several) other phase(s), typically added to enhance the mechanical properties of the final product (i.e., reinforcing phase(s)) [5]. Therefore, composites are usually characterized according to the type of the reinforcing phase. The first type is composites with continuous reinforcing elements, typically long fibers, i.e., continuous composites, whereas the second type is composites with discontinuous reinforcing elements, typically particles, whiskers, short fibers, etc., i.e., discontinuous composites. When compared to the first type, the second type is advantageous due to its easier and cheaper production and variability of the final shape of the product achievable during possible secondary treatment (forging, rolling, extrusion, severe plastic deformation (SPD) processing, etc.) [6].

Regardless of the type of composite, the grain size and distribution of the reinforcing phase within the composite have a major influence on the effectivity of enhancement of the mechanical properties. Similar to conventional materials, among the aims of the production of composite materials is to acquire homogeneous fine-grained structures with favorable distributions of the reinforcing elements within the matrices. However, simultaneous achievement of complete consolidation (high quality bonding of the phases) and an ultra-fine-grained (UFG) structure is a challenge when performed via conventional methods, such as hot isostatic pressing (HIP) or extrusion of powder-based materials, and forging of presintered semiproducts [7].

Among the promising methods for the production of UFG composites is powder metallurgy, which can advantageously be used to fabricate microcomposites and nanocomposites when using nanopowders (the size of the individual powder particles is up to hundreds of nanometers). This production method is applicable especially for tiny components for the electrotechnics [8,9]. The main advantage of such composites is negligible final porosity and high quality of bonding of the individual powder particles. Among the positive aspects of using nanopowders is also the possibility to apply lower pressures and temperatures during their consolidation/sintering [10], which enables the application of more or less conventional forming methods for the production of such materials. However, the works documenting achievement of the required structure refinement or desired redistribution of the reinforcing elements by conventional forming technologies are scarce (e.g., [11–13]). Nevertheless, compact composites with negligible porosity and a very fine structure can be fabricated via methods of intensive plastic deformation enabling significant structure refinement (formation of UFG structure) without introducing substantial changes in shapes of the processed billets. Intensive shear deformation imparted by the combination of high pressure and rotary movement processed at room/elevated temperatures also supports mutual bonding of the used materials [14–17]. The SPD technologies mostly used to prepare composite materials are the equal channel angular pressing (ECAP) [18–20] and ECAP-based methods [21,22], high pressure torsion (HPT) [23], and accumulative roll bonding (ARB) [24]. ECAP and HPT are advantageous for the consolidation of powders, as well as for their subsequent processing to the final shapes, whereas ARB is especially favorable for enhancing structure refinement of preconsolidated materials [25], or for the preparation of clad composites [26]; the latter has been successfully performed also via the repeated folding (RF) process, the fundamentals of which are similar to ARB [27].

Despite the fact that virtually all the SPD methods are applicable for the preparation of composites, the (sub)structure development is different for multi-phase materials, and single-phase ones. For single-phase materials deformed via methods such as ECAP, substructure formation involves increasing dislocation density and development of cell blocks with large misorientations resulting in the formation of dislocation cells the size of which typically decreases with increasing imposed strain. With continuing deformation, such a substructure gradually transforms into a homogeneous structure featuring fine grains with a high portion (up to 85%) of high angle grain boundaries [28]. The minimum achievable grain size for single-phase materials deformed by SPD methods is primarily controlled by the imposed strain, its character (single pass, cyclic, etc.), and processing temperature. On the other hand, the process of (sub)structure formation within multi-phase materials, i.e., composites, deformed via SPD methods is different. Generally, the decrease in the grain size is not directly proportional to the imposed strain; decreasing thickness of the reinforcing component supports a decrease in the grain size within this component during processing. However, the application of critical deformation (i.e., imposed strain, the value of which depends on the material) leads to the formation of amorphous structures and saturated solid solutions, regardless of the used technology [29–32]. Due to the interactions of the individual phases during intensive plastic processing of composites, the achievable grain size for them is generally smaller than for the original single-phase materials; SPD methods are applicable for the preparation of nanocomposites with the average grain sizes of

10 nm [33]. Therefore, the intensive plastic deformation is promising for the preparation of innovative multi-phase composites with enhanced properties.

A specific type of composite is clad composites (sometimes characterized as layered or hybrid materials) consisting of two or more different metals bond at mutual interfaces. These materials are gaining increasing interest in various industrial branches and have been subjects to numerous studies reporting the application of forming technologies for their preparation, e.g., Al/steel/Al, Ti/Cu, Ti/Al, and Mg/stainless steel were prepared using conventional rolling [34–36]. Nevertheless, rolling is usually combined with other processing technologies, as documented, e.g., by the study reporting the application of explosive bonding and subsequent rolling for the preparation of a Ti/steel clad composite, combining the excellent strength of steel and corrosion resistance of Ti, for the petrochemical industry [37]. Given by the positive effects of the different rotation rates of the rolls on the imposed shear strain, asymmetrical rolling was used to prepare clad composites combining Al and Cu [38,39].

The combination of Al and Cu is specifically favorable since both the metallic components exhibit excellent electric and heat conductivities. Replacing a portion of Cu by Al results in a material featuring lighter weight and reduced cost when compared to Cu. Optimized processing of the composites provides advantageous materials applicable, e.g., in the aerospace and automotive [40,41], or electrotechnics [42]. Given by the combination of light weight and favorable mechanical properties, Al/Cu composites are believed to possibly supplement/replace steel in particular applications in vehicles and aircrafts [41]. The studies documented favorable durability of the Al/Cu clad composites during dynamic loading (strain rates up to 100 s^{-1}), which can primarily be attributed to the development of deformation induced twins. The intensive shear strain does not only support grain refinement, but also introduces friction at mutual interfaces (given by different plastic flows of the component metals), which consequently generates heat supporting mutual bonding of the metals via atomic diffusion. Nevertheless, increasing temperature can also result in the formation of interlayers (intermetallics), which can develop within both, the continuous and discontinuous composites.

Since the conventional forming methods and heat treatments feature certain limitations, unconventional forming, i.e., via the SPD methods, has become favorable also for the preparation of clad composites. SPD is beneficial especially from the viewpoint of introducing homogeneous distribution of secondary particles, as well as grain refinement and possibly UFG structures within the composite metals. Moreover, the shear strain supports corrupting of possible oxide layers at the surfaces of the component metals and increases diffusion via introducing lattice defects [43,44]. Numerous studies have documented substantial enhancement of the final properties of clad composites when prepared using SPD [16,17,45–47]. Probably the most promising is the HPT method, which successfully suppresses the development of cracks via applying very high pressures. It also imposes very high strains, which supports redistribution of secondary particles. For the conventional ECAP technology, the strain imposed during a single pass can be increased via modifications of the dies, or by implementing back pressure.

The twist channel angular pressing (TCAP) method combines twist and bend deformation zones within a single equal channel, which enables to impart a substantial shear strain during a single pass [48,49]. TCAP has been successfully proven to impart homogeneous UFG structures, and the effectivity of single pass TCAP has been documented to be higher than double pass ECAP [22,50,51].

Since TCAP has only been used for single-phase materials so far, this study focuses on the characterization of the grains' orientations and plastic flow within an Al/Cu clad composite processed via single and double pass TCAP. Experimental texture analyses are supplemented with numerical simulations predicting the effective imposed strain values and distribution throughout the extruded composites, as well as documenting the behavior of the billet during processing, especially the behavior of the reinforcing Cu wires within the Al sheath. Last but not least, mapping of microhardness throughout the extruded billets' cross-sections was performed.

2. Materials and Methods

The aim of the presented work was to perform detailed characterization of the influences of the applied TCAP process on the behavior of grains within an Al/Cu clad composite reinforced with Cu wires. The study primarily involves the experimental double-pass extrusion, and is supplemented with the numerical prediction of the materials' behaviors. Schematics of the TCAP process acquired from the numerical simulation can be seen in Figure 1. The clad composite consisted of an Al sheath and five reinforcing Cu wires. Both the experimental and numerically simulated billets were processed via a single pass TCAP, and subsequently via the second pass, deformation route *A* for which was selected (i.e., the billet was not rotated between the individual passes—details of the selected deformation route and strain paths can be found in [52,53]).

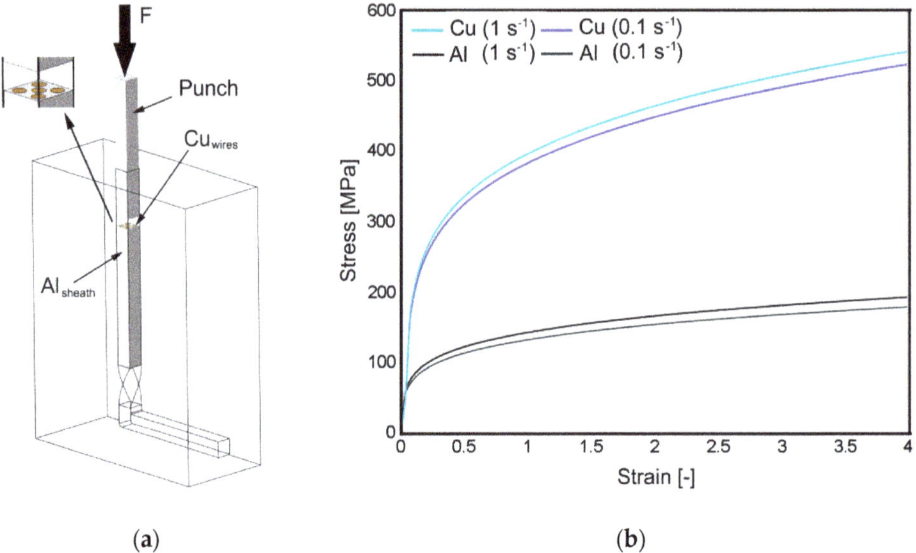

Figure 1. Schematic depiction of twist channel angular pressing (TCAP) processing of the Al/Cu clad composite (**a**) and experimental stress–strain data used for numerical simulations (**b**).

The first part of the study was focused on the practical realization of the TCAP process. The selected component metals were commercially pure (CP) Cu (99.97%) (Ferona, a.s., Prague, Czech Republic) with the composition of (in wt %): 0.0074 Ni, 0.0058 Sn, 0.0031 Fe, 0.0030 Zn, 0.0023 Si, and bal. Cu; and CP Al (99.97%) with the chemical composition of (in wt %): 0.125Fe, 0.020Cu, 0.020Zn, 0.015Mg, 0.015Mn, 0.015Ti, 0.10Si, and bal. Al. Before preparing the composite billet, all the components were annealed at 500 °C for 30 min in an electric furnace. The dimensions of the initial composite billets were 12 mm × 12 mm (square cross-section) × 130 mm (length). The processing was carried out at room temperature using a hydraulic press, MoS_2 (Ferona, a.s., Prague, Czech Republic) was applied as the lubricant. The extrusion rate was 5 mm s^{-1}.

The texture analyses of the clad composites processed via the single and double TCAP pass were performed via the SEM-EBSD method (scanning electron microscopy-electron backscatter diffraction). The samples for the analyses were prepared by manual grinding on SiC papers and subsequent manual/electrolytic polishing by a specific procedure developed by Dr. Michal Jambor, Institute of Physics of Materials, CAS. SEM-EBSD analyses were done using a Tescan Lyra 3 FIB/SEM microscope equipped with a NordlysNano EBSD detector (Oxford Instruments, Abingdon-on-Thames, Great Britain). EBSD scanning was carried out with the accelerating voltage of 20 kV and scan steps of 50 nm. The structure analyses and texture evaluations were performed using AZtecCrystal 1.1

software, and ATEX software [54]. The mechanical properties of both the extruded composites were investigated by microhardness measurements on cross-sectional cuts from the billets (perpendicular to extrusion axis) using a Zwick/Roell testing machine. The indents spacing was 1 mm and the applied parameters per indent were: load of 200 g and loading time of 10 s. The maps were assembled by the own script programmed by Adam Weiser, Institute of Physics of Materials, CAS.

The experimental study was supplemented with numerical simulations of the single and double pass extrusions. The behavior of the composite billet during single and double pass was predicted using the Forge NxT commercial software by Transvalor implementing the finite element method (FEM). The geometrical dimensions and mechanical properties in the numerical model corresponded to the real experimental parameters. The friction was determined as the Coulomb friction ($\mu = 0.02$) and the boundary conditions specified in Table 1 were defined on the basis of a previously performed study in which they were experimentally validated [52]. The geometry of the TCAP die used in both the simulation and experiment was defined by the following angles: $\omega = 90°$, $\varphi = 90°$, $\beta = 40°$, and $\psi = 20°$ (twist rotation angle, channel bending angle, twist slope angle, and arc of curvature of channels' intersection angle, respectively—the angles are described in detail, e.g., in [48,49]). The component metals were characterized using an elastic–plastic model defined by the Newton–Raphson convergent algorithm, and the die and extruder were defined as rigid parts. The clad composite billet was meshed with tetrahedral elements (215, 871 nodes in total), and automatic remeshing was activated since intensive shear deformations were expected to proceed. The room temperature stress–strain data at the strain rates of 0.1 and 1 s^{-1} (see Figure 2) imported into the database of the computational software was acquired experimentally for both the used composite metals via torsion tests performed using a servo-hydraulic torsion plastometer (SETARAM). Material behavior was finally determined via the Hansel–Spittel equation (Equation (1)),

$$\sigma_f = Ae^{m_1 T} T^{m_8} \varepsilon^{m_2} e^{m_4/\varepsilon} (1+\varepsilon)^{m_5 T} e^{m_6 \varepsilon} \dot{\varepsilon}^{m_3} \dot{\varepsilon}^{m_7 T}, \quad (1)$$

where ε is the equivalent strain, T is the temperature, $\dot{\varepsilon}$ is the equivalent strain rate, and A, m_1, m_2, m_3, m_4, m_5, m_6, m_7, m_8, and m_9 are regression coefficients. The values of the coefficients for Cu were 411.19 MPa, −0.00121, 0.21554, 0.01472, and −0.00935, respectively, and $m_5 \div m_8$ was 0. The values of the coefficients for Al were 151.323 MPa, −0.00253, 0.21142, 0.03177, and −0.00654, respectively, and $m_5 \div m_8$ was 0.

Table 1. Defined boundary conditions for component metals.

Property	Unit	Al	Cu
Temperature	°C	25	25
Young's modulus	GPa	72	111
Poisson coefficient	-	0.3	0.3
Thermal expansion	K^{-1}	24×10^{-5}	1.7×10^{-5}
Thermal conductivity	(W/(m.K))	250	394
Specific heat	($J.kg^{-1}.K^{-1}$)	1230	398
Emissivity	-	0.03	0.7
Density	($g.cm^{-3}$)	2.80	8.96

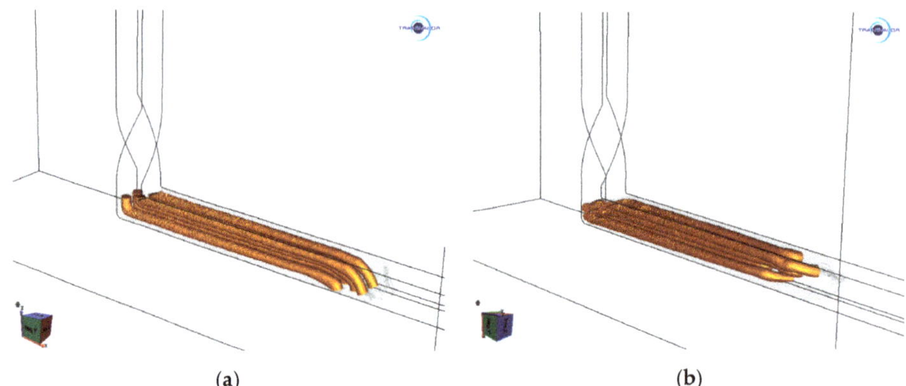

Figure 2. Deformation behavior of the extruded billet during: single pass (a) and double pass (b).

3. Results

3.1. Numerically Predicted Plastic Flow and Imposed Strain

The results of the numerical simulation showed that the billets extruded via single and double pass TCAP exhibited differences in their deformation behaviors, as did both the component metals. Figure 2a,b shows images from the simulations, the billets extruded via one and two TCAP passes in which were depicted, respectively. In Figure 2a,b, the Al sheath was visible as a contour so that the shapes of the Cu wires could clearly be seen. Both the TCAP passes imparted significant plastic deformation of both the composite components. However, differences between the single and double pass were observed, especially for the Cu wires.

As can be seen from the prediction, the ends of the Cu wires were bent downwards to the bottom of the horizontal channel after the single pass (Figure 2a), which originated when passing through the channel bending, i.e., the main deformation zone (MDZ). Additionally, the two upper wires were shifted forward after extrusion. This phenomenon can be attributed to the fact that, in the MDZ, the material plastic flow in the upper cross-sectional region of the composite billet was faster than in its bottom cross-sectional region. Additionally, the shear strain imposed to the composite in the MDZ was more significant along the inner periphery of the channel bending than along its outer periphery [55].

After the simulated double pass, the Cu wires were bent towards the side of the horizontal channel (Figure 2b). This can primarily be attributed to the selected deformation route (*A*, see Section 2 and references. [52,53]), which involved no rotation between the passes. By the effect of the presence of the twist deformation zone (TDZ), the locations of the four peripheral Cu wires changed during the second extrusion, i.e., the billet cross-section rotated by 90° compared to the cross-section of the billet extruded via a single pass. In other words, the Cu wire located in the upper left corner of the composite after the first pass shifted to the bottom left corner after the second pass, etc.

The development of the imposed effective strain during a TCAP pass is rather complicated. The TDZ primarily affects the peripheral/corner regions of the extruded billet, by the effect of which the imposed strain did not influence the reinforcing Cu wires as much as the (periphery of) Al sheath in this deformation zone. The Cu wires were not significantly affected by the shear strain during passing through the TDZ, but later on, during passing through the MDZ. As depicted in the cross-sectional cut through the simulated billet extruded via a single pass in Figure 3a, the Al sheath exhibited homogeneous imposed effective strain distribution of rather high values (up to 2.5) when compared to the Cu wires, the maximum effective strain values in which reached up to 1.5. Moreover, the imposed strain distribution through the wires' cross-sections was not homogeneous after single pass TCAP. Both phenomena were caused by the significantly lower effect of the TDZ on the wires (compared to the Al sheath).

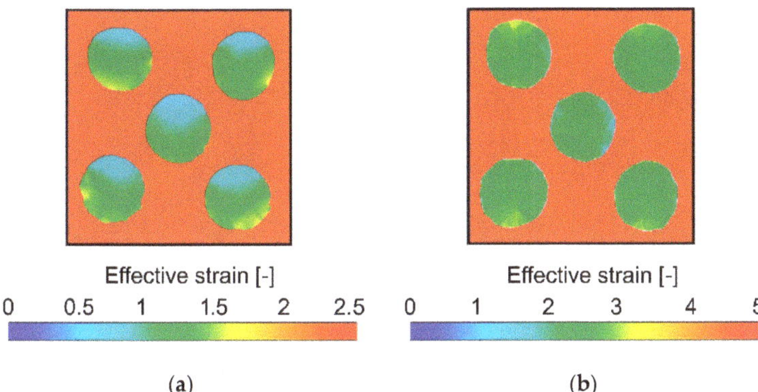

Figure 3. Effective strain distribution after: single pass (**a**) and double pass (**b**).

The prediction showed that after the double pass, the effect of the TDZ on the imposed strain was more evident than after the single pass for both the composite metals. As can be seen in Figure 3b depicting a cross-sectional cut through the billet extruded via double pass TCAP, the Al sheath exhibited the homogeneous distribution of the effective imposed strain (similar to the single pass), however, the maximum imposed strain values almost doubled (up to 5). The Cu wires exhibited not only an increase in the maximum imposed strain after the second pass, but also substantial homogenization of its distribution (please note the different scales, although the color scheme is maintained for both the Figure 3a,b—decreasing the range of the scale in Figure 3a enables better depiction of the inhomogeneities of the effective imposed strain within the Cu wires observed after the first TCAP pass).

3.2. Texture Characteristics of Extruded Composites

The effects of the TDZ and MDZ on texture developments within the experimentally extruded composite billets were evaluated via pole figures (PF), and via determination of the intensities of dominant ideal shear texture orientations [56–58]. Since the effects of the TCAP process on texture development within the Al sheath were comparable to its effects on texture development within a CP Al billet, which were characterized in the previously published study [50], this study primarily focuses on the Cu wires.

The original preannealed Cu exhibited more or less random texture with a slight tendency to form ideal cube recrystallization texture (not shown here). The texture orientations within the individual wires exhibited differences after the first pass—the individual {001}, {011}, and {111} PFs for the wires in the upper left, upper right, axial, bottom left, and bottom right positions are depicted in Figure 4a–e, respectively. As can be seen, the highest texture intensity was detected in the axial wire (Figure 4c), which exhibited dominant A fiber ideal shear texture orientation (i.e., {111}||shear plane) [56]. Regarding the peripheral wires, the pairs of the wires positioned across the cross-sectional diagonals exhibited tendencies to mirror. The bottom left (Figure 4d) and upper right (Figure 4b) wires primarily formed the C fiber ideal shear texture orientation (although by approximately 10° shifted), sometimes also denoted as the dominant α fiber defined by the Euler angles φ_1, ϕ, and φ_2 of 90°, 45°, and 0° [57,58], and the upper left (Figure 4a) and bottom right (Figure 4e) wires exhibited the tendencies to preferably form the Ab shear texture orientation (belonging to the A fiber, defined by the Euler angles φ_1, ϕ, and φ_2 of 180°, 35.26°, and 45°, respectively). However, the Ab ideal orientation was shifted towards the Bb ideal orientation (Euler angles φ_1, ϕ, and φ_2 of 180°, 54.74°, and 45°), especially within the upper left wire (Figure 4a).

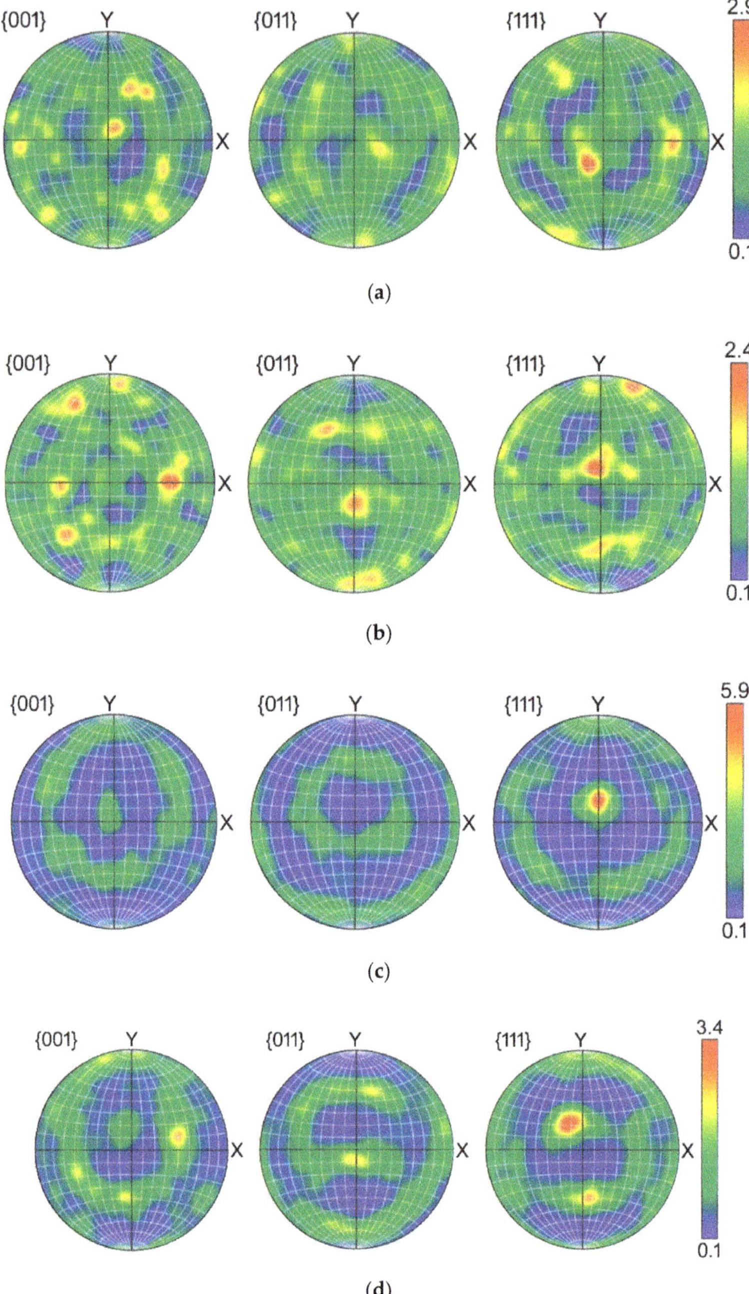

Figure 4. *Cont.*

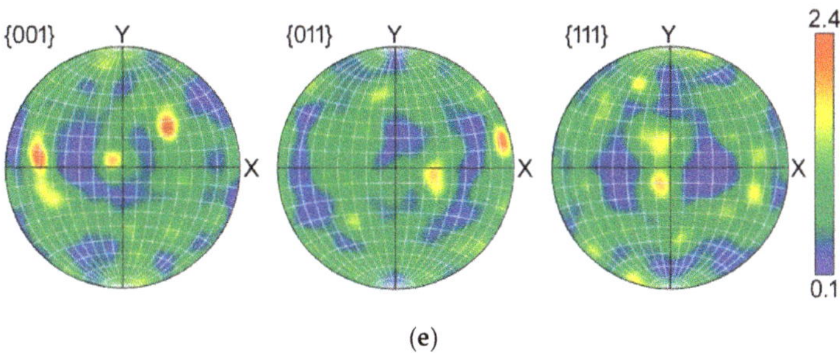

(**e**)

Figure 4. Pole figures for individual wires within composite billet extruded via single pass TCAP: upper left (**a**); upper right (**b**); axial (**c**); bottom left (**d**); and bottom right (**e**).

After the experimental double pass TCAP, homogenization of the textures within the wires occurred. Figure 5a–e depicts the distributions of the main ideal texture orientations within the individual wires after the second TCAP pass. All the wires exhibited dominant texture orientations belonging to the A fiber, however, the suborientations varied slightly (orientations belonging to the B and C fibers are not depicted since their volume fractions were neglectable). A1 and A2 ideal texture components (Euler angles φ_1, ϕ, and φ_2 of 35.26°, 45°, and 0°; and 144.74°, 45°, and 0°, respectively) were dominant in all the wires after the second TCAP pass, except the upper right wire (Figure 5b), Ab ideal orientation for which was dominant. This wire was located in the bottom right position after the first pass (after which the texture already exhibited the tendency to form the Ab ideal orientation, see Figure 4e).

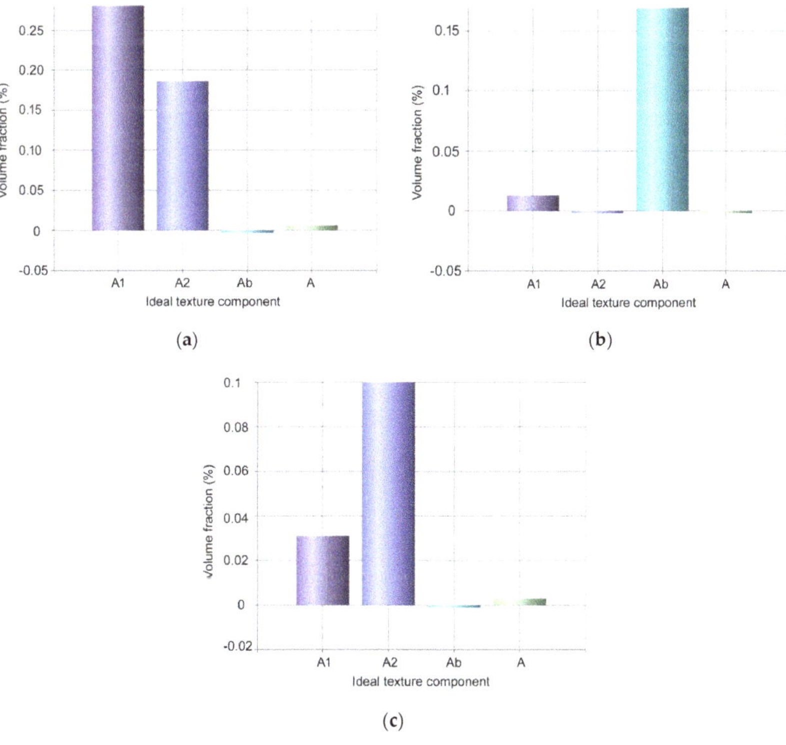

Figure 5. *Cont.*

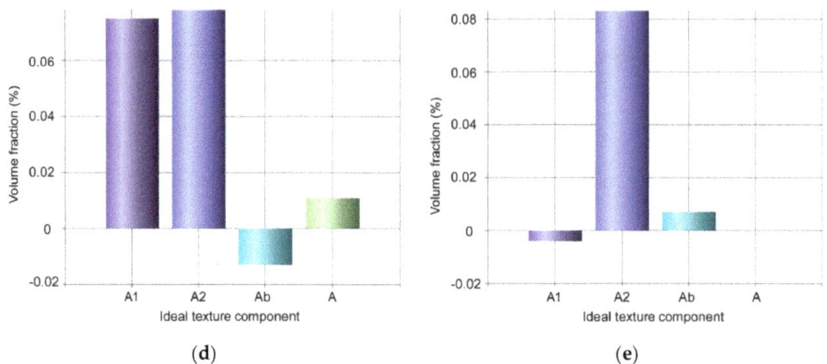

Figure 5. Volume fractions of dominant ideal texture orientations for individual wires within composite billet extruded via double pass TCAP: upper left (**a**); upper right (**b**); axial (**c**); bottom left (**d**); and bottom right (**e**).

3.3. Microhardness and Structure of Extruded Composites

The deformation strengthening introduced by the intensive shear strain and manifesting itself in macrodeformation and bending of the ends of the Cu wires (Section 3.1) contributed to increases in microhardness for both the experimentally extruded composite components. The original HV values of the initial annealed components were 58.6 for the Cu wires, and 37.4 for the Al sheath. Figure 6a,b depicts the maps of microhardness measured across the cross-sections of both the extruded billets. As can be seen, the microhardness values increased after both the single and double pass TCAP. After single pass TCAP, the average microhardness value of the Al sheath increased to 85 HV, and for the Cu wires it increased up to 107 HV. After the second pass, the average value of the Al was comparable to the first pass (86 HV), while for the Cu wires the average HV value increased to 128 HV.

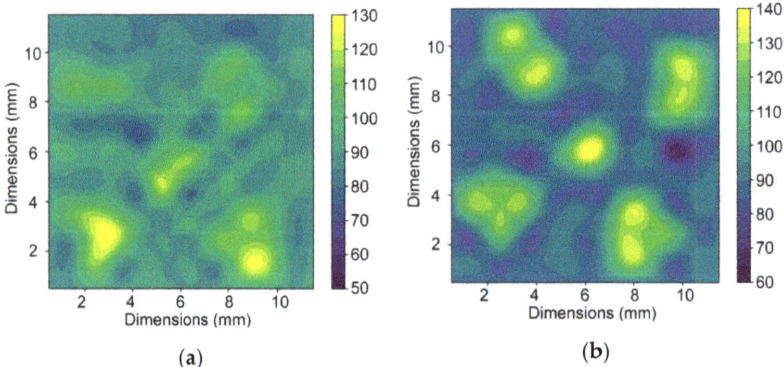

Figure 6. Maps of experimentally measured microhardness across a cross-section of billet after: single pass (**a**) and double pass (**b**).

The results point to the fact that the Cu was not strengthened to its limit after the single pass, i.e., its structure was not saturated [59], by the effect of which the microhardness still increased after the second pass. On the other hand, the Al sheath featured structure saturation already after the first pass, and then exhibited structure recovery/recrystallization at the expense of further strengthening after the second pass. This supposition is in accordance with the grain refinement of the Al sheath occurring after both the passes. After single TCAP pass, the average grain size within the Al sheath was 5.4 μm, which was comparable with the values acquired during the previous study on a CP Al billet [50,51]. However, after the second pass, the average grain size within the Al sheath decreased down to 1.75 μm (see the orientation image map—OIM—and grain size distribution in Figure 7a,b). For comparison, the OIMs for the axial wire within the billets extruded via the single and double TCAP pass are depicted in Figure 7c,d, respectively, clearly showing that the grain size of the Cu wires was larger than within the Al sheath.

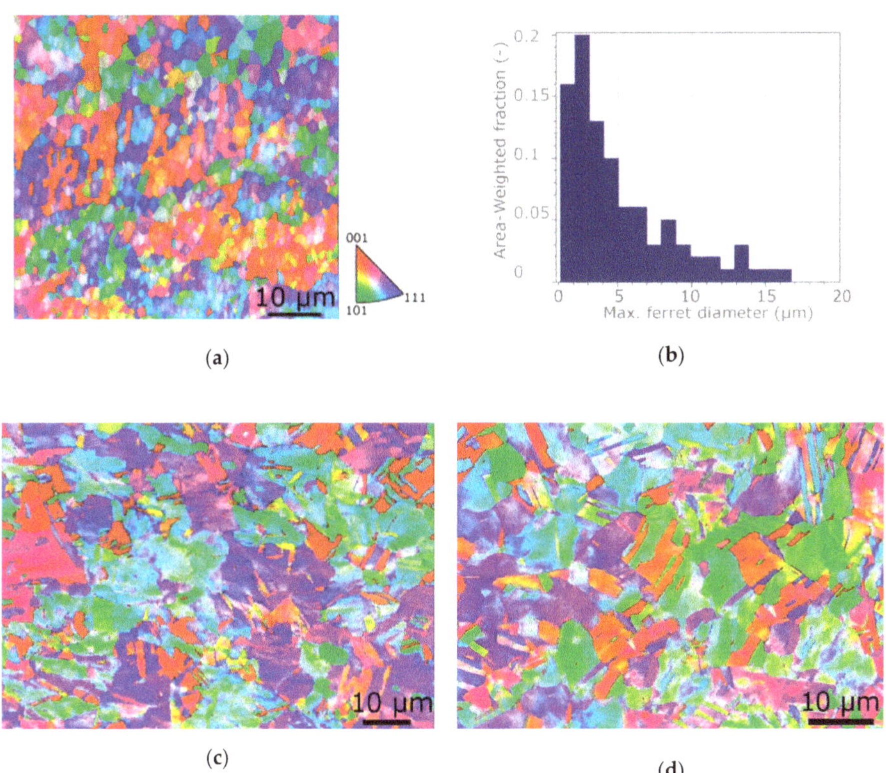

Figure 7. Structure of Al sheath after double pass TCAP: orientation image map (OIM) (**a**); grain size distribution (**b**); OIM of axial Cu wire after: Single TCAP (**c**); and double TCAP (**d**).

4. Discussion

Both the predicted and experimental results revealed that the deformation behavior of the composite billet was affected by the fact that it consisted of a combination of two different component metals. However, the individual deformation zones implemented within the die and the selected deformation route had non-negligible effects on the behavior of the billet, too. Passing through the TDZ did not generally introduce as high strains as passing through the MDZ [49]; the MDZ affects primarily the composite axial region, while the TDZ imposes high effective strain primarily to the composite (sub)surface regions. The TDZ thus imposed the shear strain primarily to the Al sheath, the plastic flow in the outer region of which was quicker than in the Cu wires. Subsequently, the MDZ contributed to the increase in the imposed shear strain also within the wires. The predicted bendings of the ends of the Cu wires after both the TCAP passes originating from different plastic flows of the individual regions of the composite billet developed during passing through the MDZ, where the upper cross-sectional region of the Al sheath adjoining to the Cu wires exhibited a tendency to flow quicker than the bottom cross-sectional region of the billet.

The development of texture during TCAP was affected by the acting strain path along which the shear strain was imposed [50,51]. As described above, the TCAP die is characterized by two independent deformation zones, twist zone (analogy to twist extrusion) two independent intersection planes in which are active, and bending zone (analogy to ECAP) another intersection plane in which is active. The TCAP die thus introduces severe shear strain into the extruded composite along three independent intersection planes, which also introduces a high amount of lattice distortions acting as obstacles for the movement of dislocations, and generation of nucleation sites during substructure formation (both these occurring phenomena were confirmed via the observed microhardness increase, and substantial grain refinement, see Figures 6 and 7). After the first TCAP pass, the axial wire exhibited dominant A fiber texture orientation, which corresponds with the supposition of the minor effect of the TDZ on this wire during the first pass, since the A fiber is an ideal shear texture orientation developing during conventional ECAP. In other words, this finding documents the dominant effect of MDZ on this wire. The texture orientations identified in the peripheral wires were primarily affected by the two intersection planes along which the shear strain was imposed to the composite in the TDZ, especially as regards their tendency to mirror across the cross-sectional diagonals. Nevertheless, double pass TCAP resulted in homogenization of texture, as all the wires exhibited the tendencies to form A fiber dominant shear texture orientation.

5. Conclusions

The study presented the results of experimental analyses, supplemented with FEM numerical prediction, of an Al/Cu clad composite prepared by room temperature extrusion via the twist channel angular pressing (TCAP) method (single and double pass). The numerical prediction showed that both the passes introduced severe shear strain to both the component metals; the maximum effective imposed strain within the Al sheath of the billet processed via the double pass reached to the value of 5. The severe imposed strain subsequently introduced significant grain refinement and deformation strengthening (especially after the second pass) resulting in the increase in microhardness up to 128 HV for the Cu wires. The results also documented that the second pass introduced homogenization of the distribution of the imposed strain within the Al sheath and Cu wires, as well as uniform tendencies to form preferential texture orientations. The Cu wires exhibited the tendency to form A fiber preferential texture orientation after the second pass, although after the first pass, one pair of the peripheral wires exhibited the formation of A fiber texture, while the second pair formed C fiber texture.

Author Contributions: The contributions of the individual authors were the following: methodology, L.K. and Z.K.; numerical prediction and evaluation, Z.K.; experimental validation, L.K.; microscopy investigation and evaluation, L.K.; writing—original draft preparation, L.K.; writing—review and editing, L.K.; project administration, L.K.; funding acquisition, Z.K. All authors have read and agreed to the published version of the manuscript.

Funding: This research was supported by the FV40286 TRIO program by Ministry of Industry and Trade.

Acknowledgments: The authors would like to appreciate the kind cooperation of Michal Jambor, Adam Weiser, and the LaPAMat team of Institute of Physics of Materials, CAS, during preparation and evaluation of experimental material.

Conflicts of Interest: The authors declare no conflict of interest. The funders had no role in the design of the study; in the collection, analyses, or interpretation of data; in the writing of the manuscript, or in the decision to publish the results.

References

1. Matthews, F.L.; Rawlings, R.D. *Composite Materials: Engineering and Science*; CRC Press: Boca Raton, FL, USA, 1999.
2. Kocich, R.; Kunčická, L.; Král, P.; Strunz, P. Characterization of innovative rotary swaged Cu-Al clad composite wire conductors. *Mater. Des.* **2018**, *160*, 828–835. [CrossRef]

3. Kunčická, L.; Kocich, R.; Dvořák, K.; Macháčková, A. Rotary swaged laminated Cu-Al composites: Effect of structure on residual stress and mechanical and electric properties. *Mater. Sci. Eng. A* **2019**, *742*, 743–750. [CrossRef]
4. Kunčická, L.; Kocich, R. Deformation behaviour of Cu-Al clad composites produced by rotary swaging. *IOP Conf. Ser. Mater. Sci. Eng.* **2018**, *369*, 012029. [CrossRef]
5. Clyne, T.W.; Withers, P.J. *An Introduction to Metal Matrix Composites*; Cambridge University Press: New York, NY, USA, 1993.
6. Tjong, S.; Ma, Z. Microstructural and mechanical characteristics of in situ metal matrix composites. *Mater. Sci. Eng. R Rep.* **2000**, *29*, 49–113. [CrossRef]
7. Viswanathan, V.; Laha, T.; Balani, K.; Agarwal, A.; Seal, S. Challenges and advances in nanocomposite processing techniques. *Mater. Sci. Eng. R Rep.* **2006**, *54*, 121–285. [CrossRef]
8. Moya, J.S.; Lopez-Esteban, S.; Pecharromán, C. The challenge of ceramic/metal microcomposites and nanocomposites. *Prog. Mater. Sci.* **2007**, *52*, 1017–1090. [CrossRef]
9. Meilakh, A.G.; Kontsevoi, Y.V.; Shubin, A.B.; Pastukhov, E.A. Steel composites based on copper-clad iron powder. *Steel Transl.* **2015**, *45*, 712–715. [CrossRef]
10. Wang, D.; Dong, X.; Zhou, P.; Sun, A.; Duan, B. The sintering behavior of ultra-fine Mo–Cu composite powders and the sintering properties of the composite compacts. *Int. J. Refract. Met. Hard Mater.* **2014**, *42*, 240–245. [CrossRef]
11. Rhee, K.Y.; Han, W.Y.; Park, H.J.; Kim, S.S. Fabrication of aluminum/copper clad composite using hot hydrostatic extrusion process and its material characteristics. *Mater. Sci. Eng. A* **2004**, *384*, 70–76. [CrossRef]
12. Kwon, H.C.; Jung, T.K.; Lim, S.C.; Kim, M.S. Fabrication of Copper Clad Aluminum Wire (CCAW) by Indirect Extrusion and Drawing. *Mater. Sci. Forum* **2004**, *449–452*, 317–320. [CrossRef]
13. Mehr, V.Y.; Toroghinejad, M.R.; Rezaeian, A. The effects of oxide film and annealing treatment on the bond strength of Al–Cu strips in cold roll bonding process. *Mater. Des.* **2014**, *53*, 174–181. [CrossRef]
14. Kunčická, L.; Kocich, R.; Strunz, P.; Macháčková, A. Texture and residual stress within rotary swaged Cu/Al clad composites. *Mater. Lett.* **2018**, *230*, 88–91. [CrossRef]
15. Sapanathan, T.; Khoddam, S.; Zahiri, S.H.; Zarei-Hanzaki, A. Strength changes and bonded interface investigations in a spiral extruded aluminum/copper composite. *Mater. Des.* **2014**, *57*, 306–314. [CrossRef]
16. Zebardast, M.; Taheri, A.K. The cold welding of copper to aluminum using equal channel angular extrusion (ECAE) process. *J. Mater. Process. Technol.* **2011**, *211*, 1034–1043. [CrossRef]
17. Lapovok, R.; Ng, H.P.; Tomus, D.; Estrin, Y. Bimetallic copper–aluminium tube by severe plastic deformation. *Scr. Mater.* **2012**, *66*, 1081–1084. [CrossRef]
18. Kocich, R.; Szurman, I.; Kursa, M.; Fiala, J. Investigation of influence of preparation and heat treatment on deformation behaviour of the alloy NiTi after ECAE. *Mater. Sci. Eng. A* **2009**, *512*, 100–104. [CrossRef]
19. Naizabekov, A.B.; Andreyachshenko, V.A.; Kocich, R. Study of deformation behavior, structure and mechanical properties of the AlSiMnFe alloy during ECAP-PBP. *Micron* **2013**, *44*, 210–217. [CrossRef]
20. Hlaváč, L.M.; Kocich, R.; Gembalová, L.; Jonšta, P.; Hlaváčová, I.M. AWJ cutting of copper processed by ECAP. *Int. J. Adv. Manuf. Technol.* **2016**, *86*, 885–894. [CrossRef]
21. Kunčická, L.; Kocich, R.; Drápala, J.; Andreyachshenko, V.A. FEM simulations and comparison of the ecap and ECAP-PBP influence on Ti6Al4V alloy's deformation behaviour. In Proceedings of the 22nd International Metallurgy and Materials Conference, Brno, Czech Republic, 15–17 May 2013; pp. 391–396.
22. Kunčická, L.; Kocich, R.; Ryukhtin, V.; Cullen, J.C.T.; Lavery, N.P. Study of structure of naturally aged aluminium after twist channel angular pressing. *Mater. Charact.* **2019**, *152*, 94–100. [CrossRef]
23. Kunčická, L.; Lowe, T.C.; Davis, C.F.; Kocich, R.; Pohludka, M. Synthesis of an Al/Al_2O_3 composite by severe plastic deformation. *Mater. Sci. Eng. A* **2015**, *646*, 234–241. [CrossRef]
24. Kocich, R.; Macháčková, A.; Fojtík, F. Comparison of strain and stress conditions in conventional and ARB rolling processes. *Int. J. Mech. Sci.* **2012**, *64*, 54–61. [CrossRef]
25. Bachmaier, A.; Pippan, R. Generation of metallic nanocomposites by severe plastic deformation. *Int. Mater. Rev.* **2013**, *58*, 41–62. [CrossRef]
26. Huang, X.; Tsuji, N.; Hansen, N.; Minamino, Y. Microstructural evolution during accumulative roll-bonding of commercial purity aluminum. *Mater. Sci. Eng. A* **2003**, *340*, 265–271. [CrossRef]
27. Dinda, G.P.; Rösner, H.; Wilde, G. Synthesis of bulk nanostructured Ni, Ti and Zr by repeated cold-rolling. *Scr. Mater.* **2005**, *52*, 577–582. [CrossRef]

28. Pippan, R.; Scheriau, S.; Taylor, A.; Hafok, M.; Hohenwarter, A.; Bachmaier, A. Saturation of Fragmentation during Severe Plastic Deformation. *Annu. Rev. Mater. Res.* **2010**. [CrossRef]
29. Tolaminejad, B.; Taheri, A.K.; Shahmiri, M.; Arabi, H. Development of Crystallographic Texture and Grain Refinement in the Aluminum Layer of Cu-Al-Cu Tri-Layer Composite Deformed by Equal Channel Angular Extrusion. *Int. J. Mod. Phys. Conf. Ser.* **2012**, *5*, 325–334. [CrossRef]
30. Suehiro, K.; Nishimura, S.; Horita, Z.; Mitani, S.; Takanashi, K.; Fujimori, H. High-pressure torsion for production of magnetoresistance in Cu–Co alloy. *J. Mater. Sci.* **2008**, *43*, 7349–7353. [CrossRef]
31. Ohsaki, S.; Kato, S.; Tsuji, N.; Ohkubo, T.; Hono, K. Bulk mechanical alloying of Cu–Ag and Cu/Zr two-phase microstructures by accumulative roll-bonding process. *Acta Mater.* **2007**, *55*, 2885–2895. [CrossRef]
32. Wilde, G.; Rösner, H. Stability aspects of bulk nanostructured metals and composites. *J. Mater. Sci.* **2007**, *42*, 1772–1781. [CrossRef]
33. Sabirov, I.; Schöberl, T.; Pippan, R. Fabrication of a W-25%Cu Nanocomposite by high pressure torsion. *Mater. Sci. Forum* **2006**, *503–504*, 561–566. [CrossRef]
34. Kim, W.N.; Hong, S.I. Interactive deformation and enhanced ductility of tri-layered Cu/Al/Cu clad composite. *Mater. Sci. Eng. A* **2016**, *651*, 976–986. [CrossRef]
35. Movahedi, M.; Kokabi, A.H.; Reihani, S.M.S. Investigation on the bond strength of Al-1100/St-12 roll bonded sheets, optimization and characterization. *Mater. Des.* **2011**, *32*, 3143–3149. [CrossRef]
36. Hosseini, M.; Manesh, H.D. Bond strength optimization of Ti/Cu/Ti clad composites produced by roll-bonding. *Mater. Des.* **2015**, *81*, 122–132. [CrossRef]
37. Polyanskii, S.N.; Kolnogorov, V.S. Cladded steel for the oil and gas industries. *Chem. Pet. Eng.* **2002**, *38*, 703–707. [CrossRef]
38. Li, X.; Zu, G.; Ding, M.; Mu, Y.; Wang, P. Interfacial microstructure and mechanical properties of Cu/Al clad sheet fabricated by asymmetrical roll bonding and annealing. *Mater. Sci. Eng. A* **2011**, *529*, 485–491. [CrossRef]
39. Li, X.; Zu, G.; Wang, P. Effect of strain rate on tensile performance of Al/Cu/Al laminated composites produced by asymmetrical roll bonding. *Mater. Sci. Eng. A* **2013**, *575*, 61–64. [CrossRef]
40. Matli, P.R.; Fareeha, U.; Shakoor, R.A.; Mohamed, A.M.A. A comparative study of structural and mechanical properties of Al–Cu composites prepared by vacuum and microwave sintering techniques. *J. Mater. Res. Technol.* **2018**, *7*, 165–172. [CrossRef]
41. Matvienko, Y.I.; Polishchuk, S.S.; Rud, A.D.; Popov, O.Y.; Demchenkov, S.A.; Fesenko, O.M. Effect of graphite additives on microstructure and mechanical properties of Al–Cu composites prepared by mechanical alloying and sintering. *Mater. Chem. Phys.* **2020**, *254*, 123437. [CrossRef]
42. Zhang, J.; Wang, B.; Chen, G.; Wang, R.; Miao, C.; Zheng, Z.; Tang, W. Formation and growth of Cu–Al IMCs and their effect on electrical property of electroplated Cu/Al laminar composites. *Trans. Nonferrous Met. Soc. Chin.* **2016**, *26*, 3283–3291. [CrossRef]
43. Bouaziz, O.; Bréchet, Y.; Embury, J.D. Heterogeneous and architectured materials: A possible strategy for design of structural materials. *Adv. Eng. Mater.* **2008**, *10*, 24–36. [CrossRef]
44. Martinsen, K.; Hu, S.J.; Carlson, B.E. Joining of dissimilar materials. *CIRP Ann.* **2015**, *64*, 679–699. [CrossRef]
45. Bouaziz, O.; Kim, H.S.; Estrin, Y. Architecturing of metal-based composites with concurrent nanostructuring: A new paradigm of materials design. *Adv. Eng. Mater.* **2013**, *15*, 336–340. [CrossRef]
46. Khoddam, S.; Estrin, Y.; Kim, H.S.; Bouaziz, O. Torsional and compressive behaviours of a hybrid material: Spiral fibre reinforced metal matrix composite. *Mater. Des.* **2015**, *85*, 404–411. [CrossRef]
47. Khoddam, S.; Sapanathan, T.; Zahiri, S.; Hodgson, P.D.; Zarei-Hanzaki, A.; Ibrahim, R. Inner architecture of bonded splats under combined high pressure and shear. *Adv. Eng. Mater.* **2016**, *18*, 501–505. [CrossRef]
48. Kocich, R.; Greger, M.; Kursa, M.; Szurman, I.; Macháčková, A. Twist channel angular pressing (TCAP) as a method for increasing the efficiency of SPD. *Mater. Sci. Eng. A* **2010**, *527*, 6386–6392. [CrossRef]
49. Kocich, R.; Kunčická, L.; Mihola, M.; Skotnicová, K. Numerical and experimental analysis of twist channel angular pressing (TCAP) as a SPD process. *Mater. Sci. Eng. A* **2013**, *563*, 86–94. [CrossRef]
50. Kocich, R.; Kunčická, L.; Král, P.; Macháčková, A. Sub-structure and mechanical properties of twist channel angular pressed aluminium. *Mater. Charact.* **2016**, *119*, 75–83. [CrossRef]
51. Kunčická, L.; Kocich, R.; Král, P.; Pohludka, M.; Marek, M. Effect of strain path on severely deformed aluminium. *Mater. Lett.* **2016**, *180*, 280–283. [CrossRef]

52. Kocich, R.; Fiala, J.; Szurman, I.; Macháčková, A.; Mihola, M. Twist-channel angular pressing: Effect of the strain path on grain refinement and mechanical properties of copper. *J. Mater. Sci.* **2011**, *46*, 7865–7876. [CrossRef]
53. Macháčková, A. Decade of twist channel angular pressing: A review. *Materials* **2020**, *13*, 1725. [CrossRef]
54. Beausir, B.; Fundenberger, J.J. Analysis Tools for Electron and X-ray Diffraction. ATEX—Software. 2017. Available online: www.atex-software.eu (accessed on 20 July 2020).
55. Kocich, R.; Greger, M.; Macháčková, A. Finite element investigation of influence of selected factors on ECAP process. In Proceedings of the METAL 2010—19th International Conference on Metallurgy and Materials, Roznov pod Radhostem, Czech Republic, 18–20 May 2010; Tanger Ltd.: Ostrava, Czech Republic, 2010; pp. 166–171.
56. Beyerlein, I.J.; Tóth, L.S. Texture evolution in equal-channel angular extrusion. *Prog. Mater. Sci.* **2009**, *54*, 427–510. [CrossRef]
57. Li, S.; Beyerlein, I.J.; Alexander, D.J.; Vogel, S.C. Texture evolution during multi-pass equal channel angular extrusion of copper: Neutron diffraction characterization and polycrystal modeling. *Acta Mater.* **2005**, *53*, 2111–2125. [CrossRef]
58. Engler, O.; Randle, V. *Introduction to Texture Analysis, Macrotexture, Microtexture, and Orientation Mapping*, 2nd ed.; Taylor and Francis Group: Milton, UK, 2010.
59. Russell, A.; Lee, K.L. *Structure-Property Relations in Nonferrous Metals*, 1st ed.; John Wiley & Sons, Inc.: Hoboken, NJ, USA, 2005.

Article

Study on Strengthening Mechanism of 9Cr-1.5Mo-1Co and 9Cr-3W-3Co Heat Resistant Steels

Long Zhao [1], Xiangru Chen [1], Tieming Wu [2] and Qijie Zhai [1,*]

1 Center for Advanced Solidification Technology (CAST), School of Materials Science and Engineering, Shanghai University, Shanghai 200444, China; zhaolongXYZ@shu.edu.cn (L.Z.); chenxr@shu.edu.cn (X.C.)
2 Shanghai Electric Group Company Limited, Shanghai 200336, China; wutm2@shanghai-electric.com
* Correspondence: qjzhai@shu.edu.cn

Received: 31 August 2020; Accepted: 24 September 2020; Published: 29 September 2020

Abstract: The strengthening mechanism of 9Cr–1.5Mo–1Co and 9Cr–3W–3Co heat resistant steel was studied by tensile test and microstructure analysis. At the same temperature, the yield strength of 9Cr–3W–3Co heat-resistant steel is higher than that of 9Cr–1.5Mo–1Co heat-resistant steel. Microstructure analysis proved that the strength of 9Cr–1.5Mo–1Co and 9Cr–3W–3Co heat-resistant steel is affected by grain boundary, dislocation, precipitation, and solid solution atoms. The excellent high temperature mechanical properties of 9Cr–3W–3Co heat-resistant steel are mainly due to the solution strengthening caused by Co and W atoms and the high-density dislocations distributed in the matrix; however, 9Cr–1.5Mo–1Co heat-resistant steel is mainly due to the martensitic lath and precipitation strengthening.

Keywords: heat-resistant steel; cast steel; microalloying; strengthening mechanism

---

1. Introduction

The consumption of coal is the main source of air pollution, especially the emission of CO_2, which leads to the increasingly serious greenhouse effect. It is a hot topic in today's society how to reduce greenhouse gas emissions and environmental pressure [1,2]. Thermal power plant is the main enterprise of coal resource consumption. In order to reduce CO_2 emissions under ensuring social and economic development, it is necessary to improve the efficiency of thermal power generation. To achieve this goal, it is necessary to the increase the temperature and pressure of the main steam engine, which puts forward higher requirements for materials [3–7]. Because of its high creep strength, good temperature oxidation resistance, excellent thermal conductivity, low thermal expansion coefficient, and low sensitivity to thermal fatigue performance, 9–12% Cr martensitic heat-resistant steel has become the heat-resistant material in thermal power plants [8–10].

The unique tempered martensite microstructure of 9–12% Cr heat-resistant steel is the key to keep good mechanical properties at a high temperature. The strengthening mechanisms are mainly grain boundary strengthening, precipitation strengthening, dislocation strengthening, and solution strengthening [11,12]. The grain boundary of 9–12% Cr martensitic heat-resistant steel is mainly composed of original austenite grain boundary, sub grain boundary, and martensitic lath boundary, whose properties are directly related to the mechanical properties of heat-resistant steel. In addition, the high-density dislocations in martensitic heat-resistant steel play an important role in improving the mechanical properties of heat-resistant steel [13]. Thirdly, the precipitates with different sizes can hinder the movement of dislocations and grain boundaries, such as $M_{23}C_6$ carbides and MX carbonitrides [14–16], which makes the material from high temperature tensile property and creep resistance [17,18]. Finally, large and small size solute atoms exist in the matrix, which plays a solid solution strengthening role in heat-resistant steel.

Lots of work has been done on the strengthening mechanism of 9–12% Cr martensitic heat-resistant steel. Yao Du et al. [19] found that the formation of MX precipitate is the main reason for the strengthening of 0.032C–9Cr–0.07Nb–0.13Mo–0.23Si–1.5Mn–0.8Ni heat-resistant steel. Peng Yan et al. [20,21] proved that dislocation and lath evolution are the main factors affecting the high-temperature strength of G115 heat-resistant steel, and the effect of precipitation strengthening is less than that of dislocation and lath strengthening, while the precipitated phase mainly hinders the movement of dislocation and lath boundary. At the same time, the decrease of supersaturation degree of interstitial atoms will also cause the decrease of mechanical properties of heat-resistant steel. S.S. Wang et al. [22] found that 9Cr–0.5Mo–1.8W heat-resistant steel was strengthened mainly by the combination of grain boundary strengthening, dislocation strengthening, and precipitation strengthening.

9Cr–1.5Mo–1Co and 9Cr–3W–3Co heat-resistant steel are two important members of 9–12% Cr martensitic heat-resistant steel. However, there are few studies in the literature comparing the strengthening mechanism of these two kinds of heat-resistant steels. 9Cr–1.5Mo–1Co heat-resistant steel is the heat-resistant material currently used in 620 °C thermal power generation units, while 9Cr–3W–3Co heat-resistant steel is a newly developed type of steel used in 650 °C thermal power generation units. The high temperature tensile properties of the two steels were compared and the mechanism was discussed. It is of great significance to the application and development of 9Cr–3W–3Co heat-resistant steel. At the same time, it is conducive to the development of new materials.

In this paper, 9Cr–1.5Mo–1Co and 9Cr–3W–3Co heat-resistant steels are taken as the research objects. The microstructure and high-temperature mechanical properties of 9Cr–1.5Mo–1Co and 9Cr–3W–3Co heat-resistant steels are studied comparatively, and their strengthening and toughening mechanism is revealed.

2. Materials and Methods

The steel used in this study was melted in an electric furnace, refining furnace, and vacuum refining furnace; then obtained by sand casting; and finally normalized at 1000 °C and tempered at 700 °C, and the chemical composition is shown in Table 1. A sample of 10 mm × 10 mm × 10 mm is cut from the sample by wire cutting. After grinding and mechanical polishing, the sample is corroded with the etchant prepared by $FeCl_3$ (5 g) + HCl (15 mL) + H_2O (80 mL) for 20 s, and the corroded sample is washed and dried with alcohol. The microstructure and the distribution of the main elements were observed and analyzed by optical microscope (OM), scanning electron microscope (SEM), and electron probe microanalyzer (EPMA). SEM was performed on JSM-6700F (JEOL, Tokyo, Japan) with an operating voltage of 15 kV. The EPMA scan was performed on EPMA-8050G (Shimadzu, Kyoto, Japan) with an operating voltage of 15 kV. The transmission sample was prepared by the tenupol-5 electrolytic double spray instrument (Struers, Ballerup, Denmark).The double spray liquid was 4% perchloric acid ethanol solution, which is cooled to −30 °C by liquid nitrogen, and the voltage was 20 V. Precipitates, laths, and dislocations were observed on thin foils using JEM-2100 emission transmission electron microscopy (JEOL, Tokyo, Japan) at 200 kV. Thermodynamic calculation was performed by Thermo-Calc software (Thermo-Calc-2015a, TCS, Solna, Sweden) and TCFe7 database was used.

Table 1. Chemical composition of the steel, wt%.

Element	C	Si	Mn	Cr	Mo	Ni	V	Nb	Co	W
9Cr-1.5Mo-1Co	0.05~0.15	0.2~0.4	0.5~1	9~12	1.5	0~0.3	0~0.3	0~0.1	1.00	0
9Cr-3W-3Co	0.05~0.15	0.2~0.4	0.5~1	9~12	0	0~0.3	0~0.3	0~0.1	3.00	2.80

Zwick Z1100TEW universal material testing machine (Zwick/Roell, Kennesaw, GA, USA) was used to measure the high temperature tensile properties. The sample size for high temperature tensile property test is shown in Figure 1 The experimental temperatures were 25 °C, 200 °C, 400 °C, 500 °C,

650 °C, and 700 °C, respectively. Three groups of experiments were carried out at each temperature to ensure the reliability of the experiment. The sample is heated to the required temperature within 60 min, which is held for 15 min at the required temperature, and then the tensile property test is started. The strain rate of the tensile test is 0.015 min^{-1}. After the tensile test, scanning electron microscope (SEM) was used to observe the morphology and analyze the fracture mechanism.

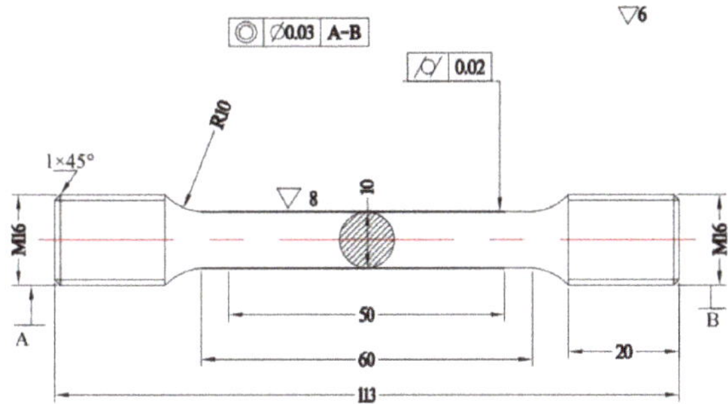

Figure 1. Sample size of high temperature tensile test.

3. Results

3.1. Evolution of High-Temperature Strength under As-Received

The high temperature tensile curves of 9Cr–1.5Mo–1Co and 9Cr–3W–3Co heat-resistant steel are shown in Figure 2. It can be seen from the figure that the tensile strength and yield strength of 9Cr–1.5Mo–1Co and 9Cr–3W–3Co heat-resistant steel decrease with the increase of temperature. However, elongation and reduction of area decreased or remained unchanged before 400 °C, and increased with the increase of tensile temperature after 400 °C. At a high temperature, the diffusion ability of metal atom is enhanced, the atom is easy to move, the pinning effect of defect on dislocation is weakened, the resistance of dislocation motion is reduced, and the macroscopic strength is reduced.

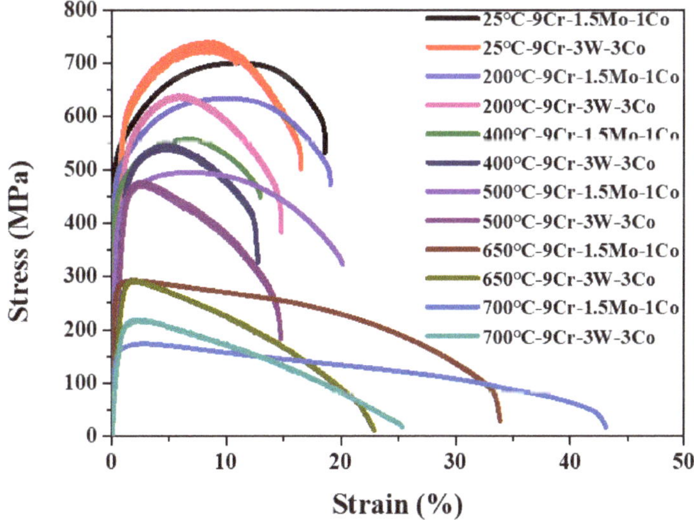

Figure 2. The stress–strain curves of 9Cr–1.5Mo–1Co and 9Cr–3W–3Co heat-resistant steel.

The relationship between the temperature and strength of 9Cr–1.5Mo–1Co and 9Cr–3W–3Co heat-resistant steel is shown in Figure 3. It can be seen from the figure that the overall strength of 9Cr–1.5Mo–1Co and 9Cr–3W–3Co heat-resistant steel decreases with the increase of tensile temperature. There is no obvious regularity difference between 9Cr–1.5Mo–1Co and 9Cr–3W–3Co heat-resistant steel of the tensile strength, and the difference between them is within 30MPa. However, compared with 9Cr–1.5Mo–1Co heat-resistant steel, the yield strength of 9Cr–3W–3Co heat-resistant steel is significantly higher than that of 9Cr–1.5Mo–1Co heat-resistant steel. At room temperature, the yield strength of 9Cr–1.5Mo–1Co heat-resistant steel is 530 MPa, which is about 100 MPa lower than that of 9Cr–3W–3Co heat-resistant steel. When the tensile temperature rises to 500 °C, the yield strength of 9Cr–1.5Mo–1Co heat-resistant steel is 401 MPa, which is about 60MPa lower than that of 9Cr–3W–3Co heat-resistant steel. When the tensile temperature rises to 700 °C, the yield strength of 9Cr–1.5Mo–1Co heat-resistant steel is 164 MPa, which is about 40MPa lower than that of 9Cr–3W–3Co heat-resistant steel.

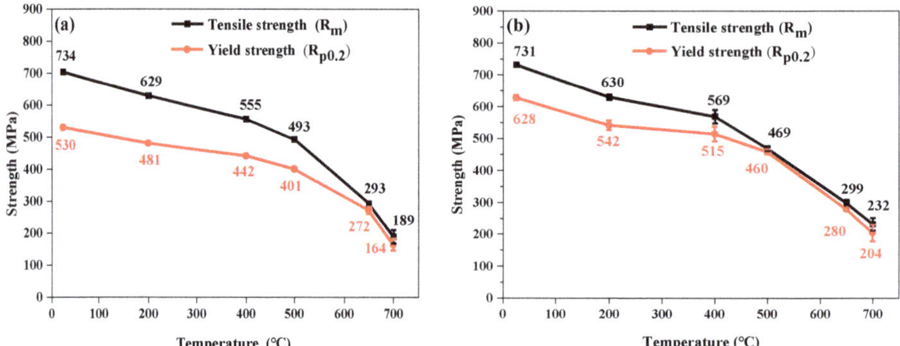

Figure 3. Relationship between the temperature and strength of heat resistant steel at different temperatures: (**a**) 9Cr–1.5Mo–1Co and (**b**) 9Cr–3W–3Co.

The relationship between the temperature and plasticity of 9Cr–1.5Mo–1Co and 9Cr–3W–3Co heat-resistant steel is shown in Figure 4. It can be seen from the figure that the plastic change rule of 9Cr–1.5Mo–1Co and 9Cr–3W–3Co heat-resistant steel is similar. When the tensile temperature is less than 400 °C, the elongation and area shrinkage of 9Cr–1.5Mo–1Co and 9Cr–3W–3Co heat-resistant steel do not change significantly with the rise of temperature. When the tensile temperature is greater than 400 °C, the elongation and reduction of area of 9Cr–1.5Mo–1Co and 9Cr–3W–3Co heat-resistant steel began to rise sharply. However, the overall elongation of 9Cr–3W–3Co heat-resistant steel is lower than that of 9Cr–1.5Mo–1Co heat-resistant steel, and there is no significant difference between the 9Cr–1.5Mo–1Co and 9Cr–3W–3Co heat-resistant steel. At room temperature, the elongation of 9Cr–1.5Mo–1Co heat-resistant steel is 19.5% and the reduction of area is 50.0%, however, the elongation and reduction of area of 9Cr–3W–3Co heat-resistant steel are 15% and 44.8% respectively, that is, the elongation and area reduction of 9Cr–1.5Mo–1Co heat-resistant steel are 4.5% and 5.2% higher than those of 9Cr–3W–3Co heat-resistant steel, respectively. When the tensile temperature is increased to 650 °C, the elongation of 9Cr–1.5Mo–1Co heat-resistant steel is increased by 30.4% and the reduction of area is increased by 58.0%, but the elongation and area reduction of 9Cr–3W–3Co heat-resistant steel are 19.1% and 86.3% respectively. In other words, the elongation and area reduction of 9Cr–1.5Mo–1Co heat-resistant steel are 11.3% higher and 2.0% lower those of 9Cr–3W–3Co heat-resistant steel, respectively. Compared with plasticity, strength is the most important index to evaluate the properties of heat-resistant steel. Under the service temperature of heat-resistant steel, the atomic diffusion speed is faster, and the plasticity of materials will rise to a certain extent. Therefore, plasticity is generally not the focus of heat-resistant steel research.

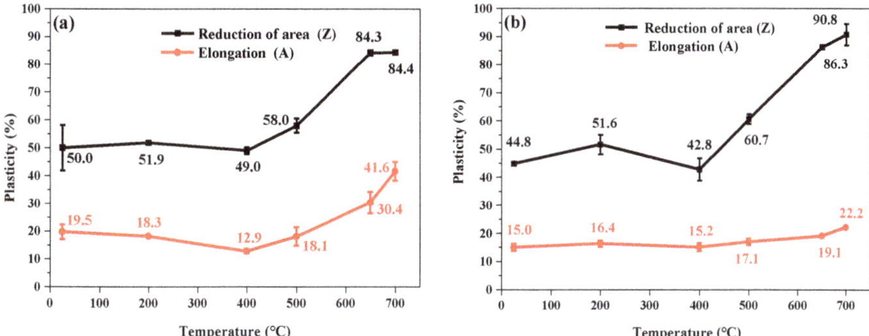

Figure 4. Relationship between temperature and plasticity of heat-resistant steel at different temperatures: (**a**) 9Cr–1.5Mo–1Co and (**b**) 9Cr–3W–3Co.

3.2. Analysis of Fracture Morphology after High Temperature Tensile

The tensile fracture of 9Cr–1.5Mo–1Co and 9Cr–3W–3Co heat-resistant steel at different temperatures is shown in Figures 5 and 6. It can be seen from the figure that 9Cr–1.5Mo–1Co and 9Cr–3W–3Co heat-resistant steel are mainly ductile fracture. With the increase of temperature, the dimple deepens and the size increases, especially at 650 °C, the dimple on the fracture surface becomes larger and deeper. Because there are many second phase particles in 9Cr–1.5Mo–1Co and 9Cr–3W–3Co that are heat-resistant, these second phase particles can effectively resist crack growth at a high temperature, which also shows that plasticity increases with the increase of tensile temperature. This is also confirmed by the results of elongation and reduction of area, that is to say, the plasticity of the material above 650 °C is obviously improved.

3.3. Microstructural Characterization of the As-Received Steel

The OM images of 9Cr–1.5Mo–1Co and 9Cr–3W–3Co heat-resistant steel are shown in Figure 7. It can be seen from the figure that the microstructures of 9Cr–1.5Mo–1Co and 9Cr–3W–3Co heat-resistant steel are typical tempered martensite. After high-temperature tempering, the martensite morphology is relatively complete, and some martensite laths are relatively wide. It is possible that during tempering, the martensite lath has merged, and at the same time, there are a large number of black dot like precipitates on the edge of the martensite lath.

SEM images of 9Cr–1.5Mo–1Co and 9Cr–3W–3Co heat-resistant steel are shown in Figure 8. It can be seen from the figure that there are precipitates with different sizes in the grain boundary, the edge of martensitic lath, and the interior of martensitic lath, and compared with the precipitates at the edge of martensitic lath and the interior of martensitic lath, the precipitates at the grain boundary have larger sizes and obvious clustering phenomenon. The obvious difference is mainly due to the higher energy at the grain boundary, which is conducive to the precipitation phase nucleation and carbide aggregation. At the same time, it can be found that the number of precipitates in 9Cr–1.5Mo–1Co heat-resistant steel is more than that in 9Cr–3W–3Co heat-resistant steel with the same area. According to the statistics of software IPP, the average size of precipitates in 9Cr–1.5Mo–1Co heat-resistant steel is 156.5 nm, and the average size of precipitates in 9Cr–3W–3Co heat-resistant steel is 270.7 nm.

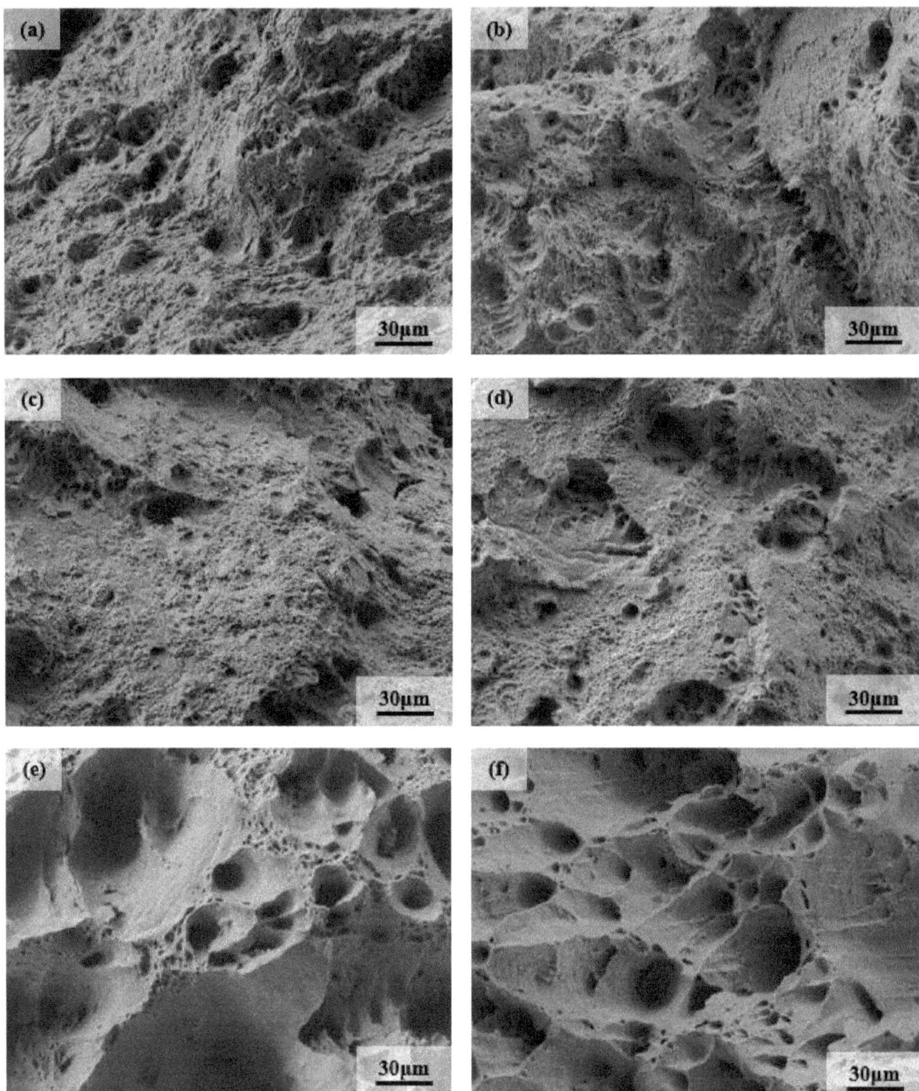

Figure 5. Tensile fracture of 9Cr–1.5Mo–1Co heat resistant steel at different temperatures: (**a**) 25 °C, (**b**) 200 °C, (**c**) 400 °C, (**d**) 500 °C, (**e**) 650 °C, and (**f**) 700 °C.

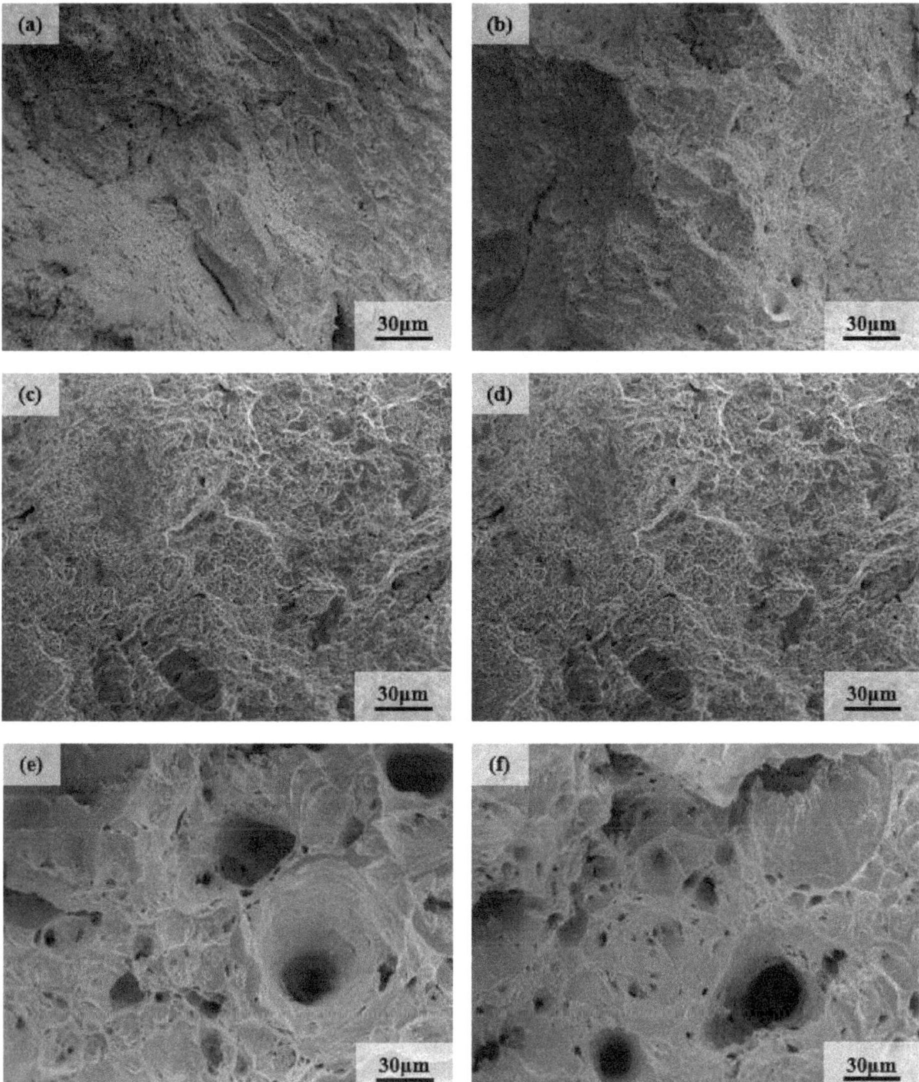

Figure 6. Tensile fracture of 9Cr–3W–3Co heat resistant steel at different temperatures: (**a**) 25 °C, (**b**) 200 °C, (**c**) 400 °C, (**d**) 500 °C, (**e**) 650 °C, and (**f**) 700 °C.

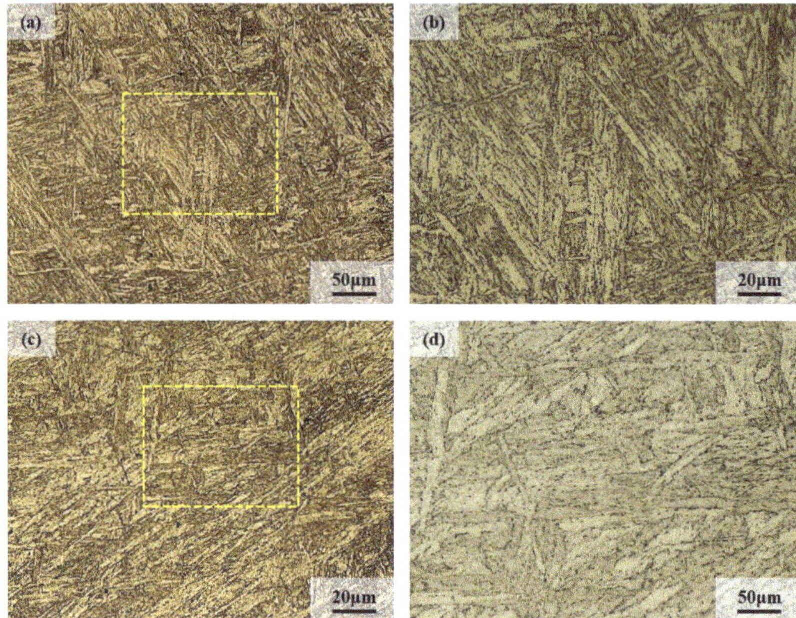

Figure 7. Optical microscope (OM) image of heat-resistant steel after heat treatment: (**a**,**b**) 9Cr–1.5Mo–1Co, (**c**,**d**) 9Cr–3W–3Co.

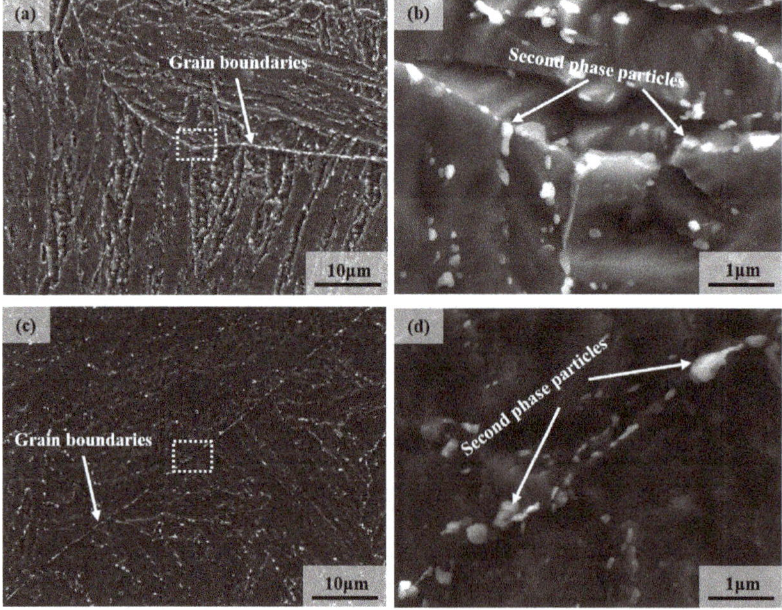

Figure 8. Scanning electron microscope (SEM) image of heat-resistant steel after heat treatment: (**a**,**b**) 9Cr–1.5Mo–1Co, (**c**,**d**) 9Cr–3W–3Co.

The results of surface element in electron probe microanalyzer(EPMA)of 9Cr–1.5Mo–1Co heat-resistant steel are shown in Figure 9. It can be seen from the figure that C, N, V, Cr, Nb,

Mo, and other elements are in the state of segregation. According to the relevant literature [23,24], a large number of carbide in the grain boundary and phase boundary may be $M_{23}C_6$, while MX precipitate may be formed inside the martensitic lath, and Co element does not participate in the formation of precipitate, which is evenly distributed in the matrix.

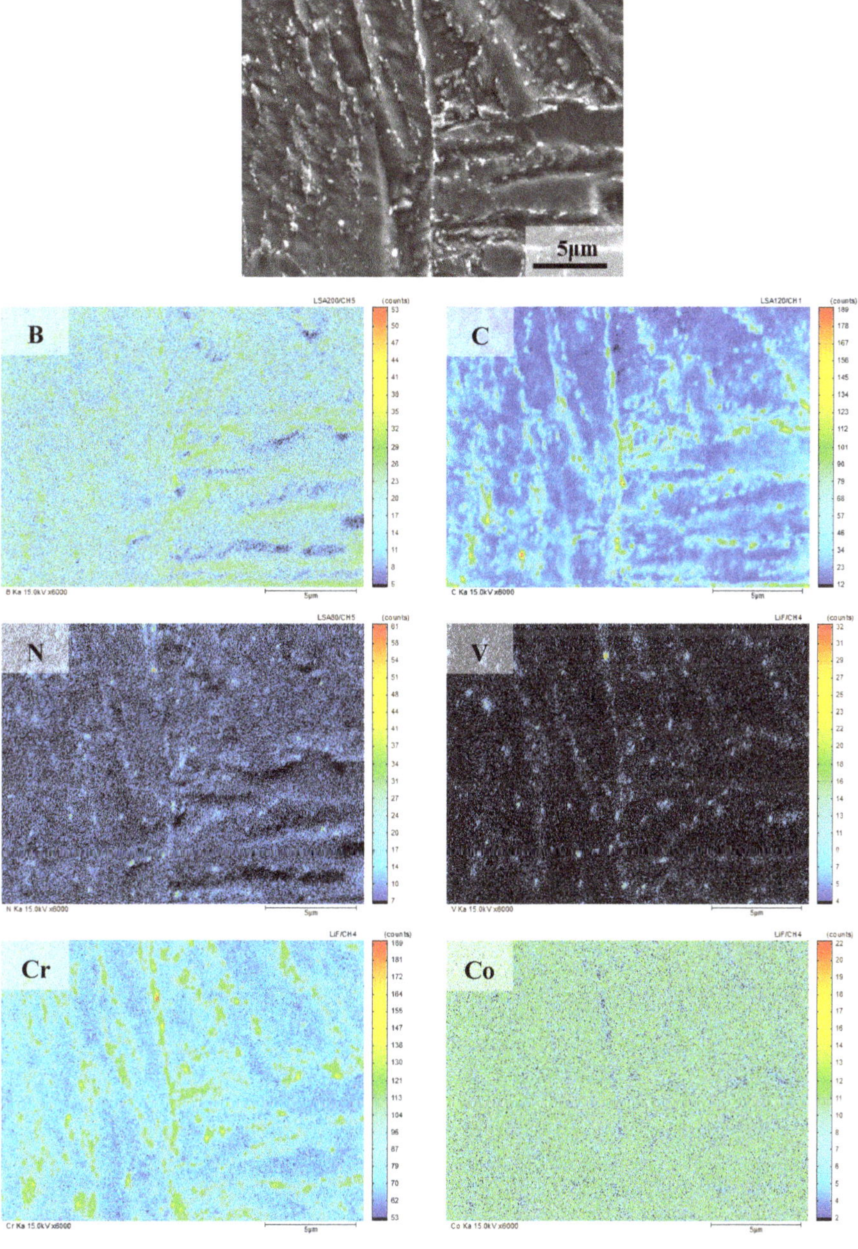

Figure 9. *Cont.*

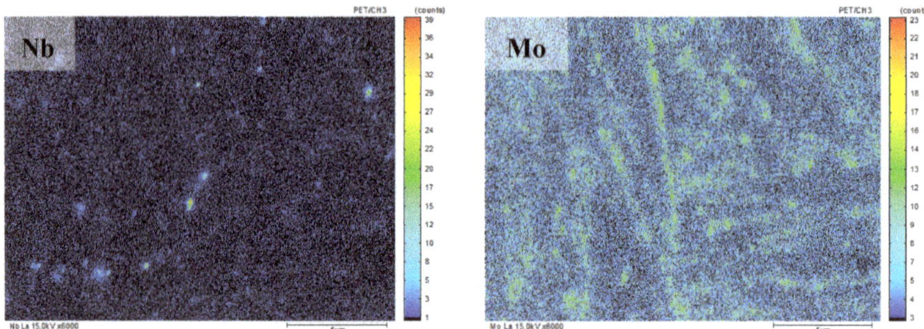

Figure 9. Analysis results of elements in 9Cr–1.5Mo–1Co heat-resistant steel.

The results of surface element in EPMA of 9Cr–3W–3Co heat-resistant steel are shown in Figure 10. Similar to 9Cr–1.5Mo–1Co, it can be seen from the figure that C, N, V, Cr, Nb, W, and other elements are in the state of segregation. According to the relevant literature [23,24], a large number of carbide in the grain boundary and phase boundary may be $M_{23}C_6$, while MX precipitate may be formed inside the martensitic lath, and Co element does not participate in the formation of precipitate, which is evenly distributed in the matrix.

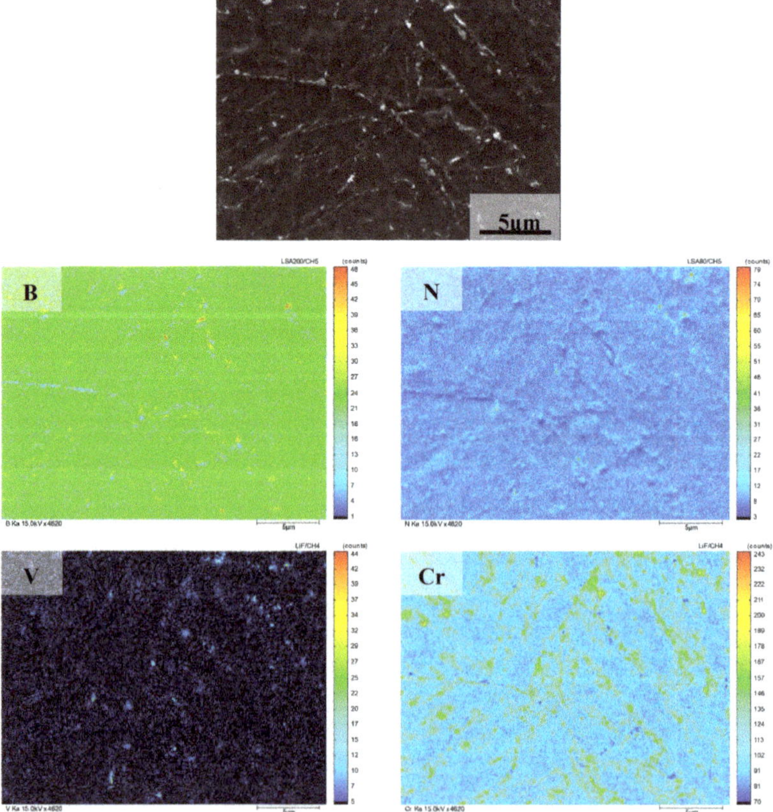

Figure 10. *Cont.*

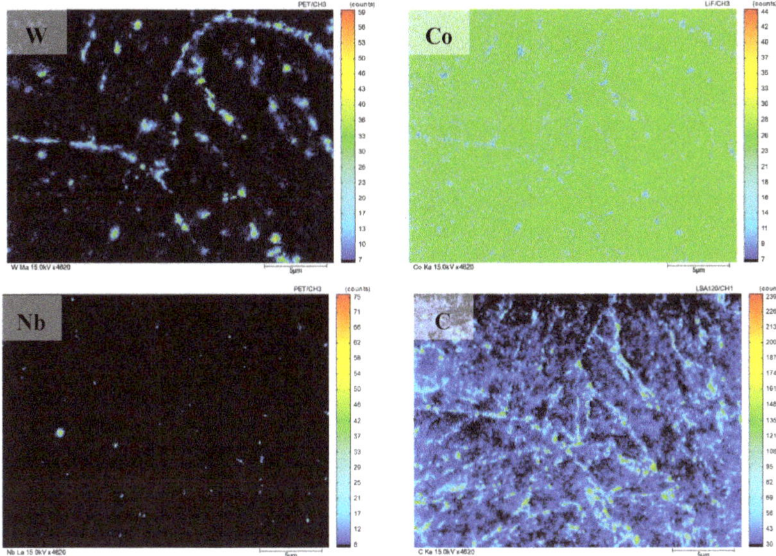

Figure 10. Analysis results of elements in 9Cr–3W–3Co heat-resistant steel.

TEM images of the matrix microstructure of 9Cr–1.5Mo–1Co and 9Cr–3W–3Co heat-resistant steel are shown in Figure 11. It can be seen from the figure that 9Cr–1.5Mo–1Co and 9Cr–3W–3Co heat-resistant steel are both typical lath-like structures with obvious precipitation at the lath boundary. Through the analysis of crystal structure, 9Cr–1.5Mo–1Co and 9Cr–3W–3Co heat-resistant steels are martensitic microstructure. At the same time, it is found that the width of the martensitic lath of 9Cr–1.5Mo–1Co heat-resistant steel is significantly smaller than that of 9Cr–3W–3Co heat-resistant steel. The width of the martensitic lath of 9Cr–1.5Mo–1Co heat-resistant steel is 385 ± 55 nm, but that of 9Cr–3W–3Co heat-resistant steel is 495 ± 55 nm.

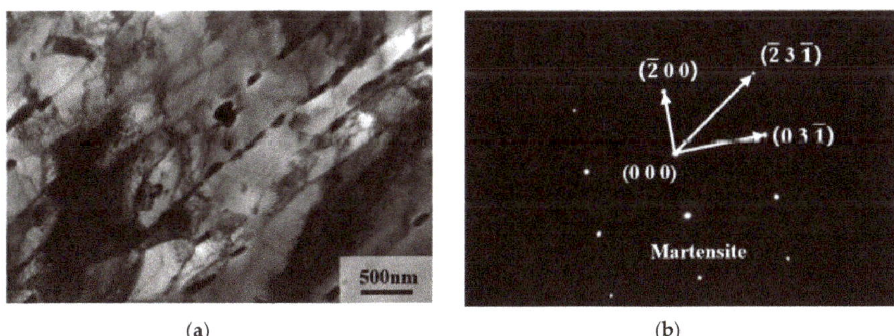

(a) (b)

Figure 11. *Cont.*

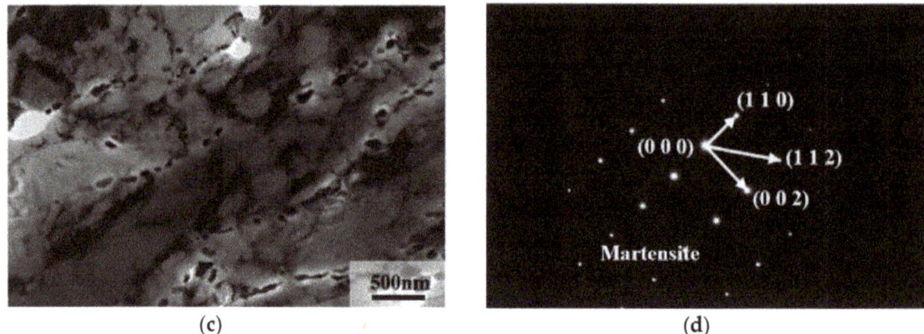

Figure 11. Transmission electron microscopy (TEM) image of heat-resistant steel after heat treatment: (**a**,**b**) 9Cr–1.5Mo–1Co, (**c**,**d**) 9Cr–3W–3Co.

TEM images of precipitated phases in 9Cr–1.5Mo–1Co and 9Cr–3W–3Co heat-resistant steel are shown in Figures 12 and 13. It can be seen from the figure that the precipitates in 9Cr–1.5Mo–1Co and 9Cr–3W–3Co heat-resistant steel are $M_{23}C_6$ precipitates and MX precipitates. The structure of these two precipitates is shown in Figure 14. For $M_{23}C_6$ precipitates, the metal atoms are distributed in 4a, 8c, 32f, and 48h of the cell, while the C atoms are mainly located in 24e. In the MX precipitate, the metal atom is located at the top of the cell and the center of the crystal surface, while the non-metal atom is located at the center of the edge and the center of the crystal cell. Because of the different positions of formation, $M_{23}C_6$ precipitates are formed at the lath boundary and sub grain boundary, MX precipitates are formed inside the martensitic lath, $M_{23}C_6$ precipitates are elliptical, and MX precipitates are spherical.

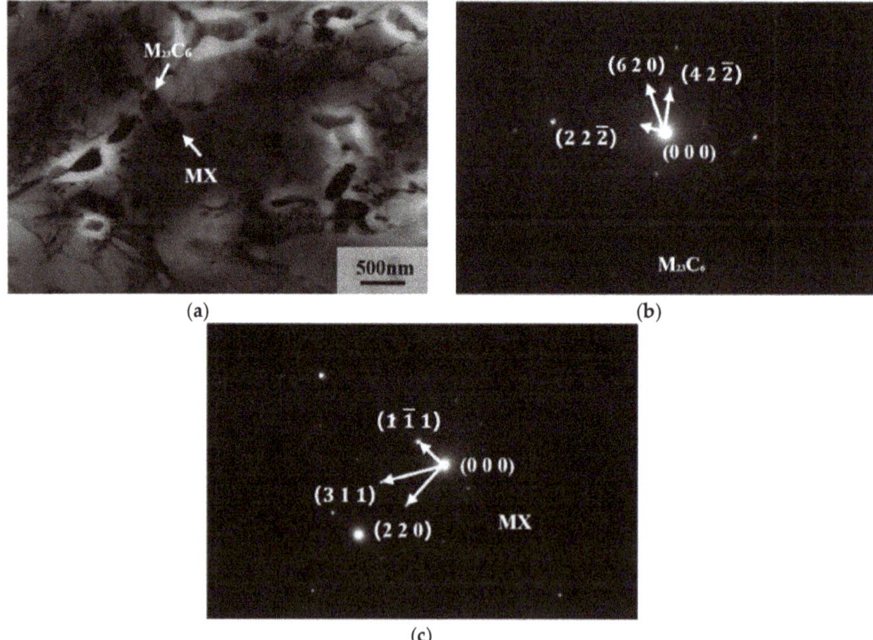

Figure 12. TEM image of precipitation phase in 9Cr–1.5Mo–1Co heat-resistant steel: (**a**) the morphology of precipitated phase, (**b**) diffraction pattern of $M_{23}C_6$ precipitates, (**c**) diffraction pattern of MX precipitates.

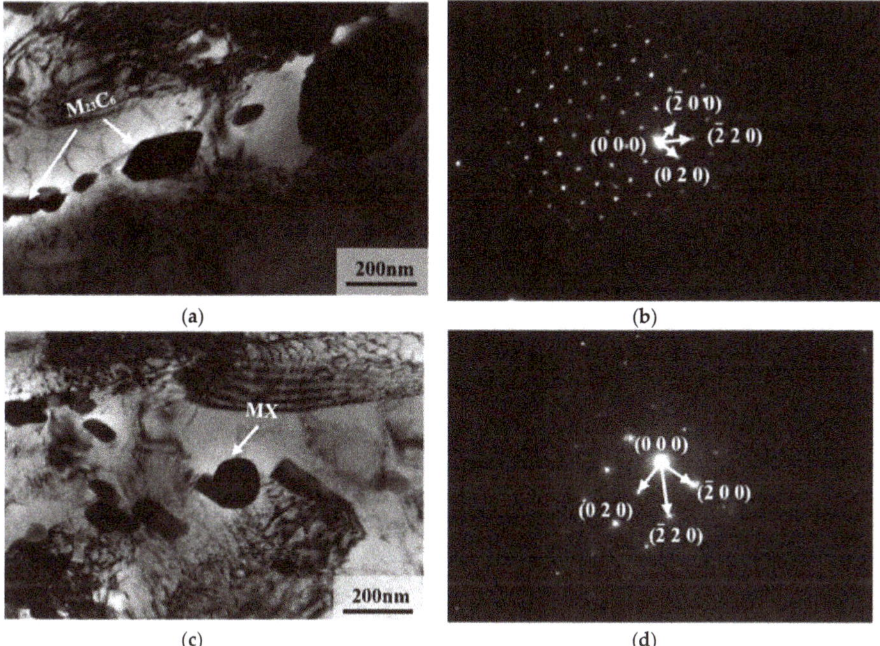

(a) (b) (c) (d)

Figure 13. TEM image of precipitation phase in 9Cr–3W–3Co heat-resistant steel: (**a**) the morphology of $M_{23}C_6$ precipitates, (**b**) diffraction pattern of $M_{23}C_6$ precipitates, (**c**) the morphology of MX precipitates, (**d**) diffraction pattern of MX precipitates

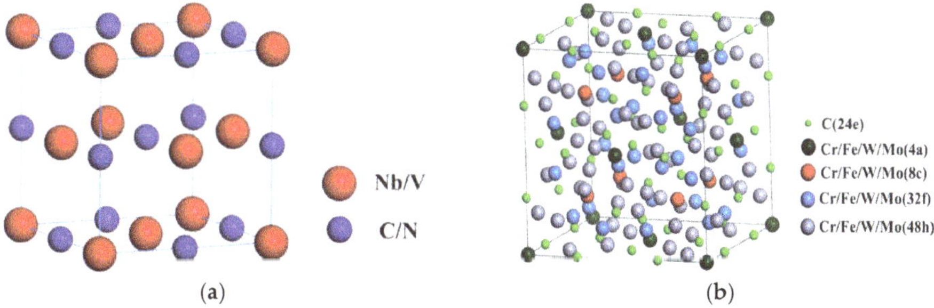

(a) (b)

Figure 14. Structure diagram of $M_{23}C_6$ precipitate and MX precipitates: (**a**) crystal structure of MX precipitate, (**b**) crystal structure of $M_{23}C_6$ precipitate.

The dislocation distribution in 9Cr–1.5Mo–1Co and 9Cr–3W–3Co heat-resistant steel is shown in Figure 15. It can be seen from the figure that a large number of simple dislocation modes, such as dislocation line and dislocation accumulation, are distributed in the martensitic lath of 9Cr–1.5Mo–1Co and 9Cr–3W–3Co heat-resistant steel, as shown by the white arrow in the figure. The existence of these dislocation structures will strengthen the heat-resistant steel [25–27]. At the same time, some dislocations are distributed around the precipitates, which will further strengthen the material. When the material is deformed as a result of the force, the dislocations will move correspondingly, but the dislocations cannot cut through these hard precipitates. In other words, the hard precipitates will hinder the wrong movement, so that the matrix can be strengthened [28]. At the same time, it is easy to see the dislocation distribution in 9Cr–1.5Mo–1Co and 9Cr–3W–3Co heat-resistant steel. The dislocations in 9Cr–3W–3Co heat-resistant steel are more concentrated, and the distribution and the number of dislocation lines around the precipitated phase are obviously more than that in 9Cr–1.5Mo–1Co heat-resistant steel.

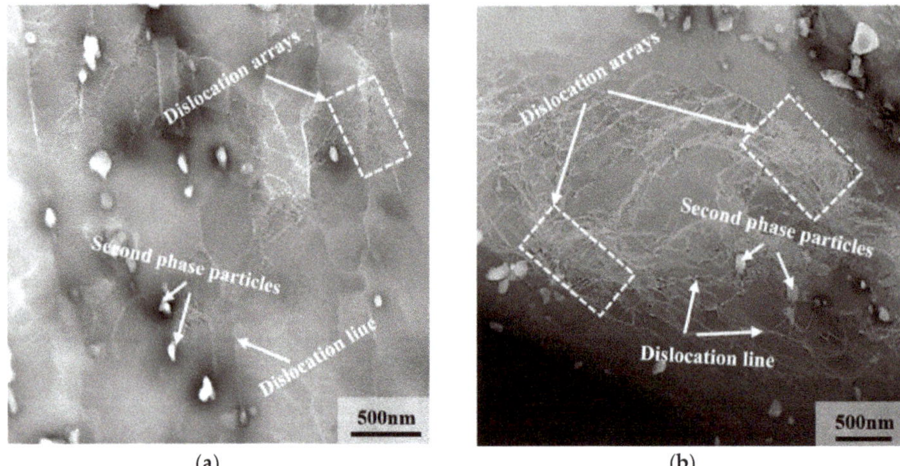

Figure 15. Dislocation distribution of heat-resistant steel under TEM after heat treatment: (**a**) 9Cr–1.5Mo–1C0 and (**b**) 9Cr–3W–3Co.

4. Discussion

Through the above results, it is not difficult to find that the overall mechanical properties of 9Cr–3W–3Co heat-resistant steel are significantly better than those of 9Cr–1.5Mo–1Co heat-resistant steel, which provides the possibility for 9Cr–3W–3Co heat-resistant steel to be widely used in 650 °C thermal power generating units. This part is mainly used to explain the reason that the properties of 9Cr–3W–3Co heat-resistant steel are better than those of 9Cr–1.5Mo–1Co heat-resistant steel.

9Cr–1.5Mo–1Co and 9Cr–3W–3Co heat-resistant steel belong to 9–12% Cr martensitic heat-resistant steel. Chromium equivalent directly affects the overall performance of heat-resistant steel. The chromium equivalent is calculated as follows [29]:

$$\begin{aligned} NCE = \%Cr + 6 \times \%Si + 4 \times \%Mo + 1.5 \times \%W + 11 \times \%V + 5 \times \%Nb + 12 \times \%Al + 8 \times \%Ti - \\ 40 \times \%C - 2 \times \%Mn - 4 \times \%Ni - 2 \times \%Co - 30 \times \%N - \%Cu \end{aligned} \quad (1)$$

After calculation, the chromium equivalent of 9Cr–1.5Mo–1Co heat-resistant steel is 8.992, while that of 9Cr–3W–3Co heat-resistant steel is 7.783. Compared with Figure 16, it is not difficult to find that 9Cr–1.5Mo–1Co and 9Cr–3W–3Co heat-resistant steels are in the austenite phase area when they are heat treated in the same austenite, which ensures that δ–Fe transformation does not occur in the process of austenitizing, that is to say, the microstructure of 9Cr–1.5Mo–1Co and 9Cr–3W–3Co heat-resistant steel is the cause of tempered martensite. However, according to the analysis of martensitic structure of 9Cr–1.5Mo–1Co and 9Cr–3W–3Co heat-resistant steel in Figure 11, the width of the martensitic lath of 9Cr–1.5Mo–1Co heat-resistant steel is obviously smaller than that of 9Cr–3W–3Co heat-resistant steel, and the smaller the martensitic lath, the better the mechanical properties. Therefore, the influence of matrix structure on the strength of 9Cr–1.5Mo–1Co heat-resistant steel is greater than that on 9Cr–3W–3Co heat-resistant steel.

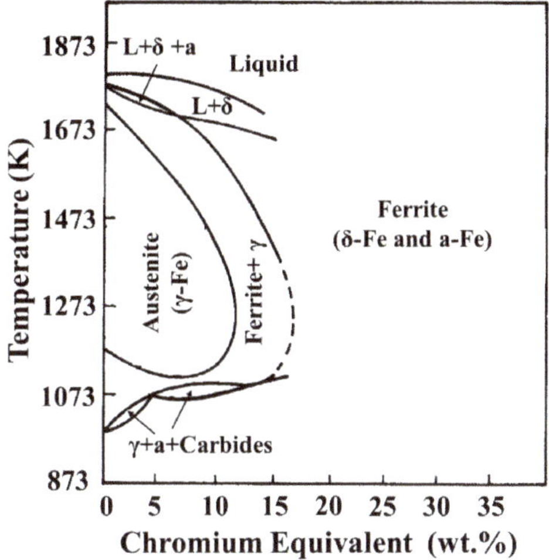

Figure 16. Pseudoequilibrium diagram showing the dependence of the existence of various phases on the chromium equivalent.

In addition to grain boundary strengthening, the strengthening methods of 9–12% Cr martensitic heat-resistant steel mainly include solution strengthening, precipitation strengthening, and dislocation strengthening. It can be seen from the above experimental results that the microstructures of 9Cr–1.5Mo–1Co and 9Cr–3W–3Co heat-resistant steel are similar, and there are four strengthening ways mentioned above. Therefore, it is necessary to distinguish the effect of the other three strengthening mechanisms on the two heat-resistant steels.

The pinning force PZ of the precipitated relative grain boundary or sub grain boundary and the pinning force PB for the martensitic lath can be expressed as follows [15]:

$$P_Z = \frac{3\gamma F_V}{d} \tag{2}$$

$$P_B = \frac{\gamma F_{VB} D}{d^2} \tag{3}$$

where γ is the boundary surface energy per unit area, F_V and F_{VB} are volume fraction, and d is the size of dispersed particles, respectively. D represents the size of structural elements, that is, the grain/subgrain size or lath thickness. It can be seen from the formula that the pinning force of precipitation relative to grain boundary or martensitic lath boundary is mainly related to the volume fraction and the size of precipitated phase. With the increase of the volume fraction of precipitates, the pinning effect of precipitates is stronger, that is, it is more favorable for the strength of heat-resistant steel. With the increase of the size of precipitates, the pinning effect of precipitates is weaker, that is, it is more unfavorable for the strength of heat-resistant steel. Figure 17 shows the precipitation amount of $M_{23}C_6$ in 9Cr–1.5Mo–1Co and 9Cr–3W–3Co heat-resistant steel calculated by Thermo-Calc software. It can be seen from the figure that the precipitation of $M_{23}C_6$ in 9Cr–1.5Mo–1Co heat-resistant steel is significantly greater than that of $M_{23}C_6$ in 9Cr–3W–3Co heat-resistant steel, which is mutually confirmed with the experimental results in Figure 8. According to the statistical results of precipitate size in Figure 8, the average size of precipitate in 9Cr–1.5Mo–1Co heat-resistant steel is smaller than that in 9Cr–3W–3Co heat-resistant steel. In conclusion, the pinning force of 9Cr–3W–3Co heat-resistant steel is smaller than that of 9Cr–1.5Mo–1Co heat-resistant steel.

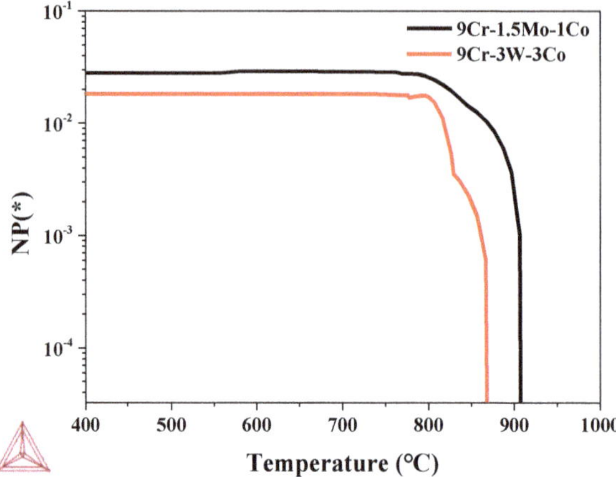

Figure 17. Change of precipitation amount of $M_{23}C_6$ precipitate with temperature.

The strengthening effect of dislocation strengthening on steel can be calculated by the following formula [30]:

$$\sigma_\rho = 0.5MGb\rho^{1/2} \tag{4}$$

where M is the Taylor coefficient, G is the shear modulus, b is the length of Burgers vector, and ρ is the dislocation density. It can be seen that the strength of the material increases with the increase of dislocation density. According to the analysis results in Figure 15, the dislocation density of 9Cr–3W–3Co heat-resistant steel is significantly higher than that of 9Cr–1.5Mo–1Co heat-resistant steel. Therefore, the contribution of dislocation to the strength of 9Cr–3W–3Co heat-resistant steel is higher than that of 9Cr–1.5Mo–1Co heat-resistant steel.

Solution strengthening is mainly related to the radius difference between the solute atom and matrix atom. The larger the radius difference between the solute atom and matrix atom, the stronger the strengthening effect on the matrix. Therefore, in 9Cr–1.5Mo–1Co and 9Cr–3W–3Co heat-resistant steels, except for the small-size non-metallic elements such as C, N, and B, which are strengthened in the form of interstitial atoms, the metal elements such as Co, W, and Mo can also strengthen 9Cr–1.5Mo–1Co or 9Cr–3W–3Co heat-resistant steel in the form of replacement atoms. According to Figures 9 and 10, Co atoms mainly exist in the matrix, while some Mo and W atoms are precipitated in the form of precipitated phase, and the rest are also solid soluble in the matrix. Combined with the element composition of 9Cr–3W–3Co or 9Cr–1.5Mo–1Co, it can be seen that the Co element content of 9Cr–3W–3Co is higher than that of 9Cr–1.5Mo–1Co. Therefore, the solid solution strengthening effect of Co on 9Cr–3W–3Co is stronger than that of 9Cr–1.5Mo–1Co. In addition, W element and Mo element belong to the same group, and the atomic number of W atom is greater than that of Mo atom, and the effect of W atom on steel solution strengthening is greater than that of Mo atom on steel. In conclusion, the strengthening effect of Co and W atoms on 9Cr–3W–3Co heat-resistant steel is higher than that of Co and Mo atoms on 9Cr–1.5Mo–1Co heat-resistant steel.

5. Conclusions

On the basis of the analysis of the tensile properties and strengthening mechanism of 9Cr–1.5Mo–1Co and 9Cr–3W–3Co heat-resistant steel at different temperatures, the following conclusions are obtained:

1. At the same temperature, the overall yield strength of 9Cr–3W–3Co heat-resistant steel is better than that of 9Cr–1.5Mo–1Co heat-resistant steel. At room temperature, the yield strength of 9Cr–3W–3Co heat-resistant steel is 628 MPa, while that of 9Cr–1.5Mo–1Co heat-resistant steel is

530 MPa. When the tensile temperature is increased to 700 °C, the yield strength of 9Cr–3W–3Co heat-resistant steel is 204 MPa, while that of 9Cr–1.5Mo–1Co heat-resistant steel is 164 MPa.

2. The strength of 9Cr–3W–3Co heat-resistant steel and 9Cr–1.5Mo–1Co heat-resistant steel is affected by grain boundary, dislocation, precipitation, and solid solution atoms, but the width of the martensitic lath of 9Cr–1.5Mo–1Co heat-resistant steel is smaller than that of 9Cr–3W–3Co heat-resistant steel, and the dislocation density of 9Cr–3W–3Co heat-resistant steel is larger than that of 9Cr–1.5Mo–1Co heat-resistant steel.
3. The excellent high-temperature mechanical properties of 9Cr–3W–3Co heat-resistant steel are mainly attributed to the solution strengthening caused by Co and W atoms and the high-density dislocations distributed in the matrix, while the precipitates are mainly pinned dislocations and hinder the movement of dislocations. However, the high temperature strength of 9Cr–1.5Mo–1Co heat-resistant steel is mainly due to the refinement and precipitation of the martensitic lath and the pinning of the grain boundary.

Author Contributions: Conceptualization, L.Z. and X.C.; Methodology, L.Z., X.C., T.W. and Q.Z.; Software, L.Z.; Validation, L.Z., X.C. and Q.Z.; Formal analysis, L.Z. and X.C.; Investigation, L.Z. and X.C.; Resources, L.Z., X.C. and Q.Z.; Data curation, L.Z. and X.C.; Writing—original draft preparation, L.Z.; Writing—review and editing, Q.Z.; visualization, L.Z.; Supervision, X.C., T.W. and Q.Z.; Project administration, T.W. and Q.Z.; funding acquisition, Q.Z. All authors have read and agreed to the published version of the manuscript.

Funding: This research received no external funding.

Acknowledgments: I would like to thank Xiaogang Lu from School of Materials Science and Engineering, Shanghai University for his great guidance in thermodynamic calculation.

Conflicts of Interest: The authors declare no conflict of interest.

References

1. Abe, F.; Kern, T.U.; Viswanathan, R. Creep-Resistant Steels, Encyclopedia of Materials: Ence and Technology (Second Edition). *Mat. Sci.* **2001**, *16*, 1840–1845. [CrossRef]
2. Viswanathan, R.; Henry, J.F.; Tanzosh, J.G.; Stanko, G.; Shingledecker, J.; Vitalis, B.; Purgert, R. US Program on Materials Technology for Ultra-Supercritical Coal Power Plants. *J. Mater. Eng. Perform.* **2013**, *22*, 2904–2915. [CrossRef]
3. Abe, F. Research and Development of Heat-Resistant Materials for Advanced USC Power Plants with Steam Temperatures of 700 degrees C and above. *Engineering* **2015**, *1*, 211–224. [CrossRef]
4. Abe, F. Progress in Creep-Resistant Steels for High Efficiency Coal-Fired Power Plants. *J. Press Vessel Technol.* **2016**, *138*, 040840. [CrossRef]
5. Masuyama, F. History of power plants and progress in heat resistant steels. *ISIJ Int.* **2001**, *41*, 612–625. [CrossRef]
6. Abe, F. Precipitate design for creep strengthening of 9% Cr tempered martensitic steel for ultra-supercritical power plants. *Sci. Technol. Adv. Mater.* **2008**, *9*, 1. [CrossRef]
7. Viswanathan, R.; Bakker, W. Materials for ultrasupercritical coal power plants-boiler materials: Part 1. *J. Mater. Eng. Perform.* **2001**, *10*, 81–95. [CrossRef]
8. Vivas, J.; Capdevila, C.; Altstadt, E.; Houska, M.; Sabirov, I.; San-Martín, D. Microstructural Degradation and Creep Fracture Behavior of Conventionally and Thermomechanically Treated 9% Chromium Heat Resistant Steel. *Met. Mater. Int.* **2019**, *25*, 343–352. [CrossRef]
9. Aghajani, A.; Somsen, C.; Eggeler, G. On the effect of long-term creep on the microstructure of a 12% chromium tempered martensite ferritic steel. *Acta Mater.* **2009**, *57*, 5093–5106. [CrossRef]
10. Fournier, B.; Dalle, F.; Sauzay, M.; Longour, J.; Salvi, M.; Caes, C.; Tournie, I.; Giroux, P.F.; Kim, S.H. Comparison of various 9–12%Cr steels under fatigue and creep-fatigue loadings at high temperature. *Mater. Sci. Eng. A-Struct. Mater. Prop. Microstruct. Process.* **2011**, *528*, 6934–6945. [CrossRef]
11. Jin, X.; Zhu, B.; Li, Y.; Zhao, Y.; Xue, F.; Zhang, G. Effect of the microstructure evolution on the high-temperature strength of P92 heat-resistant steel for different service times. *Int. J. Press. Vessel. Pip.* **2020**, *186*, 104131. [CrossRef]

12. Liu, Z.; Liu, Z.; Wang, X.; Chen, Z.; Ma, L. Evolution of the microstructure in aged G115 steels with the different concentration of tungsten. *Mater. Sci. Eng. A-Struct. Mater. Prop. Microstruct. Process.* **2018**, *729*, 161–169. [CrossRef]
13. Xiao, B.; Xu, L.; Zhao, L.; Jing, H.; Han, Y.; Song, K. Transient creep behavior of a novel tempered martensite ferritic steel G115. *Mater. Sci. Eng. A-Struct. Mater. Prop. Microstruct. Process.* **2019**, *742*, 628. [CrossRef]
14. Saini, N.; Mulik, R.S.; Mahapatra, M.M. Study on the effect of ageing on laves phase evolution and their effect on mechanical properties of P92 steel. *Mater. Sci. Eng. A-Struct. Mater. Prop. Microstruct. Process.* **2018**, *716*, 179–188. [CrossRef]
15. Dudova, N.; Plotnikova, A.; Molodov, D.; Belyakov, A.; Kaibyshev, R. Structural changes of tempered martensitic 9%Cr-2%W-3%Co steel during creep at 650 degrees C. *Mater. Sci. Eng. A-Struct. Mater. Prop. Microstruct. Process.* **2012**, *534*, 632–639. [CrossRef]
16. Jia, C.H.; Liu, Y.C.; Liu, C.X.; Li, C.; Li, H.J. Precipitates evolution during tempering of 9CrMoCoB (CB2) ferritic heat-resistant steel. *Mater. Charact.* **2019**, *152*, 12–20. [CrossRef]
17. Taneike, M.; Abe, F.; Sawada, K. Creep-strengthening of steel at high temperatures using nano-sized carbonitride dispersions. *Nature* **2003**, *424*, 294–296. [CrossRef] [PubMed]
18. Sawada, K.; Kubo, K.; Abe, F. Creep behavior and stability of MX precipitates at high temperature in 9Cr-0.5Mo-1.8W-VNb steel. *Mater. Sci. Eng. A-Struct. Mater. Prop. Microstruct. Process.* **2001**, *319*, 784–787. [CrossRef]
19. Du, Y.; Li, X.; Zhang, X.; Chung, Y.W.; Isheim, D.; Vaynman, S. Design and Characterization of a Heat-Resistant Ferritic Steel Strengthened by MX Precipitates. *Metall. Mater. Trans. A Phys. Metall. Mater. Sci.* **2020**, *51*, 638–647. [CrossRef]
20. Yan, P.; Liu, Z.; Bao, H.; Weng, Y.; Liu, W. Effect of microstructural evolution on high-temperature strength of 9Cr-3W-3Co martensitic heat resistant steel under different aging conditions. *Mater. Sci. Eng. A-Struct. Mater. Prop. Microstruct. Process.* **2013**, *588*, 22–28. [CrossRef]
21. Yan, P.; Liu, Z.; Bao, H.; Weng, Y.; Liu, W. Effect of tempering temperature on the toughness of 9Cr-3W-3Co martensitic heat resistant steel. *Materials Design.* **2014**, *54*, 874–879. [CrossRef]
22. Wang, S.S.; Peng, D.L.; Chang, L.; Hui, X.D. Enhanced mechanical properties induced by refined heat treatment for 9Cr-0.5Mo-1.8W martensitic heat resistant steel. *Mater. Des* **2013**, *50*, 174–180. [CrossRef]
23. Nikitin, I.; Fedoseeva, A.; Kaibyshev, R. Strengthening mechanisms of creep-resistant 12%Cr-3%Co steel with low N and high B contents. *J. Mater. Sci.* **2020**, *55*, 7530–7545. [CrossRef]
24. Kostka, A.; Tak, K.G.; Hellmig, R.; Estrin, Y.; Eggeler, G. On the contribution of carbides and micrograin boundaries to the creep strength of tempered martensite ferritic steels. *Acta Mater.* **2007**, *55*, 539–550. [CrossRef]
25. Rojas, D.; Garcia, J.; Prat, O.; Carrasco, C.; Sauthoff, G.; Kaysser-Pyzalla, A.R. Design and characterization of microstructure evolution during creep of 12% Cr heat resistant steels. *Mater. Sci. Eng. A-Struct. Mater. Prop. Microstruct. Process.* **2010**, *527*, 3864–3876. [CrossRef]
26. Guguloth, K.; Roy, N. Study on the creep deformation behavior and characterization of 9Cr-1Mo-V-Nb steel at elevated temperatures. *Mater. Charact.* **2018**, *146*, 279–298. [CrossRef]
27. Wang, S.; Chang, L.; Lin, D.; Chen, X.; Hui, X. High Temperature Strengthening in 12Cr-W-Mo Steels by Controlling the Formation of Delta Ferrite. *Met. Mater. Trans. A* **2014**, *45*, 4371–4385. [CrossRef]
28. Kipelova, A.; Kaibyshev, R.; Belyakov, A.; Molodov, D. Microstructure evolution in a 3% Co modified P911 heat resistant steel under tempering and creep conditions. *Mater. Sci. Eng. A-Struct. Mater. Prop. Microstruct. Process.* **2011**, *528*, 1280–1286. [CrossRef]
29. George, G.; Shaikh, H.; Parvathavarthini, N.; George, R.P.; Khatak, H.S. On the microstructure-polarization behavior correlation of a 9Cr-1Mo steel weld joint. *J. Mater. Eng. Perform.* **2001**, *10*, 460–467. [CrossRef]
30. Maruyama, K.; Sawada, K.; Koike, J.I. Strengthening mechanisms of creep resistant tempered martensitic steel. *ISIJ Int.* **2001**, *41*, 641–653. [CrossRef]

Article

Assessment of Retained Austenite in Fine Grained Inductive Heat Treated Spring Steel

Anna Olina [1,*], Miroslav Píška [1], Martin Petrenec [1], Charles Hervoches [2], Přemysl Beran [2,3], Jiří Pechoušek [4] and Petr Král [5]

1 Department of Manufacturing Technology, Faculty of Mechanical Engineering, Brno University of Technology, Technicka 2, 61669 Brno, Czech Republic; piska@fme.vutbr.cz (M.P.); mpetrenec@gmail.com (M.P.)
2 Nuclear Physics Institute of the CAS, Řež 130, 25068 Řež, Czech Republic; hervoches@ujf.cas.cz (C.H.); premysl.beran@esss.se (P.B.)
3 European Spallation Source ERIC, Box 176, 22100 Lund, Sweden
4 Department of Experimental Physics, Faculty of Science, Palacky University in Olomouc, 17. Listopadu 12, 77900 Olomouc, Czech Republic; jiri.pechousek@upol.cz
5 Institute of Physics of Materials, Czech Academy of Sciences, Zizkova 22, 61662 Brno, Czech Republic; pkral@ipm.cz
* Correspondence: y121510@fme.vutbr.cz; Tel.: +420-773-782-177

Received: 21 October 2019; Accepted: 1 December 2019; Published: 5 December 2019

Abstract: Advanced thermomechanical hot rolling is becoming a widely used technology for the production of fine-grained spring steel. Different rapid phase transformations during the inductive heat treatment of such steel causes the inhomogeneous mixture of martensitic, bainitic, and austenitic phases that affects the service properties of the steel. An important task is to assess the amount of retained austenite and its distribution over the cross-section of the inductive quenched and tempered wire in order to evaluate the mechanical properties of the material. Three different analytical methods were used for the comparative quantitative assessment of the amount of retained austenite in both the core and rim areas of the sample cross-section: neutron diffraction—for the bulk of the material, Mössbauer spectroscopy—for measurement in a surface layer, and the metallographic investigations carried by the EBSD. The methods confirmed the excessive amount of retained austenite in the core area that could negatively affect the plasticity of the material.

Keywords: spring steel; heat treatment; retained austenite; Mössbauer spectroscopy; neutron diffraction

1. Introduction

The modern production of environmentally friendly vehicles has imparted progress in the automotive industry, leading to the development of many innovative features. The main impacts are fuel savings, which can be provided by reducing the weight of vehicles, for example, by replacing the contemporary wiring with innovative copper-aluminum clad composite wires [1,2], by using modern lightweight construction materials [3], or by introducing electric automobiles [4].

Due to the electric car strategy, new requirements are also given to the chassis components in terms of higher strength (above 2100 N/mm^2) and sufficient toughness. Several approaches could be taken in order to obtain the required results. For example, an application of the optimized treatment with the implementation of a new processing technology, which can preferably be done via the methods for applying severe plastic deformation (SPD), such as equal channel angular pressing (ECAP) [5,6], accumulative roll bonding (ARB) [7], or rotary swaging [8,9]. The advanced treatments contribute to the optimization of the final mechanical properties via mechanical mixing and substantial grain refinement [10].

Another approach is to develop thermo-mechanically rolled fine-grained steel with a mixed microstructure consisting of martensite and/or bainite and a considerable amount of stabilized retained austenite (RA) [11,12]. In the last few years, many reports about the development of quenching and partitioning (Q&P) processes for the production of heat-treated steel with the optimal combination of high strength and ductility [13–19] can be found. All studies emphasize the key role of RA and its morphology, size, and distribution. The stability of RA also plays a key role in the prevention of tempered martensite embrittlement caused by the decomposition of RA to cementite and ferrite [20].

Grain refinement of steel microstructures was also reported to improve the stability of retained austenite by decreasing the size of blocks of RA [21]. Thermo-mechanical rolling is used to obtain the fine grain structure of the hot rolled wire products [22], along with the vanadium and/or niobium micro alloying of steel [23,24]. Vanadium micro-alloying also suppresses the growth of austenite grains under higher temperatures up to 1000 °C during austenitization in comparison with V-free spring steel [25–27], which contributes to higher strength of the steel after heat treatment.

Based on the above mentioned, the vanadium micro-alloying spring steel is supposed to provide an excellent combination of strength and ductility under optimal parameters of heat treatment. At the same time, the amount of RA and its distribution in the structure should be carefully studied because RA could be a key factor in the production of the heat-treated steel with advanced properties. Hence, the precise and reliable methods of RA assessment should be implemented.

2. Experimental Material and Methods

2.1. Material

A wire rod of vanadium micro-alloyed 0.6C-0.6Mn-1.4Si-0.6Cr (wt.%) spring steel was investigated in this study. The chemical composition of the steel is presented in Table 1. A wire rod with a diameter of 17.00 mm was produced by thermo-mechanical hot rolling and was cold drawn to a diameter of 15.50 mm preceding inductive heat treatment. The microstructure of hot rolled wire rod consisted of a fine grain perlite-ferrite structure with a cementite lamellae of 17 nm and former austenite grain size of G13.0 (an average grain diameter of 3.3 μm) according to ASTM E112 (Figure 1). The tensile mechanical properties of hot rolled wire were: ultimate tensile strength (UTS) 1350 N/mm^2 with a reduction of area value of (Z) 46%.

Table 1. Chemical composition of the steel (in wt.%).

Element	C	Mn	Si	Cr	V	Ni	Mo	Fe
	0.560	0.580	1.400	0.570	0.150	0.024	0.002	balanced

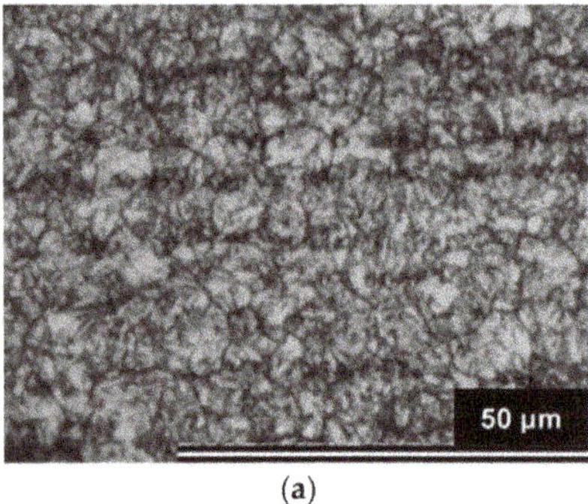

(a)

(b)

Figure 1. Austenite grain size and microstructure of hot rolled wire: (**a**) austenite grain size G13.0 according to ASTM E112; (**b**) cementite lamellae (transmission electron microscopy).

Parameters of the induction heat treatment are presented in Table 2. All specimens were inductively heated up to an austenitization temperature 850 °C. Then specimens were quenched by water spraying (Q1–Q4 specimens) to the specified temperatures. Specimens Q1 and Q2 were quenched to 40 °C and specimens Q3 and Q4 were quenched 100 °C below calculated martensite start temperature $M_S = 280\ ^{\circ}C$ (Equation (1)) [28]. Temperature M_f was estimated based on [29] as −50 °C with respect to the carbon content in the steel.

Table 2. Parameters of the induction heat treatment.

Specimen	Austenitization Temperature, °C	Temperature after Quenching, °C	Tempering Temperature, °C
Q1	850	40	-
Q2	850	40	-
Q3	850	180	-
Q4	850	180	-
QT1	850	40	460
QT2	850	40	420
QT3	850	180	460
QT4	850	180	300

Finally, specimens were inductively tempered at three different temperatures (Table 2) and cooled by water spraying until the ambient temperature (QT1–QT4 specimens). Tempering temperatures 460 °C and 300 °C were chosen as the limit values for the heat treatment of silicon alloying steel grades [20], and tempering temperature 420 °C was set in order to improve the strength characteristics of the heat-treated spring steel.

It should be additionally mentioned that measurements of quenched specimens Q3 and Q4 were considered as informative ones. Structural characteristics of these samples were affected by the slow air cooling until ambient temperature after the interrupted quenching by water spraying to 180 °C before the measurements, as the direct investigation under 180 °C required comprehensive in situ study. At the same time, measurements of QT3 and QT4 specimens fully corresponded to the investigated heat treatment parameters as they were tempered immediately after reducing to 180 °C.

$$M_S(^{\circ}C) = 521 - 353C - 225Si - 24.3Mn - 27.4Ni - 17.7Cr - 25.8Mo \quad (1)$$

Mechanical properties of the specimens were measured by means of the tensile test machine WPM UPC 1200 for quenched and tempered conditions. Hardness maps for all specimens were obtained by means of semi-automatic micro hardness equipment Durascan (Struers, Ballerup, Denmark), method HV1.

Then, three different analytical methods were used in order to characterize the obtained microstructures of specimens after heat treatment. The main focus was given to the amount of RA, its distribution, and its effect on the mechanical properties of steel.

2.2. Neutron Diffraction

The first method was the neutron powder diffraction (ND), which provided the average information from the whole bulk of the material. The room temperature diffraction patterns were collected on the MEREDIT instrument (Nuclear Physics Institute, Řež, Czech Republic) [30] at the Nuclear Physics Institute in Rez near Prague [31]. The mosaic copper monochromator (planes (220)) (Nuclear Physics Institute, Řež, Czech Republic) was used to provide neutrons with a wavelength of 1.46 Å. Data were collected between 4° and 144° of 2θ with a step of 0.08° 2θ. The samples were rotated along the vertical axis during measurement to minimize the influence of the texture and the preferred orientation on the phase fraction analysis. The full pattern structural refinements were performed using the program FullProf (Version 6.30 - Sep2018, The FullProf Suite, France) [32]. For the microstructural analysis, the instrument profile was obtained by fitting the diffraction pattern of the standard SiO_2 powder sample collected at the same conditions. Two sets of samples were investigated. The first set contained the initial cylinders of as-prepared heat-treated specimens with a diameter of 15.5 mm. The second set was manufactured from the first one after the ND measurements by removing the rim part. Only the core of the original cylindrical specimens with a square profile and dimensions of 7.5 × 7.5 mm^2 was used. The 15 mm length of the specimens was kept constant for all the measurements.

2.3. Mössbauer Spectroscopy

Mössbauer spectroscopy is a nuclear resonance spectroscopic technique based on the physical phenomenon of recoilless nuclear emission and resonant absorption of gamma rays [33]. This experimental technique provides qualitative and quantitative analysis of materials (e.g., structural, phase, and magnetic information) containing specific elements. The ^{57}Fe isotope shows the most favorable parameters for Mössbauer spectroscopy. The backscattering geometry allows us to analyze surfaces of bulk materials. Hence, ^{57}Fe Mössbauer spectroscopy has become a very important experimental method in steel characterization. In this study, scattering method utilized the conversion X-rays registration (the conversion X-rays Mössbauer spectroscopy—CXMS), which analyzes material surface up to depths of 1–20 μm.

Nuclear hyperfine interactions can be analyzed by means of Mössbauer spectroscopy [34]. The hyperfine parameters are isomer shift, quadrupole splitting, and hyperfine magnetic splitting. The isomer shift is a result of the Coulomb interaction between the nuclear/nuclei charge and the electron charge [35]. The charges distributed asymmetrically around the atomic nucleus (electrons, ions, and dipoles) increase the electric field gradient, which differs from zero on the site of nucleus. These electric quadrupole interactions cause a splitting of the excited nuclear level [35] and provide the information about bond properties and the local symmetry of iron site. [33,34,36]. The third hyperfine parameter is magnetic splitting. This magnetic field can originate within the atom itself, within crystals via exchange interactions, or it can be external one. The magnetic field (nuclear Zeeman effect) splits the nuclear states [35].

The iron-phases analysis of the specimens was established by means of Mössbauer Spectroscopy [37]. Polished non-etched metallographic cross-sections were prepared for all specimens. The measurements were performed in a backscattering geometry at room temperature with a $^{57}Co(Rh)$ source, and spectra were recorded up to 512 channels. The isomer shift referred to the calibration of the alpha-iron sample at room temperature. The Mössbauer spectra fitting procedure was carried out using MossWinn (version 4.0, Author Dr. Zoltán Klencsár, Budapest, Hungary) [38] software. Paramagnetic, γ-Fe, and RA phases were fitted by one singlet. Magnetically ordered phases (ferrites, α-Fe, and those in the martensitic structure) represented by sextet components were fitted with a magnetic hyperfine field distribution [37]. The fitting process was set as free for all parameters of the Mössbauer spectra.

2.4. Electron Backscatter Diffraction (EBSD) Analysis

The same cross-sections then were additionally used for EBSD analysis of the specimens Q1–Q4 in order to obtain the information about phase distribution within the rim and core areas of the specimens. Samples for EBSD were mechanically ground using SiC papers up to 4000 grit, and were polished using colloidal silica with a 0.06 μm particle size. Finally, the specimens were electropolished using 900 mL acetic acid and 100 mL perchloric acid at 20 °C for 10 s in order to remove the strains induced by mechanical preparation. The microstructures were investigated by scanning electron microscope Tescan Lyra 3 equipped with a NordlysNano detector operating at an accelerating voltage of 20 kV with the specimen tilted at 70°. The EBSD data were analyzed using HKL Channel 5 (version 5.11.10405.0, Oxford Instruments plc, Abingdon, United Kingdom) software developed by Oxford Instruments.

3. Results and Discussion

3.1. Mechanical Properties

In order to visualize the inhomogeneity of the microstructure after fast induction heating, the hardness maps were measured for all specimens (Figure 2). Average hardness values for all specimens Q1–Q4 and QT1–QT4, along with the results of the tensile test, are shown in Table 3.

Table 3. Mechanical properties of the specimens.

Specimen	Ultimate Tensile Strength, N/mm^2	Reduction of Area, %	Average Hardness Value, Method HV1
Q1	-	-	864
Q2	-	-	862
Q3	-	-	780
Q4	-	-	788
QT1	2114	35	635
QT2	2176	19	653
QT3	1815	36	580
QT4	1699	0	692

Based on the results of the tensile test, only samples QT1 and QT3 had the cup and cone fracture pattern after the tensile test (Figure 3a,c) and could be used for the production of chassis components. For specimen QT2 (Figure 3b), it had a non-round cup and a cone pattern, and specimen QT4 had a brittle fracture pattern (Figure 3d), which indicates the low plasticity of these specimens; the suggested schemes of the heat treatment (QT2 and QT4) with lower tempering temperatures are not appropriate for the production of the chassis components with advanced properties.

3.2. Neutron Diffraction

The neutron powder diffraction technique has an advantage for when the necessity of the average structural and microstructural information from a large volume is needed. All collected neutron diffraction patterns showed the presence of two crystallographic phases. An example of a measured and calculated neutron diffraction pattern of the specimen QT3 is presented in Figure 4. The minor phase reflections can be indexed as a face-centered cubic (FCC) lattice with a cell parameter of about 3.58 Å. This phase was recognized as RA. The major phase reflections could be within the first approximation indexed as a body-centered cubic (BCC) lattice with a cell parameter of about 2.87 Å, which looked like a fingerprint of the ferrite phase. From the crystallographic point of view, there was no possibility to distinguish a difference between ferrite and low-carbon α-martensite phase. α-martensite had the same lattice as ferrite, but due to the high dislocation and defect density, the strong microstructural reflection broadening or asymmetry could be observed, but the quantitative evaluation of the phase ration was not possible.

The detailed analysis of the neutron diffraction profile showed the significant reflection broadening of both phases surpassing the instrument broadening (significant specimen contribution). It indicates the presence of an increased microstrain in the specimens. In addition, in the case of the specimens Q1–Q4, the shape of the reflections of the ferrite phase was found to be strongly asymmetric, indicating the split of the cell parameters and the decrease of the symmetry from cubic to tetragonal. The appearance of the tetragonality indicates the presence of the martensitic phase. The results of the structural and microstructural analysis are depicted below.

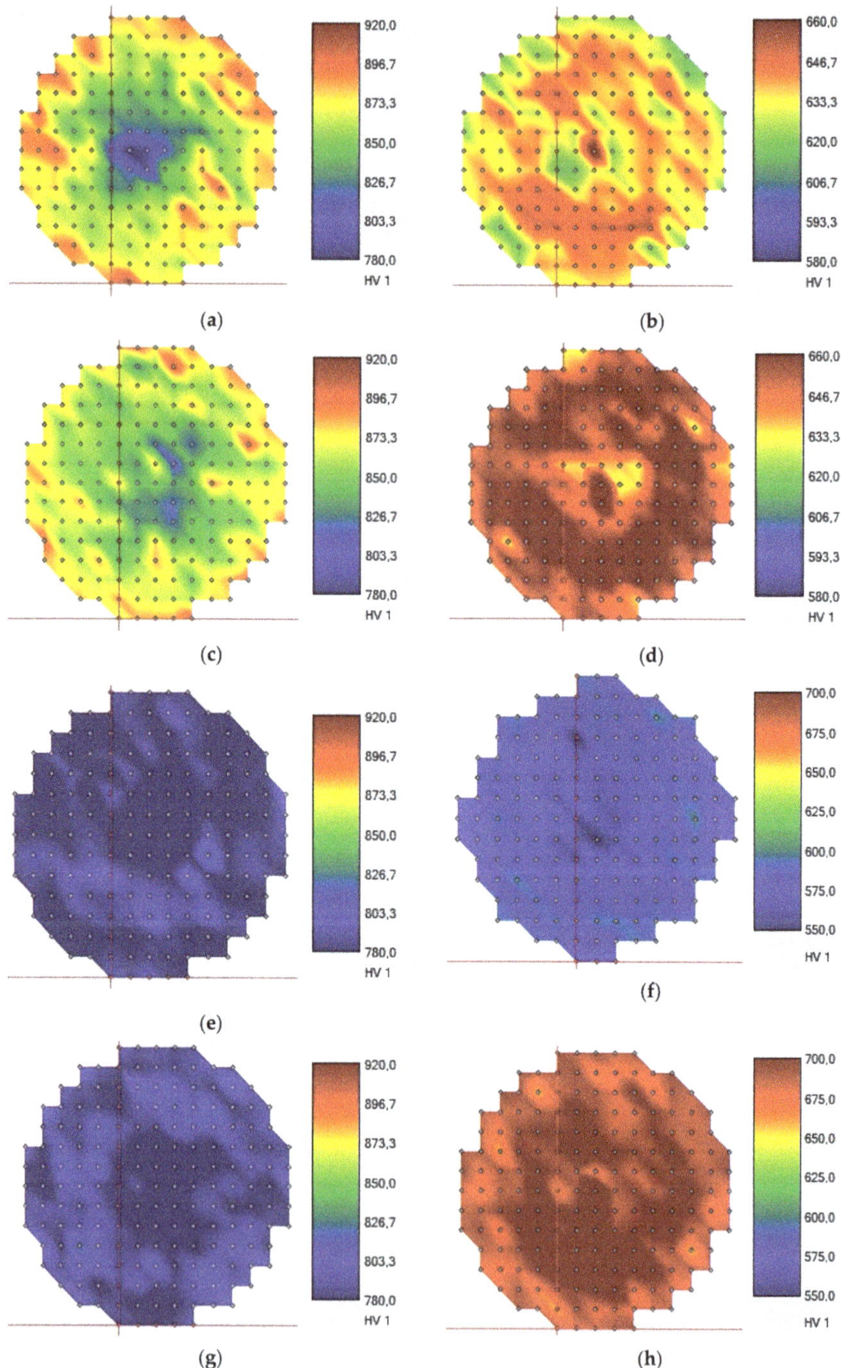

Figure 2. Hardness mapping of the specimens, method HV1: (**a**) Q1; (**b**) QT1; (**c**) Q2; (**d**) QT2; (**e**) Q3; (**f**) QT3; (**g**) Q4; (**h**) QT4.

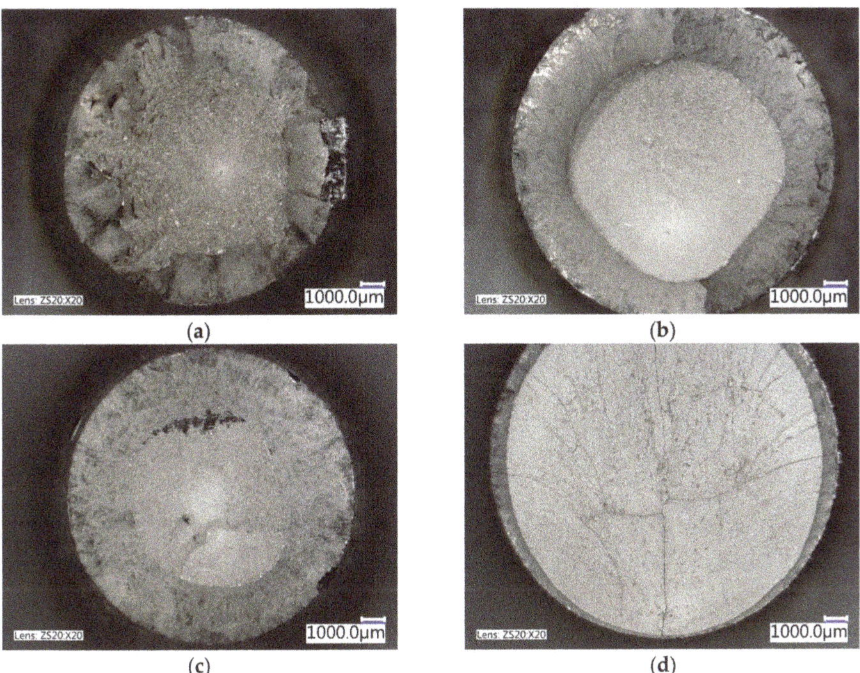

Figure 3. Fracture surface of the QT specimens: (**a**) QT1—cup and cone pattern; (**b**) QT2—non-round cup and cone pattern; (**c**) QT3—cup and cone pattern; (**d**) QT4—brittle fracture.

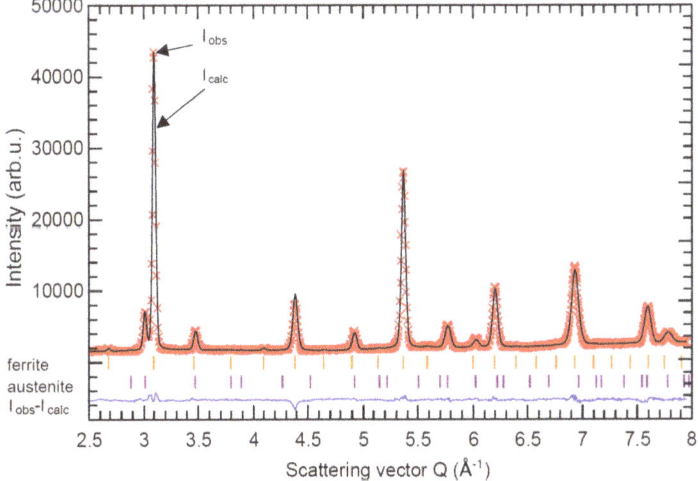

Figure 4. Measured (red crosses) and calculated (black line) neutron diffraction pattern of the specimen QT3 together with their difference (blue bottom line). Small vertical bars represent the Bragg positions for ferrite and austenite, respectively.

The amount of RA calculated from the full pattern fitting is presented in Figure 5 as a function of quenching and tempering temperatures, i.e., parameters of the heat treatment. The comparison of the measurements of the whole specimen and the core area confirmed that the excessive amount of

RA was located at the specimen center. The reason for the higher amount of RA in the core area was possibly the relatively slow heat transfer from the surface to the core during the quenching process, which leads to the significantly slower and delayed transformation within the specimen core.

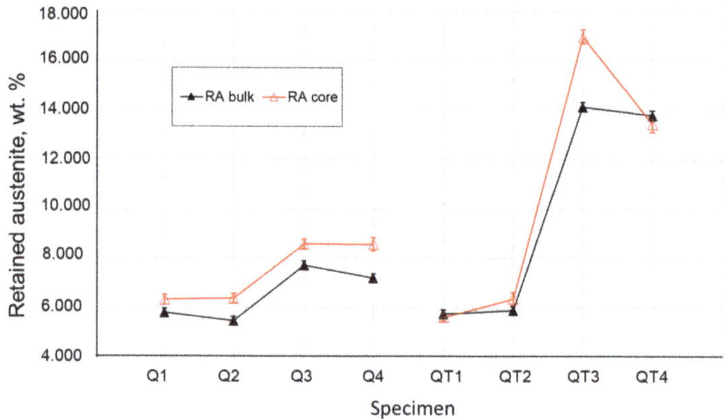

Figure 5. Weight fraction of retained austenite in (%).

In order to evaluate the influence of RA on the mechanical properties of the material, specimens QT1 and QT3 should be compared. The amount of RA in the bulk of the specimen QT1 is 2.4 times lower than in QT3, but both samples showed the reduction of area ≥35%, which is acceptable for the production of chassis components. It indicates that ductility is not affected by the higher amount of stabilized RA in the QT3 specimen, and it reaches similar values for both QT1 and QT3 specimens. Contrarily, UTS and hardness show a significant decrease for the QT3 specimen with the higher amount of the soft FCC (RA) phase.

The change in lattice parameter for ferrite/α-martensite and austenite, which perfectly describes its evolution with respect to the different parameters of heat treatment, can be seen in Figure 6. As the speed of induction heating was considered to be constant for all samples, it can be concluded that the evolution of the lattice parameter was attributed to the different temperatures and, therefore, to the carbon activity and its diffusion under such temperatures. The highest value of tetragonality, i.e., the c/a parameter, was observed for quenched specimens Q1 and Q2 (1.008 in the bulk), which confirmed the high amount of martensitic phase in the structure (Figure 6) [39]. On the contrary, the tetragonality of all QT specimens was zero, which suggests that the tempering temperature was sufficient for carbon diffusion from BCC lattice.

The lattice parameter for the FCC phase was the opposite of the BCC evolution. Obtained results also correspond with the description of phase transformation in the TRIP C-Mn-Si sheet steel studied by Yu et al. [40] by means of in situ neutron diffraction. Yu et al. attributed the expansion of the lattice parameter of FCC to the carbon enrichment of RA, and expected the shrinkage of FCC lattice after further tempering above 470 °C due to diffusion of the carbon from RA to ferrite.

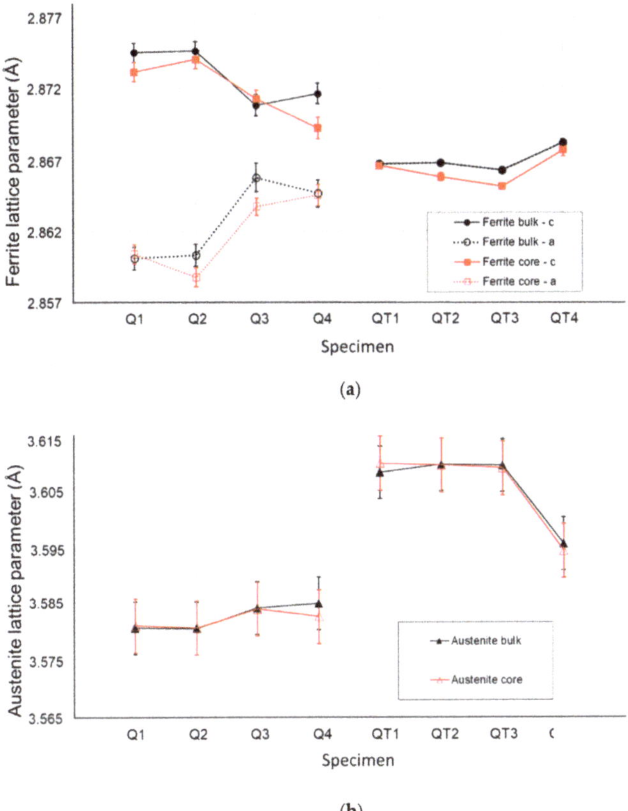

(**a**)

(**b**)

Figure 6. Lattice parameters: (**a**) body-centered cubic – ferrite; (**b**) face-centered cubic – austenite.

The evaluation of the reflection profile using an anisotropic strain-broadening model incorporated in the FullProf program (Version 6.30 - Sep2018, The *FullProf* Suite, France) revealed an average and directional μ-strain (combination of strain of type-II and type-III) of the individual phases. Figures 7 and 8 show the values of the μ-strain for all specimens, both for the bulk and the core areas for the ferrite and austenite phase, respectively. A strong directional asymmetry was found for the ferrite phase (μ-strain along 100 is significantly higher than the one along 111, see Figure 7) in comparison with the austenite phase where the asymmetry is minimal (represented by error bars in Figure 8).

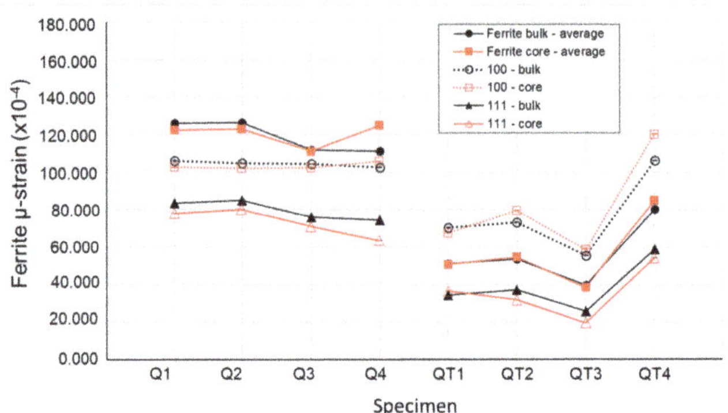

Figure 7. Distribution of µ-strain in BCC.

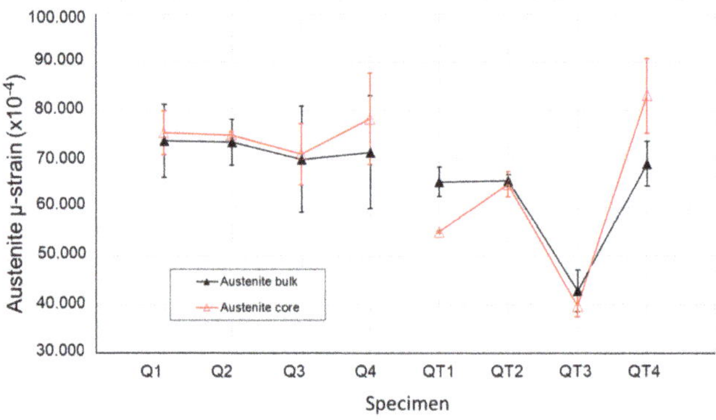

Figure 8. Distribution of µ-strain in FCC.

Specimens Q1 and Q2, which were quenched under the same conditions until ambient temperature, had comparable values of the ferrite µ-strain. A small decrease of the ferrite µ-strain was observed for Q3 and Q4 specimens, which were quenched to the higher temperature. All correlated tempered specimens QT showed a significant decrease of the µ-strain. An increase for specimen QT4 is due to the lowest tempering temperature 300 °C from the QT specimens.

The austenite µ-strain is smaller than in the ferrite phase. The values were comparable for all heat treatment samples, except the QT3 specimen where it was significantly smaller. It is in good agreement with the fact that the QT3 specimen had the highest RA content and also the highest hardness (see Table 3) in comparison to other QT specimens. The effect of the tempering temperature on the austenite µ-strain could be described by a comparison of µ-strain values in the core areas: the higher the tempering temperature (460 °C), the lower µ-strain in the core area was observed. Tempering reduced dislocation density and the µ-strain of phases depending on the tempering temperature. The quenched steel has a high dislocation density of about 10^{15} m^{-2}, which correlated with the µ-strain values that was found by Harjo et al. [41]. Lower tempering temperature (420 °C) contributed to higher values of µ-strain, i.e., internal stresses due to the dislocations and other intergranular defects, which could be the reason for the lower plasticity of the QT2 specimen in comparison with specimen QT1 (Table 3). An additional reason for the lower plasticity of the QT2 specimen could be tempering embrittlement, which was studied in [20].

Specimen QT3 had the lowest value of austenite µ-strain for both core and rim areas, which could be explained by the combination of initial low µ-strain after interrupted quenching and by following the µ-strain release after sufficient tempering under 460 °C. Sample Specimen QT4, which was similarly quenched but tempered under 300 °C, had higher values of the μ-strain that is linked with the carbon diffusion within the phases. As the carbon activity after heating up to 300 °C was low, the μ-strain value of the sample QT4 could be considered to be very close to the value in the quenched Q4 and in the Q3 specimen, respectively.

3.3. Mössbauer Spectroscopy

Backscattering the ^{57}Fe Mössbauer spectroscopy results indicates the same shape for all the Mössbauer spectra for the specimens. Figure 9 shows the recorded Mössbauer spectra for specimens Q1 and QT1, as well as Q3 and QT3, (both core and rim) after implemented fitting. The intensity of the austenite singlet lines was decreased for the QT1 specimen in comparison with Q1, which showed the decrease in the amount of RA in both bulk and rim areas.

Selected parameters that resulted from the fitting of the obtained spectra are listed in Table 4. With the amount of RA that is determined by spectral area of the phase in the spectrum (relative to all iron-bearing phases in specimen), the following parameters were obtained and are presented: δ—isomer shift and LW—spectral line width.

Table 4. Numerical results of the Mössbauer spectroscopy in comparison with neutron diffraction.

Specimen	RA ± 1.00*, %	δ ± 0.02, mm/s	LW ± 0.05, mm/s	RA Measured by Neutron Diffraction, %
Q1 bulk	7.10	−0.07	0.35	5.81 ± 0.16
Q1 rim	6.70	−0.09	0.33	
Q2 bulk	6.80	−0.09	0.40	5.48 ± 0.16
Q2 rim	6.20	−0.10	0.37	
Q3 bulk	8.60	−0.07	0.35	7.73 ± 0.16
Q3 rim	8.40	−0.05	0.31	
Q4 bulk	8.60	0.02	0.39	7.22 ± 0.15
Q4 rim	8.40	−0.06	0.33	
QT1 bulk	6.80	−0.14	0.49	5.76 ± 0.17
QT1 rim	6.10	−0.13	0.55	
QT2 bulk	6.80	−0.12	0.41	5.91 ± 0.15
QT2 rim	3.40	−0.15	0.26	
QT3 bulk	12.70	−0.12	0.44	14.13 ± 0.18
QT3 rim	9.40	−0.11	0.39	
QT4 bulk	14.10	−0.12	0.44	13.77 ± 0.19
QT4 rim	12.60	−0.13	0.37	

* The uncertainty of retained austenite measurement is 1%; however, we present the results with two decimal places to highlight differences in measurement and fitting results.

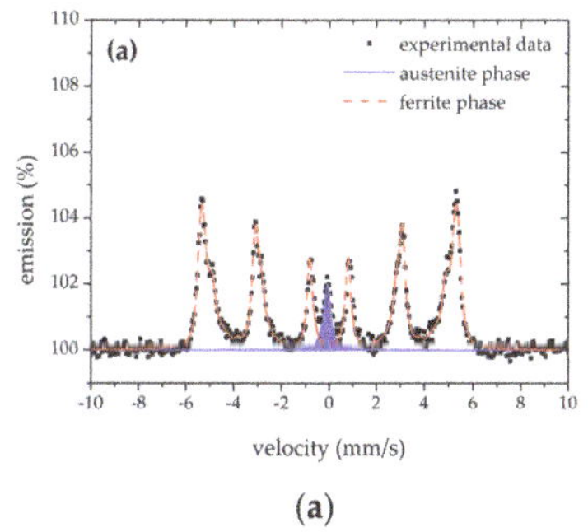

(a)

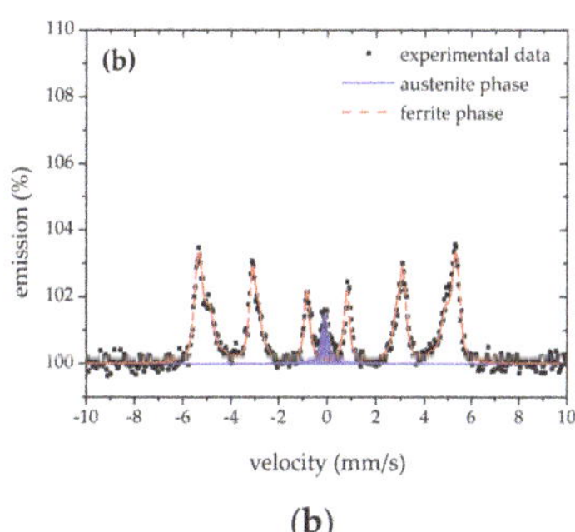

(b)

Figure 9. *Cont.*

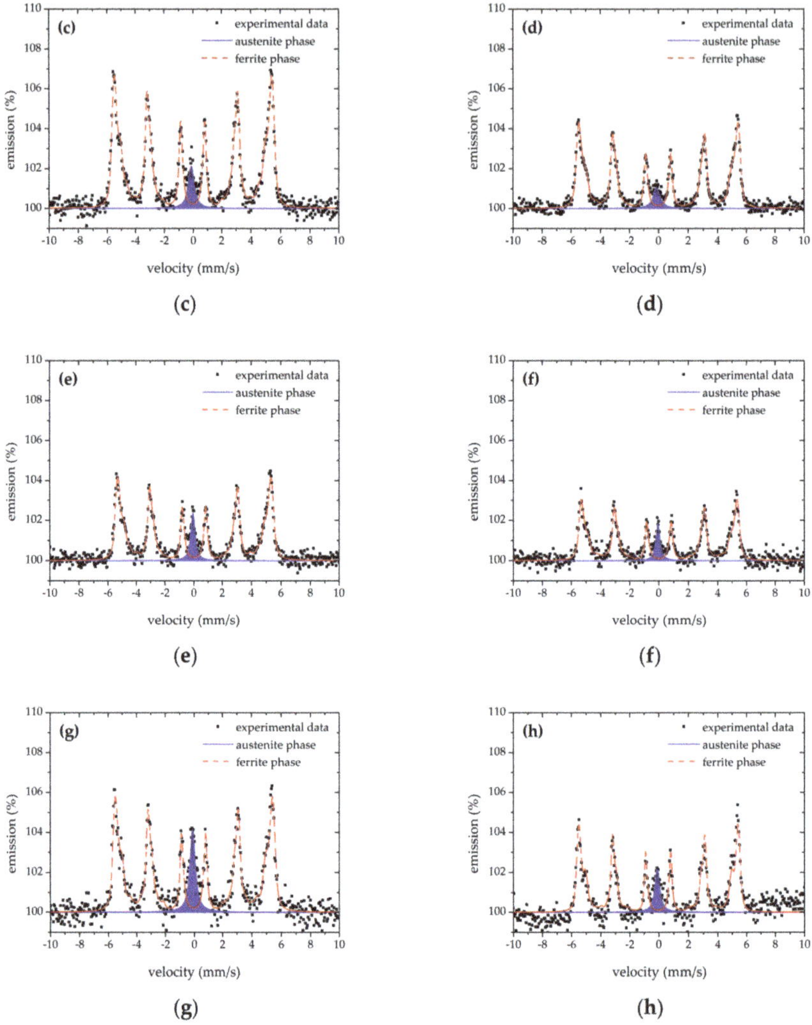

Figure 9. Mössbauer spectra: (**a**) specimen Q1—bulk; (**b**) specimen Q1—rim area; (**c**) specimen QT1—bulk; (**d**) specimen QT1—rim area; (**e**) specimen Q3—bulk; (**f**) specimen Q3—rim area; (**g**) specimen QT3—bulk; (**h**) specimen QT3—rim area.

Isomer shifts had values of around 0.00 ± 0.07 mm/s. Q samples exhibited the average hyperfine magnetic field B^*_{hf} in the interval 31.5 ± 0.5 T, while QT samples exhibited the B^*_{hf} in the interval 32.5 ± 0.5 T. No iron carbides (cementite) were distinguished in the spectra, which would exhibit a sextet component with a low B_{hf} value of ~20 T.

Only the Q4 bulk Mössbauer measurement exhibited a positive value for the RA isomer shift $\delta = 0.02$ mm/s, so the lowest value of the average hyperfine magnetic field $B^*_{hf} = 31.0$ T for the ferrite was calculated. Oppositely, the QT2 rim Mössbauer measurement exhibited the lowest (negative) value of RA isomer shift $\delta = -0.15$ mm/s, and the highest value of ferrite average hyperfine magnetic field $B^*_{hf} = 32.9$ T was also calculated. The QT2 rim specimen exhibited the lowest line width value LW = 0.26 mm/s of an RA singlet pattern. Contrarily, the highest LW = 0.55 mm/s of RA singlet was found for the QT1 rim specimen, which could inspire ideas of fitting the RA by a more complex way as a combination of a few singlets and/or doublets [42–47]. Hence, the structure of RA in such a case could be discussed. The same recoilless fraction was taking into account for both austenite and ferrite phases [43,45].

3.4. Electron Backscatter Diffraction (EBSD) Analysis

EBSD was the third method applied in this study for the quantitative analysis of the microstructure. The method was based on the collecting of the diffracted patters (the Kikuchi bands) from the focus electron beam in a scanning electron microscope with their following indexing in order to obtain the complex information about microstructure, phase distribution, crystal orientation, and strain within the polycrystalline material.

For all analyzed samples, the fine grain structure was observed. The results of the EBSD analysis for specimen Q1 (both rim and core areas) are presented in Figure 10. The amount of indexed points was on average 70–80% (Figure 10c,d) and the amount of the observed FCC fraction was about 0.01% (Figure 10e,f), which was significantly low in comparison with the results obtained by the means of neutron diffraction and the Mössbauer spectroscopy.

Based on obtained results, it can be suggested that the EBSD technique is limited in the evaluation of the amount of RA. This method cannot be applied in order to evaluate the distribution of RA in high carbon fine grain steels due to the relatively small size of RA films and also due to the high dislocation density in phase boundaries. The same limitations were also observed by Hofer et al. [12] due to their inability to resolve the ultra fine austenite films in the structure.

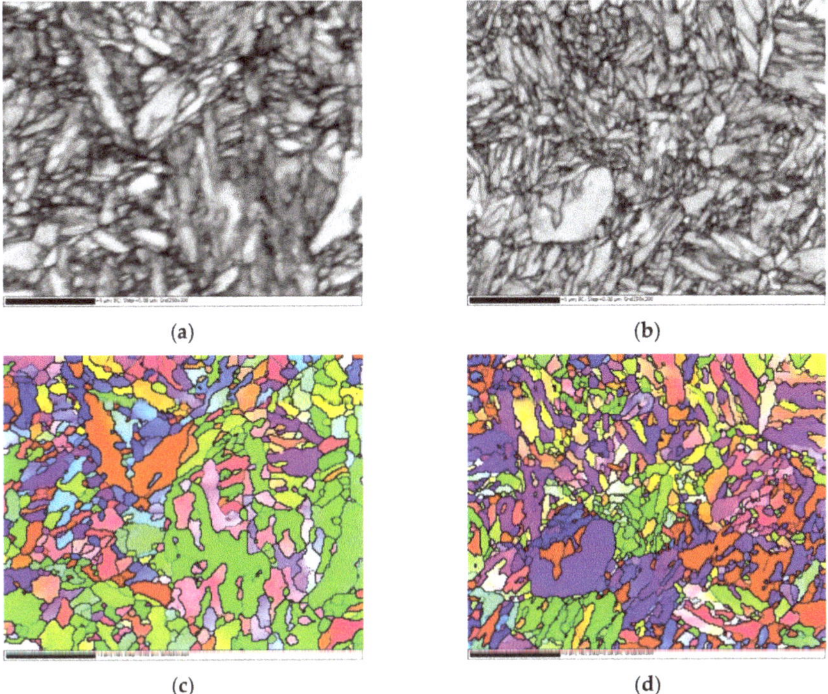

Figure 10. *Cont.*

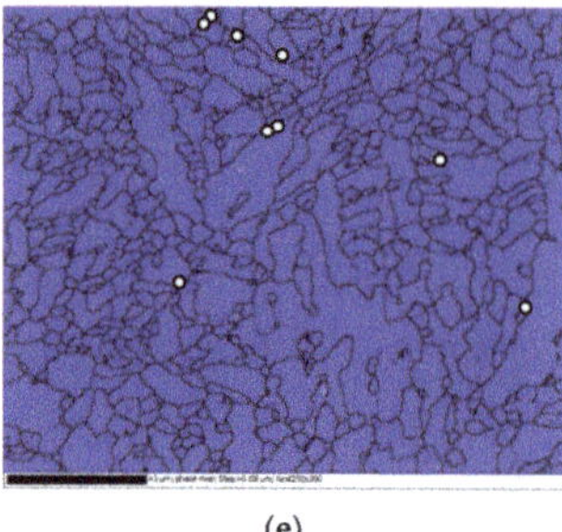

(e)

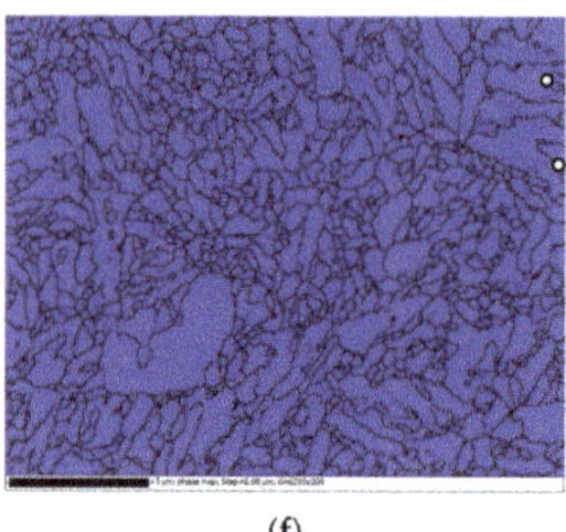

(f)

Figure 10. Results of the electron backscatter diffraction for specimen Q1: (**a**) quality image for rim area; (**b**) quality image for core area; (**c**) crystal orientation map for rim area; (**d**) crystal orientation map for core area; (**e**) phase map for rim area (blue—Fe BCC, white dots—Fe FCC); (**f**) phase map for core area (blue—Fe BCC, white dots—Fe FCC).

4. Conclusions

In this study, four schemes of the inductive heat treatment were applied for the inductive heat treatment of fine grain vanadium micro-alloyed spring steel in order to obtain the best combination of high strength (above 2100 N/mm^2) and sufficient ductility. The microstructures were studied by means of three analytical methods, and the main focus was given to the assessment of RA. The analytical techniques of neutron powder diffraction, Mössbauer spectroscopy, and EBSD analysis were applied. and the results can be concluded as following:

- Neutron diffraction provided the results from the bulk of the material. Depending on the quenching and tempering parameters, the measured amount of the RA was 5.8 (2) % for specimens QT1 and QT2, which did not change significantly from the measured Q1 and Q2 specimens, respectively. The amount of RA for specimens QT3 and QT4 was on average 14.0 (2) % because of the interrupted quenching to 180 °C, in comparison with the quenching to 40 °C for specimens QT1 and QT2. Additionally, RA was proven to distribute in the core area of the specimens due to the fast inductive heat treatment, which provides a lower transformation speed in the core area of the specimens in comparison with the surface layer.
- Backscattering the Mössbauer spectroscopy, a commercial newly developed method of RA assessment, provided the results from a polished cross-section surface (depth up to 1–20 µm from the surface) and showed a correlation with the neutron diffraction within the scatter range (see Table 4). Mössbauer spectroscopy also confirmed the excessive amount of RA in the core area of the specimens.
- The EBSD analysis is usually applied in order to describe the phase distribution in the microstructure, and this study deals with the limitations of this method. Although, the average amount of indexed points was 70–80%, the FCC phase was not resolved, which allowed the prediction of the ultra-fine RA films in the microstructure.
- Due to higher content of silicon, RA was stabilized in the microstructure after tempering, which was confirmed by the results of both the neutron diffraction and Mössbauer spectroscopy, and the ultra-fine RA films around the martensitic phase were predicted by the EBSD method.
- Although stabilized RA did not deteriorate the ductility and plasticity of the inductive quenched and tempered specimens QT1 and QT3, a high amount of the soft FCC phase reduced the tensile strength and hardness of the heat-treated steel. Hence, the suggested interrupted quenching for specimen QT3 was found to not be reliable for the production of inductive heat-treated spring steel with advanced properties and a UTS above 2100 N/mm^2.
- The differences in mechanical properties of the specimens QT after the heat treatment were explained not only based on the amount of RA, but also based on the µ-strain in the BCC and

FCC phases measured by neutron diffraction. Specimen QT1 had the best combination of strength and plasticity due to a higher amount of martensite in the structure (after quenching to 40 °C) and sufficient tempering (460 °C), which contributed to the increase of the tensile strength and decrease of the μ-strain, especially in the core area, respectively. Lower tempering temperatures lead to higher μ-strain values and they support the brittle fracture.

Author Contributions: Conceptualization, A.O. and M.P. (Martin Petrenec); Funding acquisition, M.P. (Miroslav PISKA); Investigation, C.H., P.B. and J.P.; Validation, P.K.; Writing—original draft, A.O.; Writing—review & editing, M.P. (Miroslav PISKA), M.P. (Martin Petrenec), C.H., P.B., J.P. and P.K.

Funding: This research was funded by FME BUT Brno, Czech Republic as the "Research of perspective production technologies", FSI-S-19-6014.

Acknowledgments: This research work was supported by the FME BUT Brno, Czech Republic as the "Research of perspective production technologies", FSI-S-19-6014 and measurements of neutron powder diffraction were carried out at the CANAM infrastructure of the NPI CAS Rez supported through MŠMT project No. LM2015056. Presented results were obtained with the use of infrastructure Reactors LVR-15 and LR-0, which is financially supported by the Ministry of Education, Youth and Sports - project LM2015074". Lukas Kouril is gratefully acknowledged for his help with Mössbauer spectroscopy measurements.

Conflicts of Interest: The authors declare no conflict of interest.

References

1. Kocich, R.; Kunčická, L.; Král, P.; Strunz, P. Characterization of innovative rotary swaged Cu-Al clad composite wire conductors. *Mater. Design.* **2018**, *160*, 828–835. [CrossRef]
2. Kunčická, L.; Kocich, R.; Dvořák, K.; Macháčková, A. Rotary swaged laminated Cu-Al composites: Effect of structure on residual stress and mechanical and electric properties. *Mater. Sci. Eng. A.* **2019**, *742*, 743–750. [CrossRef]
3. Naizabekov, A.B.; Andreyachshenko, V.A.; Kocich, R. Study of deformation behavior, structure and mechanical properties of the AlSiMnFe alloy during ECAP-PBP. *Micron.* **2013**, *44*, 210–217. [CrossRef] [PubMed]
4. Bobeth, S.; Matthies, E. New opportunities for electric car adoption: the case of range myths, new forms of subsidies, and social norms. *Energy Effic.* **2018**, *11*, 1763–1782. [CrossRef]
5. Kocich, R.; Macháčková, A.; Kursab, M. FEA of plastic flow in AZ63 alloy during ECAP process. *Acta. Phys. Pol. A.* **2012**, *122*, 581–587. [CrossRef]
6. Kunčická, L.; Kocich, R.; Ryukhtin, V.; Cullen, J.C.T.; Lavery, N.P. Study of structure of naturally aged aluminium after twist channel angular pressing. *Mater. Charact.* **2019**, *152*, 94–100. [CrossRef]
7. Kocich, R.; Macháčková, A.; Fojtík, F. Comparison of strain and stress conditions in conventional and ARB rolling processes. *Int. J. Mech. Sci.* **2012**, *64*, 54–61. [CrossRef]
8. Kunčická, L.; Kocich, R.; Hervoches, C.; Macháčková, A. Study of structure and residual stresses in cold rotary swaged tungsten heavy alloy. *Mater. Sci. Eng. A.* **2017**, *704*, 25–31. [CrossRef]
9. Kocich, R.; Kunčická, L.; Dohnalík, D.; Macháčková, A.; Šofer, M. Cold rotary swaging of a tungsten heavy alloy: Numerical and experimental investigations. *Int. J. Refract. Met. Hard Mater.* **2016**, *61*, 264–272. [CrossRef]
10. Kocich, R.; Kunčická, L.; Macháčková, A.; Šofer, M. Improvement of mechanical and electrical properties of rotary swaged Al-Cu clad composites. *Mater. Design.* **2017**, *123*, 137–146. [CrossRef]
11. Varshney, A.; Sangal, S.; Kundu, S.; Mondal, K. Super strong and highly ductile low alloy multiphase steels consisting of bainite, ferrite and retained austenite. *Mater. Design.* **2016**, *95*, 75–88. [CrossRef]
12. Hofer, C.; Winkelhofer, F.; Clemens, H.; Primig, S. Morphology change of retained austenite during austempering of carbide-free bainitic steel. *Mater. Sci. Eng. A.* **2016**, *664*, 236–246. [CrossRef]
13. Paravicini Bagliani, E.; Santofimia, M.J.; Zhao, L.; Sietsma, J.; Anelli, E. Microstructure, tensile and toughness properties after quenching and partitioning treatments of a medium-carbon steel. *Mater. Sci. Eng. A.* **2013**, *559*, 486–495. [CrossRef]
14. De Diego-Calderón, I.; Sabirov, I.; Molina-Aldareguia, J.M.; Föjer, C.; Thiessen, R.; Petrov, R.H. Microstructural design in quenched and partitioned (Q&P) steels to improve their fracture properties. *Mater. Sci. Eng. A.* **2016**, *657*, 136–146.

15. Gao, G.; Zhang, B.; Cheng, C.; Zhao, P.; Zhang, H.; Bai, B. Very high cycle fatigue behaviors of bainite/martensite multiphase steel treated by quenching-partitioning-tempering process. *Int. J. Fatigue.* **2016**, *92*, 203–210. [CrossRef]
16. Luo, P.; Gao, G.; Zhang, H.; Tan, Z.; Misra, R.D.K.; Bai, B. On structure-property relationship in nanostructured bainitic steel subjected to the quenching and partitioning process. *Mater. Sci. Eng. A.* **2016**, *661*, 1–8. [CrossRef]
17. Guia, X.; Gaoa, G.; Guoa, H.; Zhaoa, F.; Tana, Z.; Bai, B. Effect of bainitic transformation during BQ & P process on the mechanical properties in an ultrahigh strength Mn-Si-Cr-C steel. *Mater. Sci. Eng. A.* **2017**, *684*, 598–605.
18. HajyAkbary, F.; Sietsma, J.; Miyamoto, G.; Kamikawa, N.; Petrov, R.H.; Furuhara, T.; Santofimia, M.J. Analysis of the mechanical behavior of a 0.3C-1.6Si-3.5Mn (wt%) quenching and partitioning steel. *Mater. Sci. Eng. A.* **2016**, *667*, 505–514. [CrossRef]
19. Hao, Q.; Qin, S.; Liu, Y.; Zuo, X.; Chen, N.; Huang, W.; Rong, Y. Effect of retained austenite on the dynamic tensile behavior of a novel quenching-partitioning-tempering martensitic steel. *Mater. Sci. Eng. A.* **2016**, *662*, 16–25. [CrossRef]
20. Horn, R.M.; Ritchie, R.O. Mechanism of tempered martensite embrittlement in low alloy steels. *Metal. Trans. A* **1978**, *9*, 1039–1053. [CrossRef]
21. Lan, H.F.; Du, L.X.; Li, Q.; Qiu, C.L.; Li, J.P.; Misra, R.D.K. Improvement of strength-toughness combination in austempered low carbon bainitic steel: The key role of refining prior austenite grain size. *J. Alloy. Compd.* **2017**, *710*, 702–710. [CrossRef]
22. DeArdo, A.J.; Garcia, C.I.; Palmiere, E.J. Thermomechanical Processing of Steels. In *ASM Handbook, Vol 4: Heat Treating*, 3rd ed.; ASM International: Geauga County, OH, USA, 1991; pp. 237–255.
23. Zahng, C.; Liu, Y.; Jiang, C.; Xiao, J. Effect of niobium and vanadium on hydrogen-induced delayed fracture in high strength spring steel. *J. Iron Steel Res. Int.* **2011**, *18*, 49–53. [CrossRef]
24. Yu, Q.; Sun, Y. Abnormal growth of austenite grain of low-carbon steel. *Mater. Sci. Eng. A* **2006**, *420*, 34–38. [CrossRef]
25. Yang, G.; Sun, X.; Li, Z.; Li, X.; Yong, O. Effects of vanadium on the microstructure and mechanical properties of a high strength low alloy martensite steel. *Mater. Design.* **2013**, *50*, 102–107. [CrossRef]
26. Moropoulos, S.; Karagiannis, S.; Ridley, N. The effect of austenitising temperature on prior austenite grain size in low-alloyed steel. *Mater. Sci. Eng. A.* **2008**, *483–484*, 735–739. [CrossRef]
27. Muszka, K.; Hodgson, P.D.; Majta, J. Study of the effect of grain size on the dynamic mechanical properties of microalloyed steels. *Mater. Sci. Eng. A* **2009**, *500*, 25–33. [CrossRef]
28. Canale, L.C.F.; Totten, G.E. Steel heat treatment failures due to quenching. In *Failure Analysis of Heat Treated Components*, 1st ed.; Canale, L.C.F., Mesquita, R.A., Totten, G.E., Eds.; ASM International: Geauga County, OH, USA, 2008; pp. 255–284.
29. Kobasko, N.; Aronov, M.; Powell, J.; Vanas, J. Intensive Quenching of Steel Parts: Equipment and Method. In Proceedings of the 7th IASME/WSEAS International Conference on Health Transfer, Thermal Engineering and Environment, Moscow, Russia, 20–22 August 2009.
30. Neutron Diffraction Experiments on MEREDIT Instrument. Available online: http://www.xray.cz/xray/csca/kol2010/abst/beran.htm (accessed on 16 October 2019).
31. Mikula, P. Past and Present Status of Neutron Scattering at the Research Reactor in Řež. *Mater. Struct.* **2006**, *13*, 51–62.
32. Rodrıguez-Carvajal, J. Recent advances in magnetic structure determination by neutron powder diffraction. *Phys. B* **1993**, *192*, 55–69. [CrossRef]
33. Greenwood, N.N.; Gibb, T.C. *Mössbauer Spectroscopy*; Springer: Amsterdam, The Netherlands, 1971; pp. 1–76.
34. Vértes, A.; Korecz, L.; Burger, K. *Mössbauer Spectroscopy*; Elsevier Scientific Pub. Co.: Amsterdam, The Netherlands, 1979; pp. 34–52.
35. Gütlich, P.; Bill, E.; Trautwein, A.X. *Mössbauer Spectroscopy and Transition Metal Chemistry: Fundamentals and Applications*; Springer: Berlin, Germany, 2011; pp. 73–135.
36. Danon, J. *Lectures on the Mössbauer Effect*; Gordon and Breach: New York, NY, USA, 1968; pp. 66–104.
37. Pechousek, J.; Kouril, L.; Novak, P.; Kaslik, J.; Navarik, J. Austenitemeter – spectrometer for rapid determination of residual austenite in steels. *Measurement* **2019**, *131*, 671–676. [CrossRef]
38. Klencsár, Z.; Kuzmann, E.; Vértes, A. User-friendly software for Mössbauer spectrum analysis. *J. Radioanal. Nucl. Chem.* **1996**, *210*, 105–118. [CrossRef]

39. Vatavuk, J.; Canale, L.C.F. Steel failures due to tempering and isothermal heat treatment. In *Failure Analysis of Heat Treated Components*, 1st ed.; Canale, L.C.F., Mesquita, R.A., Totten, G.E., Eds.; ASM International: Geauga County, OH, USA, 2008; pp. 285–309.
40. Yu, D.; Chen, Y.; Huang, L.; An, K. Tracing phase transformation and lattice evolution in a TRIP sheet steel under high-temperature annealing by real-time in situ neutron diffraction. *Crystals* **2018**, *131*, 360. [CrossRef]
41. Harjo, S.; Kawasaki, T.; Gong, W.; Aizawa, K. Dislocation characteristics in lath martensitic steel by neutron diffraction. *J. Phys. Conf. Ser.* **2016**, *476*, 1–7. [CrossRef]
42. Varga, I.; Kuzmann, E.; Vértes, A. Kinetics of a-c phase transition of Fe-12Cr-4Ni Alloy aged between 500–650 °C. *Hyperfine Interact.* **1998**, *112*, 169–174. [CrossRef]
43. Schwartzendruber, L.J.; Bennett, L.H.; Schoefer, E.A.; Delong, W.T.; Campbell, H.C. Mössbauer-effect examination of ferrite in stainless steel welds and castings. *Welding Res. Suppl.* **1974**, *53*, 1–12.
44. Jha, B.K.; Mishra, N.S. Microstructural evolution during tempering of a multiphase steel containing retained austenite. *Mater. Sci. Eng. A* **1999**, *263*, 42–55. [CrossRef]
45. Besoky, J.I.; Danon, C.A.; Ramos, C.P. Retained austenite phase detected by Mössbauer spectroscopy in ASTM A335 P91 steel submitted to continuous cooling cycles. *J. Mater. Res. Technol.* **2019**, *8*, 1888–1896. [CrossRef]
46. Pierce, D.T.; Coughlin, D.R.; Williamson, D.L.; Clarke, K.D.; Clarke, A.J.; Speer, J.G.; De Moor, E. Characterization of transition carbides in quench and partitioned steel microstructures by Mössbauer spectroscopy and complementary techniques. *Acta. Mater.* **2015**, *90*, 417–430. [CrossRef]
47. Uwakweh, O.N.C.; Bauer, J.P.; Génin, J.-M.R. Mössbauer Study of the Distribution of Carbon Interstitials in Iron Alloys and the Isochronal Kinetics of the Aging of Martensite: The Clustering-Ordering Synergy. *Metal. Trans. A* **1900**, *21*, 589–602. [CrossRef]

Article

Strain Range Dependent Cyclic Hardening of 08Ch18N10T Stainless Steel—Experiments and Simulations

Jaromír Fumfera [1], Radim Halama [2,*], Radek Procházka [3] and Petr Gál [2,4] and Miroslav Španiel [1]

1 Department of Mechanics, Biomechanics and Mechatronics, Faculty of Mechanical Engineering, Czech Technical University in Prague, Technicka 4, 16000 Prague 6, Czech Republic; jaromir.fumfera@fs.cvut.cz (J.F.); miroslav.spaniel@fs.cvut.cz (M.Š.)

2 Department of Applied Mechanics, Faculty of Mechanical Engineering, VSB-Technical University of Ostrava, 17.listopadu 2172/15, 70800 Ostrava, Czech Republic; Petr.Gal@ujv.cz

3 Comtes FHT, a. s., Průmyslová 995, 334 41 Dobřany, Czech Republic; rprochazka@comtesfht.cz

4 ÚJV Řež, a. s., Hlavní 130, Řež, 250 68 Husinec, Czech Republic

* Correspondence: radim.halama@vsb.cz; Tel.: +420-597-321-288; Fax: +420-596-916-490

Received: 18 November 2019; Accepted: 12 December 2019; Published: 17 December 2019

Abstract: This paper describes and presents an experimental program of low-cycle fatigue tests of austenitic stainless steel 08Ch18N10T at room temperature. The low-cycle tests include uniaxial and torsional tests for various specimen geometries and for a vast range of strain amplitude. The experimental data was used to validate the proposed cyclic plasticity model for predicting the strain-range dependent behavior of austenitic steels. The proposed model uses a virtual back-stress variable corresponding to a cyclically stable material under strain control. This internal variable is defined by means of a memory surface introduced in the stress space. The linear isotropic hardening rule is also superposed. A modification is presented that enables the cyclic hardening response of 08Ch18N10T to be simulated correctly under torsional loading conditions. A comparison is made between the real experimental results and the numerical simulation results, demonstrating the robustness of the proposed cyclic plasticity model.

Keywords: austenitic steel 08Ch18N10T; cyclic plasticity; cyclic hardening; experiments; finite element method; low-cycle fatigue

---

1. Introduction

Austenitic stainless steels, for example, 316L in PWR (pressurized water reactor) and 08Ch18N10T in the Russian VVER concept (water–water power reactor), are usually used for components in primary circuit reactor internals (a block consisting of guided tubes, a core barrel, a core barrel bottom and a core shroud), in main primary pipes, and so forth. During their design life, these components must withstand mechanical operational loads (e.g., pressure pulses and vibrations), thermal loads (regimes such as heating up and shut-downs), corrosive loads and also irradiation. These regimes subject the reactor internals to cyclic loading.

When designing or assessing the long term operation of existing structural components, it is necessary to include fatigue evaluations. In the last decade, the finite element method (FEM) with phenomenological models has mainly been used in practical applications [1]. A description and a short history of the development of constitutive models of cyclic plasticity has been provided by the authors in a previous publication [2]. Their goal is to describe as accurately as possible the stress-strain behavior of the material, which is found on the basis of experiments under cyclic loading conditions [3]. A small deviation in the stress-strain prediction can lead to a major fatigue error, especially in low cycle fatigue. In this case, stainless steels show cyclic hardening in the initial stage, followed by cyclic softening [4,5]. This phenomenon depends on the strain range and also on the type of loading. Non-proportional loading induces more cyclic hardening than proportional loadings. The most sensitive materials are materials with low stacking fault energy, for example, austenitic stainless steels [6]. Low-cycle fatigue tests of this type were presented for example, by Jin et al. in Reference [7]. They presented results for 316L stainless steel under proportional and non-proportional loadings. In another study, Xing et al. [8] presented the results of experimental testing on 316L stainless steel under proportional and non-proportional loadings with various strain amplitudes. The authors also presented a numerical study and compared the numerical results with the experimental data. They used the visco-plastic numerical model, based on the Ohno-Wang kinematic hardening rule.

The temperature in VVER concept reactor usually does not exceed 350 °C in most components. Temperature effect have significant influence, which is presented in Reference [9]. The additional hardening due to non-proportional loading has been investigated by many authors. The basic concept involves modifying the isotropic or kinematic hardening rule with a non-proportional parameter. For example, Benallal and Marquis [10] introduced the non-proportional angle, which is defined as the angle between the direction of the increment in plastic deformation and the direction of the deviatoric stress. Another approach was introduced by Tanaka [11]. He introduced the fourth rank tensor, which characterizes the internal dislocation structure of the material. This parameter is dependent on the loading path.

The goal of all the studies mentioned above was to understand the behavior of the material under specific cyclic loading conditions and to provide the material data for a better fatigue and lifetime assessment of the structural parts. This paper follows up on the main author's previous paper [2] which presents some results of uniaxial low-cycle fatigue tests of austenitic stainless steel 08Ch18N10T at room temperature. The experimental program includes uniaxial tests of hourglass-type specimens and is now extended by new results for notched specimens with 3 different notch geometries considering strain amplitudes up to 3%. Torsional loading tests of notched-tube specimens are also newly presented.

In a previous paper [2], the authors presented a new constitutive material model that is used for finite element (FE) simulations of experiments on 08Ch18N10T material. The constitutive material model is based on the Chaboche model. The proposed material model is in very good agreement with uniaxial loading condition results. In this paper, the model has been modified to provide a better description of the torsional loading. This modification also enables the cyclic hardening response of 08Ch18N10T steel to be simulated correctly under torsional loading conditions. The constitutive material model is based on the memory surface introduced in the stress space, which is analogous to the theory of Jiang and Sehitoglu [12] for treating the impact of the strain amplitude on the stress response of the material. The new theory is shown on the kinematic hardening rule based on Chaboche's model with three backstress parts. Recently, an approach has been introduced that takes into account a new internal variable called virtual backstress, corresponding to a cyclically stable material. This provides an easy way to identify the parameters and to reduce the number of material parameters. A comparison between the real experimental results and the numerical simulation results demonstrates the robustness of the constitutive plasticity model.

2. Experiments

The experimental section describes the low-cycle fatigue test measurements of specimens in pure tension/compression mode and in torsion mode.

2.1. Experimental Setup

Pure axial tension-compression tests were carried out using a MAYES electromechanical testing machine with a loading capacity of 100 kN. The test specimens were placed in MTS 646 hydraulic collet grips to ensure repeatability of the alignment conditions in tensile/compression mode. The axial deformation of the specimens was controlled by an MTS 634.25 extensometer with an initial gage length of 10 mm with a 50% measuring range, for uniform gage specimens and with an initial length of 20 mm with a 20% measuring range, for elliptically-shaped specimens.

Pure torsion tests were conducted on an MTS Bionix servo-hydraulic testing machine with an axial load capacity of 25 kN and 250 Nm in torsion. The test specimens were carefully mounted in MTS 647 hydraulic wedge grips and were tested with the axial load control set to zero. An EPSILON 3550 axial/torsional extensometer was employed to measure and control the torsional shear angle with a range of $\pm 2^{\circ}$. The initial gauge length of the extensometer was 25 mm. The whole test setup is shown in Figure 1.

These tests were conducted at room temperature and were loaded using a triangular waveform at a strain rate of 0.002 s^{-1}. During the experimental measurements all channels were recorded, for example, time, force/torque, displacement/angle, axial/torsional extensometer, with a recording frequency of 20 Hz.

The digital image correlation (DIC) was used for an analysis of the 3D deformation on the surface of some specimens, see Figure 1. During cyclic loading, the frame rate was set to cover at least 20 fps per one loading cycle. The MERCURY RT optical measuring system was used to capture and analyze the 3D images. The configuration of the system consists of two 5 Mpx CMOS BONITO cameras with circular polarizing filters to reduce the glare from the reflected surface of the specimen.

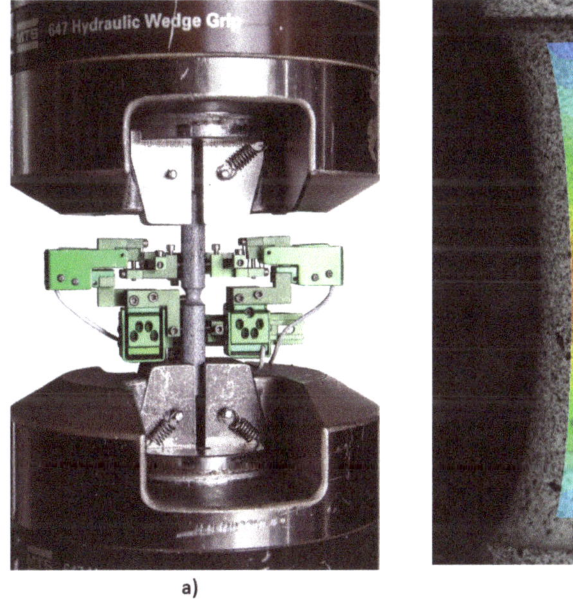

a)

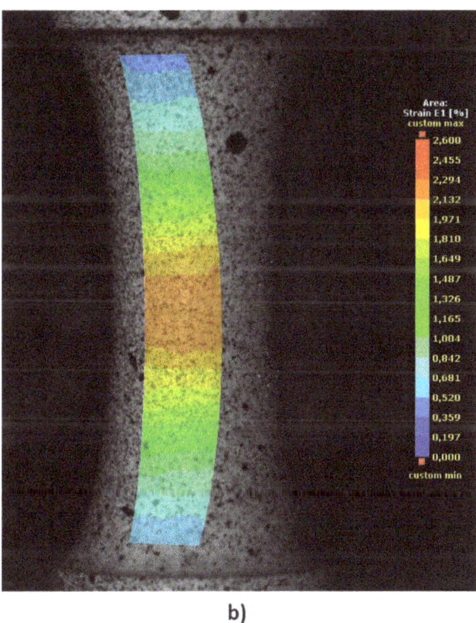

b)

Figure 1. Experiment: (**a**) Experimental Setup, (**b**) digital image correlation (DIC) Snapshot of Specimen IDF-6.

The test setup (see Figure 1) on the MTS servo-hydraulic testing machine consists of hydraulic wedge grips, a notched specimen and an EPSILON axial/torsional extensometer and a snapshot of a notched specimen (see Figure 1) under a loading with a random contrast pattern, which the DIC algorithm requires and a strain map on the surface.

The stochastic pattern on the surface of the specimen and two digital video cameras allows 3D strain measurements throughout the fatigue life until fracture, with resolution of 1100 DPI (1 px = 0.22 μm). In addition, the DIC system can continuously store all captured images in the computer memory. The fatigue life of each loading condition takes at least several dozen of cycles, even tens of thousands of cycles, which can generate up to hundreds of thousands of images to be processed. All captured images were processed later in post processing to prevent data loss. This loss occurs when the bitrate increases while real-time processing is being used.

2.2. Experimental Program

The experimental program consists of 6 series of specimens. The first series is used for the material parameter identification process. According to the ASTM E606 standard [13], the classic uniform-gage geometry of a specimen is limited to a total strain amplitude of ϵ_a = 0.5%. For higher strain levels, non-uniform hour-glass type geometry is required in order to prevent buckling. The material parameters identification series (IDF) was therefore compiled from uniform-gage (UG) specimens (see Figure 2) and non-uniform-gage specimens with an elliptical longitudinal section (E9, see Figure 2). To identify the material parameters (described in detail in Reference [2]), it is necessary to know the stress-strain curves in the cycles. For UG specimen geometry, tested according to Reference [13], this can be calculated directly from the elongation of the extensometer and from the force measured during the experiment. For E9 specimen geometry, the strain was measured by the DIC (due to the experimental setup, the strain cannot be calculated directly from the elongation of the extensometer for non-uniform gage geometries).

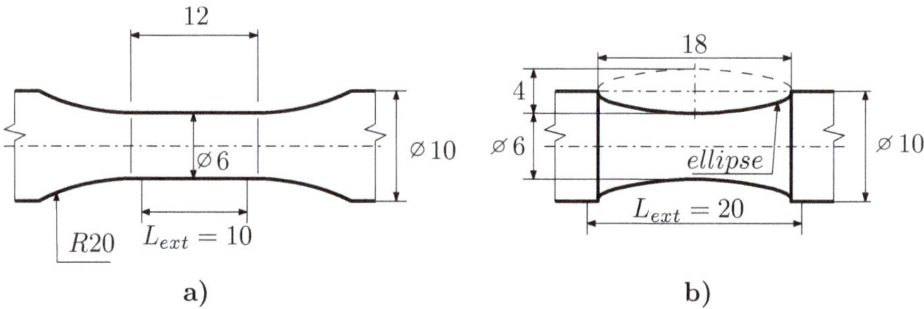

Figure 2. IDF Specimen Geometry: (**a**) UG, (**b**) E9.

The next series consists of E9 geometry (see Figure 2), notch geometry with an $R = 1.2$ mm (R1.2, see Figure 3), geometry with an $R = 2.5$ mm notch (R2.5, see Figure 3) and geometry with an $R = 5$ mm notch (R5, see Figure 4). The last series is the notched tube geometry (NT, see Figure 4), which was exposed to torsional loading.

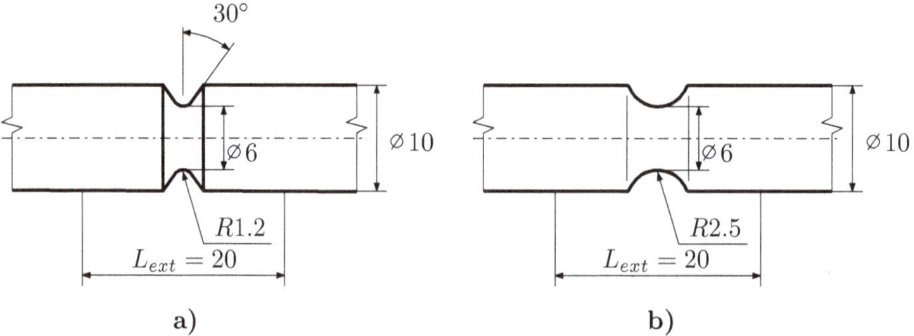

Figure 3. Notched Specimens: (**a**) R1.5, (**b**) R2.5.

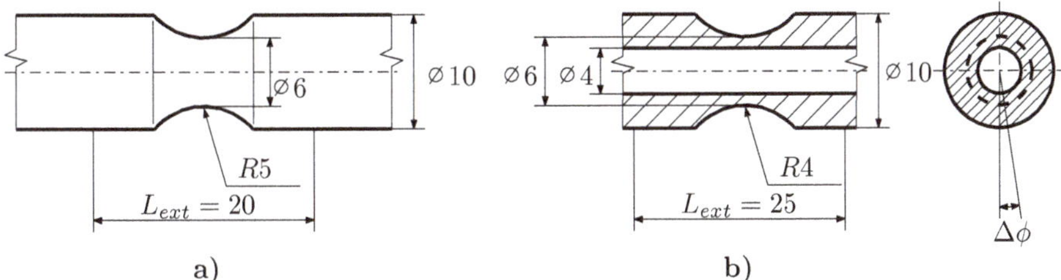

Figure 4. Notched and Notched Tube Specimens: (**a**) R5, (**b**) NT.

All boundary conditions of the experiments and their simulations are together with resulting experimental lifetimes reported in Appendix A.

3. Constitutive Model with Strain Range Dependency

The concept of single yield surface plasticity with strain range dependency is used. Isothermal conditions are considered, since the influence of the strain rate is not taken into account. The constitutive model is described in detail in Reference [2], so just a brief recapitulation of some key equations is presented here.

3.1. Cyclic Plasticity and Memory Surface

The plasticity function is defined as

$$f = \sqrt{\frac{2}{3}(\boldsymbol{s} - \boldsymbol{a}) : (\boldsymbol{s} - \boldsymbol{a})} - Y = 0, \tag{1}$$

where $\boldsymbol{s}$ is the deviatoric part of stress tensor $\boldsymbol{\sigma}$, $\boldsymbol{a}$ is the deviatoric part of back-stress $\boldsymbol{\alpha}$. The actual yield surface size Y is defined as

$$Y = \sigma_y + R, \tag{2}$$

where R is the isotropic variable and σ_Y is the initial size of the yield surface. The accumulated plastic strain increment dp is defined as

$$dp = \sqrt{\frac{2}{3} d\epsilon^p : d\epsilon^p}. \tag{3}$$

The superposition of the virtual back stress parts is defined as

$$\boldsymbol{\alpha}_{virt} = \sum_{i=1}^{M} \boldsymbol{\alpha}^i_{virt}, \tag{4}$$

and for each part

$$d\boldsymbol{\alpha}^i_{virt} = \frac{2}{3} C_i d\boldsymbol{\epsilon}_p - \gamma_i \boldsymbol{\alpha}^i_{virt} dp. \tag{5}$$

For 08Ch18N10T material, three backstress parts are taken into consideration, so $M = 3$.

The evolution of the memory surface size R_M is directed by the following rule

$$dR_M = H(g) \langle \mathbf{L} : d\boldsymbol{\alpha}_{virt} \rangle \tag{6}$$

where

$$g = \|\boldsymbol{\alpha}_{virt}\| - R_M <= 0 \tag{7}$$

and

$$\boldsymbol{L} = \frac{\boldsymbol{\alpha}_{virt}}{\|\boldsymbol{\alpha}_{virt}\|}. \tag{8}$$

3.2. Isotropic Hardening

The cyclic isotropic hardening is linear in p, defined incrementally as

$$dR = R_0(R_M)dp, \tag{9}$$

where

$$R_0(R_M) = A_R R_M^2 + BR_M + C_R \text{ for } R_M \geq R_{M0} \tag{10}$$

$$R_0(R_M) = A_R R_{M0}^2 + BR_{M0} + C_R \text{ otherwise,} \tag{11}$$

where A_R, B_R, C_R and R_{M0} are material parameters.

3.3. Kinematic Hardening

Chaboche's kinematic hardening rule is used in this study. The backstress is composed of M parts

$$\boldsymbol{\alpha} = \sum_{i=1}^{M} \boldsymbol{\alpha}_i, \tag{12}$$

the memory term is a function of memory surface R_M and accumulated plastic strain p

$$d\boldsymbol{\alpha}_i = \frac{2}{3} C_i d\boldsymbol{\epsilon}_p - \gamma_i \phi(p, R_M) \boldsymbol{\alpha}_i dp, \tag{13}$$

where M, C_i and γ_i are the same as in Equation (5). Function ϕ is defined as

$$\phi(p, R_M) = \phi_0 + \phi_{cyc}(p, R_M), \tag{14}$$

where ϕ_0 is a material parameter. ϕ_{cyc} is a function defined as follows

$$d\phi_{cyc} = \omega(R_M) \cdot \left(\phi_\infty + \phi_{cyc}(p, R_M)\right) dp \tag{15}$$

$$\phi_\infty(R_M) = A_\infty R_M^4 + B_\infty R_M^3 + C_\infty R_M^2 + D_\infty R_M + E_\infty \tag{16}$$

$$\omega(R_M) = A_\omega + B_\omega R_M^{-C_\omega} \text{ for } R_M \geq R_{M\omega} \tag{17}$$

$$\omega(R_M) = A_\omega + B_\omega R_{M\omega}^{-C_\omega} \text{ otherwise} \tag{18}$$

where A_∞, B_∞, C_∞, D_∞, E_∞, A_ω, B_ω, C_ω and $R_{M\omega}$ are material parameters.

3.4. Modification for Torsional Loading

The original plasticity model shows very good prediction under uniaxial loading conditions [2]. The original model also predicts well for notched specimen geometries but produces an error of up to about 15 % under shear stress loading conditions, as will be shown in Section 6. For a low loading level (see Figure 5a)), where there is limited cyclic hardening, the prediction of the original model [2] is satisfactory. For a high loading level (see Figure 5b)), the model overpredicts the cyclic hardening under dominant shear stress loading conditions and the formulation of the material model needs to be modified.

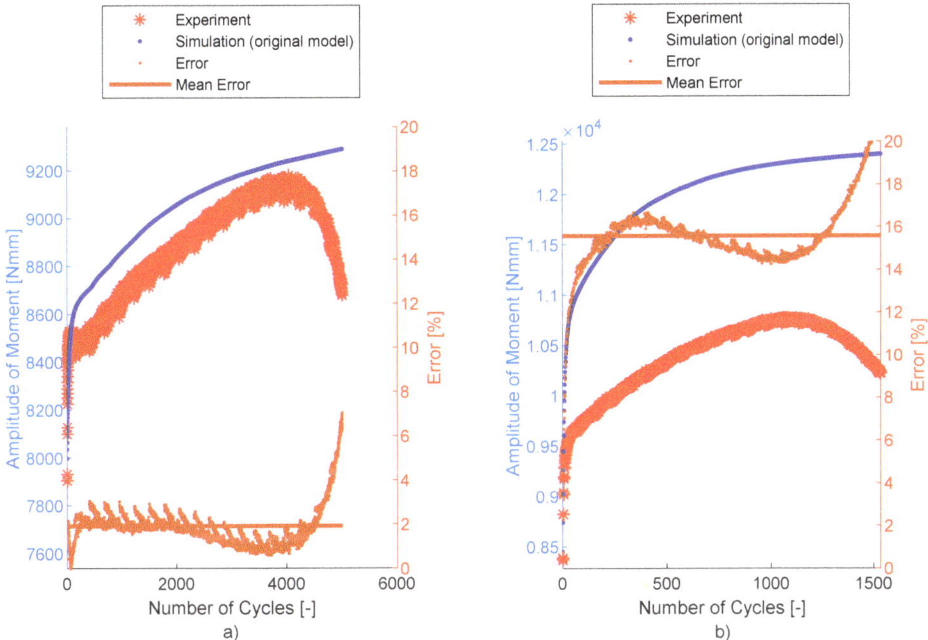

Figure 5. Original model under torsional loading: (**a**) specimen NT-1 (low loading level), (**b**) specimen NT-6 (high loading level).

The first modification of the original model [2] is to separate the memory surface function into two memory surfaces. Memory surface R_M for the isotropic hardening part remains the same as in the original model defined by the set of Equations (4)–(8). The new memory surface $R_{M\phi}$ for the kinematic hardening part is modified and is defined by analogy as

$$\boldsymbol{\alpha}_{virt\phi} = \sum_{i=1}^{M} \boldsymbol{\alpha}^{i}_{virt\phi} \tag{19}$$

$$d\boldsymbol{\alpha}^{i}_{virt\phi} = \frac{2}{3} C_i d\boldsymbol{\varepsilon}_p - \gamma_i K \boldsymbol{\alpha}^{i}_{virt\phi} dp, \tag{20}$$

where

$$K = (\delta_{IJ} + (1 - \delta_{IJ}) K_{shear}), \tag{21}$$

where δ_{IJ} is Kronecker delta, I, J are indexes of stress tensor σ and K_{shear} is a new material parameter. The rest of the equations for defining the memory surface of the kinematic hardening part remain analogous to the original model [2]:

$$dR_{M\phi} = H(g_{\phi}) \left\langle \mathbf{L}_{\phi} : d\boldsymbol{\alpha}_{virt\phi} \right\rangle \tag{22}$$

where

$$g_{\phi} = \left\| \boldsymbol{\alpha}_{virt\phi} \right\| - R_{M\phi} <= 0 \tag{23}$$

and

$$\mathbf{L}_{\phi} = \frac{\boldsymbol{\alpha}_{virt\phi}}{\left\| \boldsymbol{\alpha}_{virt\phi} \right\|}. \tag{24}$$

A quick analysis of this modified formulation shows that it provides practically the same prediction in uniaxial loading conditions (because $R_{M\phi} \simeq R_M$) as the original formulation. However, depending on the value of K_{shear}, it can give a different prediction under shear loading conditions: it is more effective for higher loading levels than for lower loading levels and it can reduce the over prediction of the model for $K_{shear} > 1$.

The second modification to the original model [2], also associated with the memory surface, is to omit limits $R_{M\omega}$ and R_{M0} and to set boundaries of the memory surfaces instead: R_M^{min} and R_M^{max}. The value of the memory surface R_M and $R_{M\phi}$ used for controlling the isotropic and kinematic hardening part can lie only between these two bounds. For simplification and for mathematically correct expression, the memory surface size that is used, R_M^{used}, is defined as

$$R_M^{used} = R_M^{min} \text{ for } R_M < R_M^{min} \tag{25}$$

$$R_M^{used} = R_M \text{ for } R_M^{min} < R_M < R_M^{max} \tag{26}$$

$$R_M^{used} = R_M^{max} \text{ for } R_M > R_M^{max} \tag{27}$$

and analogously for $R_{M\phi}^{used}$. The variable R_M in Equations (13)–(18) of the original model is simply replaced by variable $R_{M\phi}^{used}$. The modified form of the kinematic hardening equations is now

$$d\boldsymbol{\alpha}_i = \frac{2}{3} C_i d\boldsymbol{\epsilon}_p - \gamma_i \phi(p, R_{M\phi}^{used}) \boldsymbol{\alpha}_i dp \tag{28}$$

$$\phi(p, R_{M\phi}^{used}) = \phi_0 + \phi_{cyc}(p, R_{M\phi}^{used}) \tag{29}$$

$$d\phi_{cyc} = \omega(R_{M\phi}^{used}) \cdot \left(\phi_{\infty} + \phi_{cyc}(p, R_{M\phi}^{used}) \right) dp \tag{30}$$

$$\phi_{\infty}(R_{M\phi}^{used}) = A_{\infty}(R_{M\phi}^{used})^4 + B_{\infty}(R_{M\phi}^{used})^3 + C_{\infty}(R_{M\phi}^{used})^2 + D_{\infty} R_{M\phi}^{used} + E_{\infty} \tag{31}$$

$$\omega(R_{M\phi}^{used}) = A_{\omega} + B_{\omega}(R_{M\phi}^{used})^{-C_{\omega}}. \tag{32}$$

The third modification of the original model [2] is the definition of the formulation of isotropic hardening as a non-linear formulation in p as

$$dR = A_R \cdot \exp(B_R \cdot R_M^{used}) \cdot p^{C_R}, \tag{33}$$

where A_R, B_R and C_R are material parameters. This very important modification deserve a short analysis. In the original model [2], for cyclic loading, the actual yield stress Y increases practically linearly with the number of cycles. This means that with many cycles, the actual yield stress Y can theoretically go higher than the total stress amplitude and the computed deformation becomes only elastic.

4. Identification of Material Parameters

The material parameter identification process for 08Ch18N10T is based on knowing the shape of the stress-strain hysteresis loops during the fatigue life. A total of twelve uniaxial specimens and eight torsional specimens are used for the identification process. This is described in detail in References [14] and [2], so just a brief recapitulation of the key steps updated by the unique features of the proposed modification to the material model is done here.

The Young modulus E, the Poisson ratio μ and the yield strength σ_y are obtained from a tensile test. The actual yield strength evolution during the fatigue life is determined using the root mean square error method. Chaboche material parameters C_1, γ_1, C_2, γ_2, C_3, γ_3 are identified from two selected hysteresis loops (the bigger loop and the smaller loop).

The first guess of the memory surface size R_M for each specimen is computed. It is assumed here that $R_{M\phi} \simeq R_M$. Boundary parameters R_M^{min} and R_M^{max} are simply the maximum and minimum values of R_M computed in the identification process.

The actual yield stress is fitted as a function of R_M^{used} and parameters A_R, B_R, C_R are found from Equation (33).

Using the experimental data from the tensile test and performing a simulation of this test, parameter ϕ_0 is found based on the Equation (13) as an optimal value of ϕ. The value of function ϕ from Equation (13) is found, using a similar optimization process as for determining the Chaboche material parameters. ϕ_∞ is the value of ϕ for $n = N_d$, where N_d is the number of cycles after which the crack occurs on the specimen and the force starts to drop during the experiment. From Equation (16), ϕ_∞ is then set as a function of R_M^{used} by finding material parameters A_∞, B_∞, C_∞, D_∞, E_∞.

For each NT geometry specimen tested, the *Error* value in each cycle between the experimental amplitude of torque $T_{a\,exp}$ and the simulation amplitude of torque $T_{a\,sim}$ can be defined as

$$Error = (T_{a\,exp} - T_{a\,sim})/T_{a\,exp} \cdot 100\,[\%]. \tag{34}$$

The *MeanError* over all cycles is calculated as

$$MeanError = \frac{1}{N_d} \sum_{n=1}^{N_d} Error_n, \tag{35}$$

where index n is the number of cycles. The total error over all NT geometry specimens tested is defined as

$$TotalError = \frac{1}{S} \sum_{s=1}^{S} MeanError_s, \tag{36}$$

where s is the NT specimen index and $S = 8$ is the total number of NT specimens tested (see Table A3 in Appendix A for details).

For the different K_{shear} from Equation (21), the *TotalError* value is captured in Figure 6. The final K_{shear} material parameter is identified as the optimal value of K_{shear} where the *TotalError* is minimal.

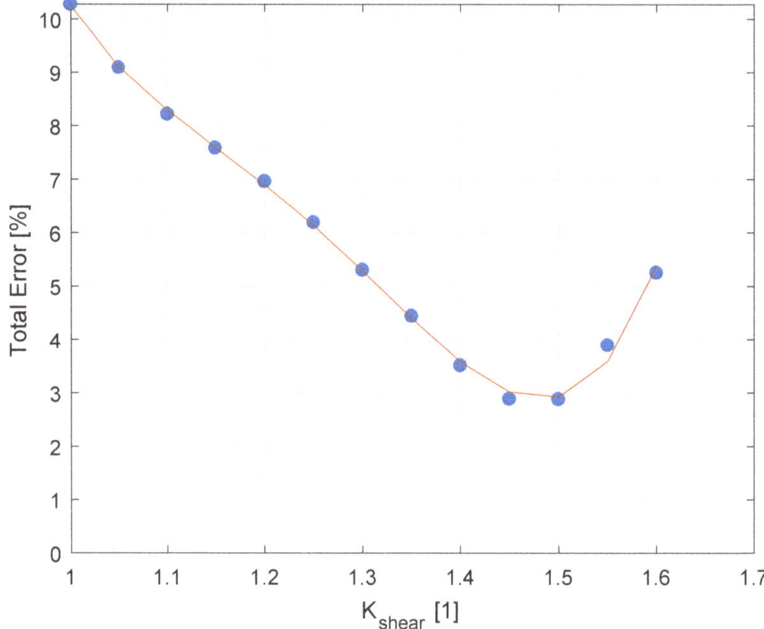

Figure 6. Identification of material parameter K_{shear}.

The material value parameters are presented in Table 1.

Table 1. Material parameters of the new proposed model for 08Ch18N10T.

E [MPa]	ν	σ_y [MPa]	C_1 [MPa]	γ_1	C_2 [MPa]
210,000	0.3	150	63,400	148.6	10,000
γ_2	C_3 [MPa]	γ_3	A_∞	B_∞	C_∞
911.4	2000	0	-1.3127×10^{-9}	1.7981×10^{-6}	-8.6705×10^{-4}
D_∞	F_∞	A_R [MPa^{-1}]	B_R	C_R [MPa]	R_M^{min} [MPa]
1.6678×10^{-1}	-10.600	3.0113×10^{-1}	1.4865×10^{-1}	1.1818×10^{-2}	130.54
A_ω	B_ω	C_ω	R_M^{max} [MPa]	ϕ_0	K_{shear}
0	2.0024×10^{-13}	-4.8591	506.59	2.3178	1.5

The experimental data from the IDF series of experiments can also be plotted into fatigue diagram ϵ_a-N_f, where N_f is the number of cycles to failure and ϵ_a is the amplitude of the total strain. Due to the experimental setup, ϵ_a is not completely constant during the experiment in the case of E9 geometry (during the experiments, the amplitude of extensometer elongation $\frac{\Delta L_{ext}}{2}$ is controlled to be constant, so for UG geometry the ϵ_a is also constant but it is not completely constant for E9 geometry), so the mean value during the experiment is plotted. Fatigue data are shown in Figure 7. Other lifetimes are reported in the form of tabular data in Appendix A.

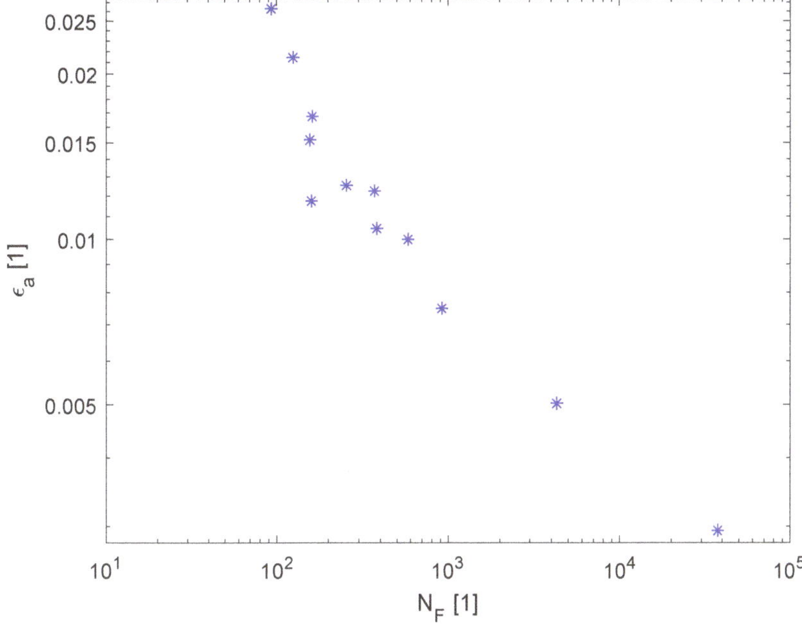

Figure 7. Fatigue data of the IDF series of experiments.

5. FE Simulations

The geometry of most specimens is not uniform, so the non-uniform stress and strain field in their cross-section are expected and FEA must be used for simulations. The constitutive model is implemented into Abaqus FE software using the USDFLD subroutine. FE models of each of the tested geometries were created, see Figures 8–10. The symmetry boundary condition is defined on the right edge of the model. The left edge of the model always corresponds with the cross-section where the extensometer is attached to the body of the specimen during the experiment. The displacement boundary condition on the upper edge of the FE model is created with the same amplitude value as was recorded from the extensometer during the experiment. Abaqus CAX8R mesh elements are used for the axisymmetric models and C3D8R elements are used for the NT geometry, which is a 3D model with cyclic symmetry. The element size in fine mesh areas has been determined using sensitivity study to 0.1 mm.

The Abaqus Chaboche plasticity material model with combined hardening and the USDFLD subroutine is used. The equations of the constitutive model are coded into the USDFLD subroutine for calculating the actual memory surfaces size R_M^{used} and $R_{M\phi}^{used}$, which, combined with the accumulated plastic strain p, determines the actual yield stress Y, the value of function ϕ and the memory term of the Chaboche model $\phi \cdot \gamma_i$. The full Abaqus USDFLD subroutine code written in Fortran is available in Appendix B.

This subroutine makes possible to use the material model presented here in engineering computations. Combined with the material parameters identification process described in Section 4, it can also be used for other materials.

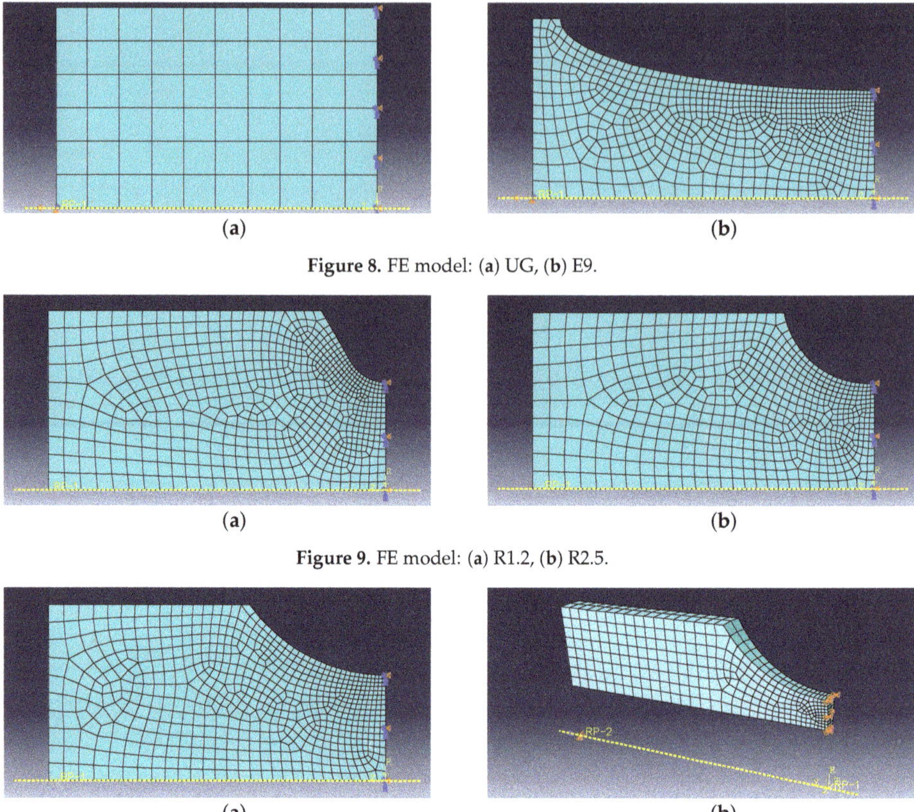

Figure 8. FE model: (**a**) UG, (**b**) E9.

Figure 9. FE model: (**a**) R1.2, (**b**) R2.5.

Figure 10. FE model: (**a**) R5, (**b**) NT.

6. Experimental and Simulation Results

The implementation of the plasticity model presented here (including the presented modification) into FE code was verified using FE simulations of all experiments mentioned in Section 2.2. The following figures show some results of experiments and their FE simulations. Due to the large scale of the experimental program, only two representative specimens with low and high load levels were selected for demonstration in this section. The results of remaining specimens are presented in the form of error values in following tables. The compared variables in each figure are the amplitudes of the force measured during the experiment ($F_{a\ exp}$) and computed by the FE simulations ($F_{a\ sim}$). Two constitutive models are shown—the original model [2] and the modified model presented in this paper. The actual error between each FE simulation and the experiment and the mean error value, are also displayed.

The error between the experiment and the FE simulation in each cycle n is calculated simply as

$$Error = \frac{F_{a\ exp} - F_{a\ sim}}{F_{a\ exp}} \times 100\,\%. \tag{37}$$

The mean error and the total error are calculated using Equations (35) and (36) considering corresponding number of specimens in the series.

The Figure 11 and Table 2 show the experimental and simulation results of E9 geometry series representing the uniaxial loading conditions. The prediction capability of these two models is comparable.

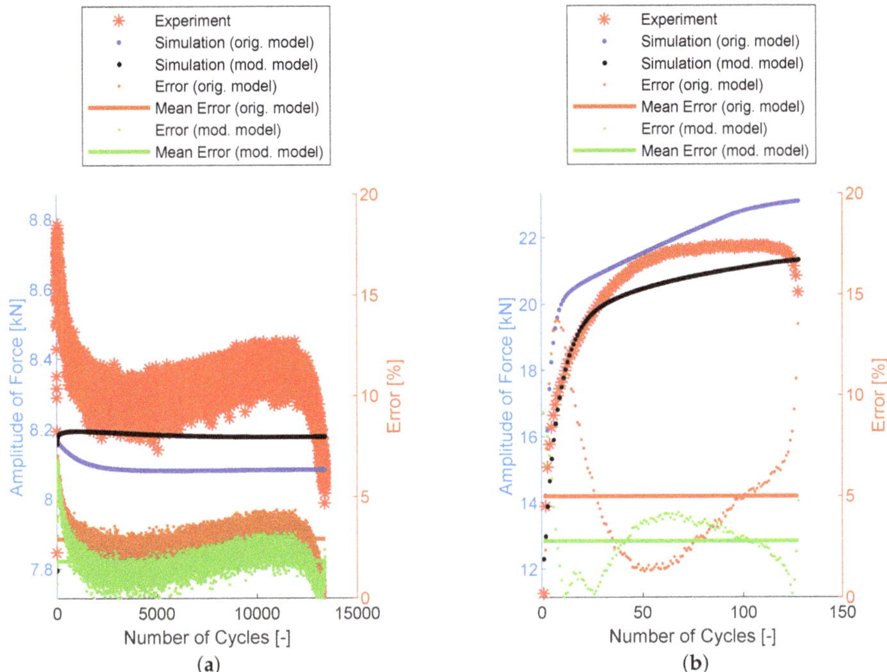

Figure 11. Amplitude of force—experiment vs. simulations [2]: (**a**) E9-1, (**b**) E9-17.

Table 2. Mean error of all E9 specimens tested—experiment vs. simulations [2].

Specimen Name	Orig. Model Mean Err. [%]	Mod. Model Mean Err. [%]	Specimen Name	Orig. Model Mean Err. [%]	Mod. Model Mean Err. [%]
E9-1	2.9226	1.8207	E9-10	7.8144	8.8757
E9-2	2.3311	1.2756	E9-11	2.5028	3.9003
E9-3	2.4027	1.1938	E9-12	4.3523	6.5915
E9-4	1.6977	0.7773	E9-13	4.0929	3.4343
E9-5	8.0687	7.0447	E9-14	2.1610	3.8515
E9-6	8.8658	7.4521	E9-15	2.9195	2.9485
E9-7	11.7310	10.4229	E9-16	1.8601	2.7524
E9-8	3.8241	3.9171	E9-17	4.9766	2.7579
E9-9	9.8245	9.5508			

The NT geometry series results are in Figure 12 and Table 3. In this case, the compared variables are the amplitudes of the torque measured during the experiment ($T_{a\ exp}$) and computed by the FE simulations ($T_{a\ sim}$). The errors are calculated using Equations (34)–(36). For this geometry, the difference in the prediction capability of the original model and the modified model is not the same—the modified model provides a better prediction of the cyclic hardening of the material under torsional loading for high loading levels.

Finally, the notched specimen geometry series R1.2, R2.5 and R5 follows on Figures 13–15 and Tables 4–6. The stress field in the cross-section of these specimens is no longer uniaxial and the prediction capabilities of both models are also comparable.

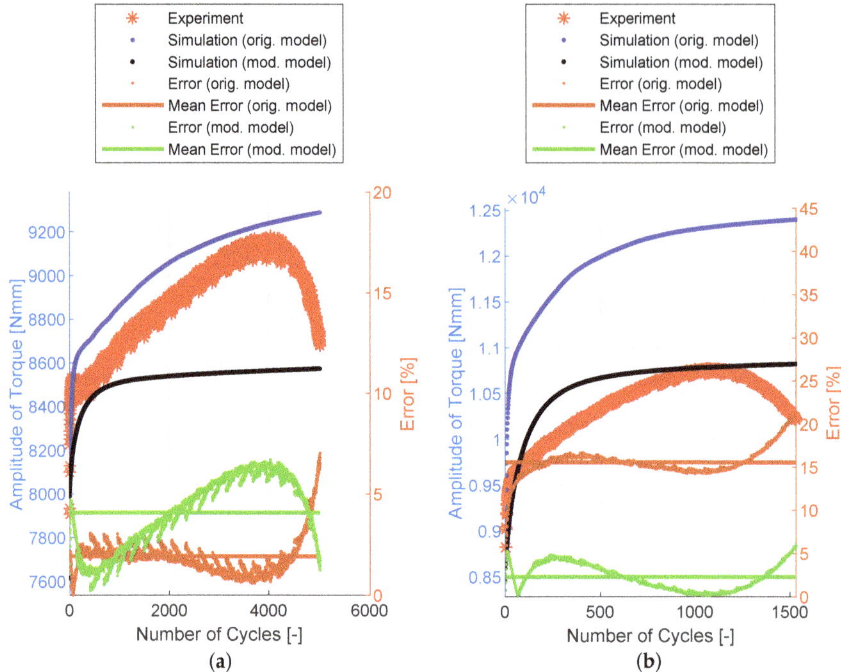

Figure 12. Amplitude of force—experiment vs. simulations: (**a**) NT-1, (**b**) NT-6.

Table 3. Mean error of all NT specimens tested—experiment vs. simulations.

Specimen Name	Orig. Model Mean Err. [%]	Mod. Model Mean Err. [%]	Specimen Name	Orig. Model Mean Err. [%]	Mod. Model Mean Err. [%]
NT-1	1.9100	4.0682	NT-5	14.2137	1.3947
NT-2	0.8367	5.9823	NT-6	15.5549	2.2815
NT-3	11.2048	1.3797	NT-7	13.1168	1.5014
NT-4	11.1021	1.0934	NT-8	8.8054	4.7887

Table 4. Mean error of all R1.2 specimens tested—experiment vs. simulations.

Specimen Name	Orig. Model Mean Err. [%]	Mod. Model Mean Err. [%]	Specimen Name	Orig. Model Mean Err. [%]	Mod. Model Mean Err. [%]
R1.2-1	2.8075	2.4172	R1.2-10	1.6518	1.7538
R1.2-2	3.7011	3.1679	R1.2-11	2.0827	2.2332
R1.2-3	2.2438	2.2027	R1.2-12	3.9411	3.2028
R1.2-4	2.8530	2.7056	R1.2-13	2.5308	3.1540
R1.2-5	2.8984	2.7105	R1.2-14	1.4521	1.8444
R1.2-6	4.7877	4.4405	R1.2-15	3.6781	2.6435
R1.2-7	1.4888	1.4897	R1.2-16	1.5820	1.9106
R1.2-8	7.1382	6.7943	R1.2-17	1.6089	2.5930
R1.2-9	2.4171	2.2355	R1.2-18	1.2789	2.2219

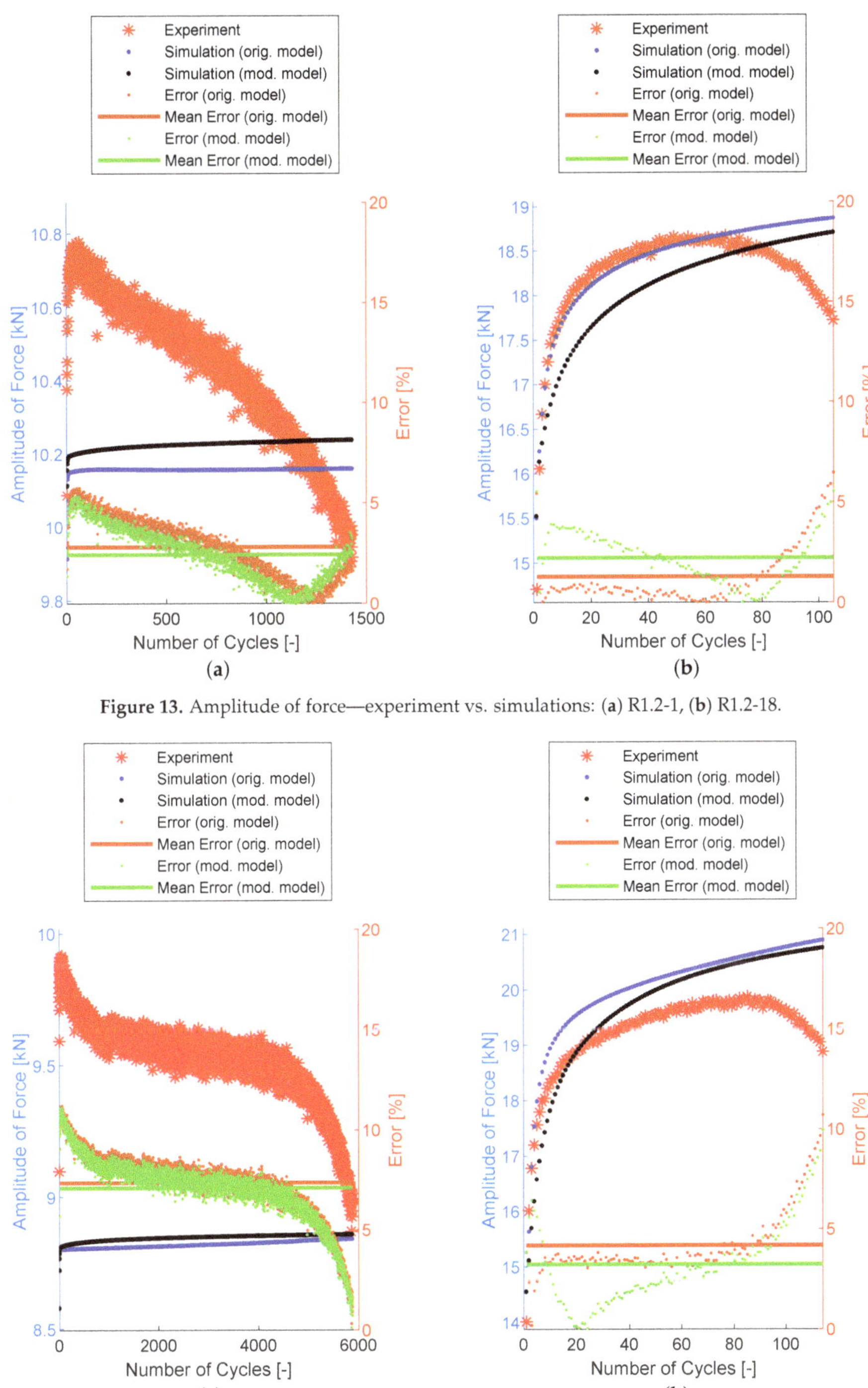

Figure 13. Amplitude of force—experiment vs. simulations: (**a**) R1.2-1, (**b**) R1.2-18.

Figure 14. Amplitude of force—experiment vs. simulations: (**a**) R2.5-1, (**b**) R2.5-21.

Table 5. Mean error of all R2.5 specimens tested—experiment vs. simulations.

Specimen Name	Orig. Model Mean Err. [%]	Mod. Model Mean Err. [%]	Specimen Name	Orig. Model Mean Err. [%]	Mod. Model Mean Err. [%]
R2.5-1	7.3714	7.1025	R2.5-12	2.1944	1.6489
R2.5-2	8.1586	7.6327	R2.5-13	1.2466	1.0057
R2.5-3	9.1468	8.6587	R2.5-14	8.7778	9.1473
R2.5-4	6.8139	6.8130	R2.5-15	2.6624	3.0678
R2.5-5	6.6714	6.6118	R2.5-16	1.4643	1.3563
R2.5-6	9.9838	9.1708	R2.5-17	0.9873	1.5697
R2.5-7	4.3249	3.4860	R2.5-18	1.4020	1.4515
R2.5-8	3.8551	3.8250	R2.5-19	1.6099	2.6423
R2.5-9	1.0034	0.9027	R2.5-20	0.9634	2.4069
R2.5-10	4.7921	4.9816	R2.5-21	4.1944	3.2605
R2.5-11	1.9673	2.1464			

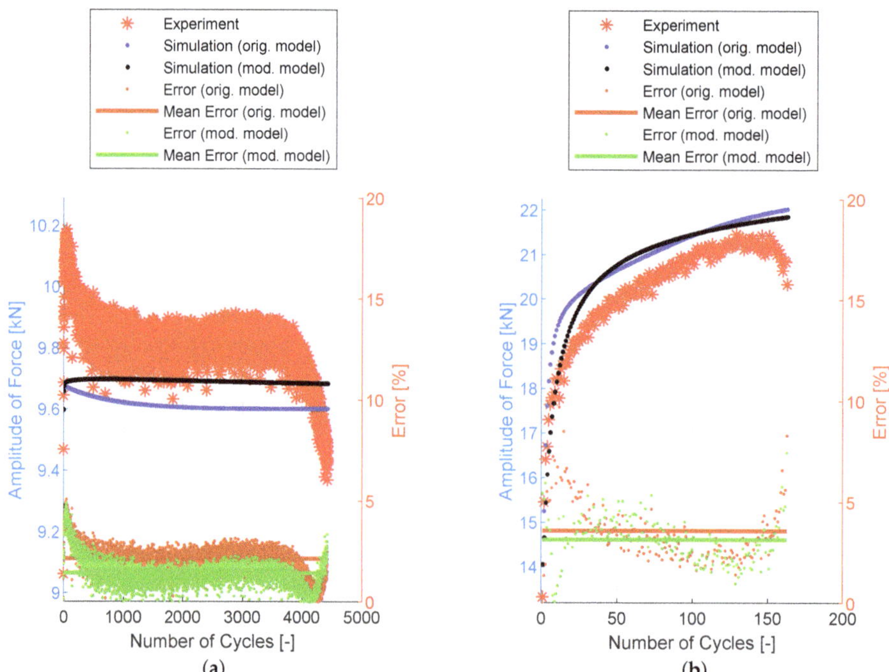

Figure 15. Amplitude of force—experiment vs. simulations: (**a**) R5-1, (**b**) R5-24.

Table 6. Mean error of all R5 specimens tested—experiment vs. simulations.

Specimen Name	Orig. Model Mean Err. [%]	Mod. Model Mean Err. [%]	Specimen Name	Orig. Model Mean Err. [%]	Mod. Model Mean Err. [%]
R5-1	2.1303	1.4186	R5-13	6.7479	6.8700
R5-2	2.0673	1.8112	R5-14	5.1055	5.4414
R5-3	0.7021	0.8284	R5-15	1.3043	1.4251
R5-4	0.9757	0.9284	R5-16	1.1829	1.3661
R5-5	1.4847	1.4209	R5-17	3.6903	3.6048
R5-6	1.7435	1.6993	R5-18	3.1399	2.9518
R5-7	2.9066	2.7548	R5-19	6.1649	6.1226
R5-8	5.3372	5.4106	R5-20	2.8263	2.6683
R5-9	4.9004	4.5530	R5-21	1.0485	1.2882
R5-10	2.3623	2.6227	R5-22	8.2167	7.6119
R5-11	7.0110	6.8065	R5-23	2.2011	1.6441
R5-12	2.3912	3.1025	R5-24	3.5803	3.1425

7. Discussion

As has been shown in the previous sections, the model proposed in Reference [2] can capture very well the static and cyclic stress-strain curve for uniaxial loading conditions with a reasonable number of material parameters. Using the modification proposed in this paper, with only two extra material parameters, the error under torsional loading conditions can be reduced significantly, without any harm under uniaxial loading conditions, as is shown in Table 7, where the total errors (defined in Equation (36)) are summarized. Both models also produce very good predictions for notched specimen geometries, where the stress-strain field is not uniaxial.

In Figures 11–15, the range of response quantity axis has been chosen to make visible the difference between experimental values and predicted ones. That is why the error seems to be higher than actually is. This is true especially in case of the lowest strain amplitude.

In the calibration process dealing with torsional loading, the K_{shear} value is found as a compromise between all loading levels, so the proposed modification improves prediction for most, but not all, specimens tested (see Tables 3 and 7).

Table 7. Total error comparison between the original model and the modified model.

Geometry	The Original Nodel [2] Total Error [%]	The Modified Model Total Error [%]
E9	4.84	4.61
NT	9.60	2.85
R1.2	2.79	2.76
R2.5	4.27	4.23
R5	3.30	3.23

It should be pointed out that presented tests consist only of single loading modes. Combinations of these modes, for example, proportional and also non-proportional combination of tension and torsion probably induces cyclic non-proportional hardening and another cyclic phenomenons. These conditions are also limiting for eventual FE analysis on the real components. Combined loading conditions considering proportional as well as non-proportional loading are potential topics for future investigation.

8. Conclusions

This paper has described the experimental setup and the experimental program for a low-cycle fatigue test of 08Ch18N10T austenitic stainless steel. Using FE simulations, material model [2] capable of capturing the strain range dependent cyclic hardening has been newly verified on notched specimens, where the stress-strain field is non-uniform and for torsional loading. With a newly proposed modification, model can correctly simulate cyclic hardening also for shear stress loading conditions.

The extensive experimental program was subsequently completely simulated. The Chaboche plasticity model combined with non-linear isotropic hardening has already been implemented into Abaqus commercial FE software. The model presented here can easily be implemented into Abaqus using the USDFLD subroutine as a simple extension of the Abaqus default cyclic plasticity model. The full Fortran code of subroutine can be found in Appendix B. This implementation makes the proposed model ready to use for some engineering computations. The usage limitations are given by the conditions under which the model has been tested (simple, uncombined loading).

The original cyclic plasticity model presented in [2] provides a good prediction of the cyclic response of uniaxial and notched specimens. With the modification for torsional loading that has been presented here, it can also provide a good prediction of cyclic hardening under torsional loading conditions. It can also easily be applied to the Abdel-Karim-Ohno model or to a modified version with promised ratcheting prediction [8]. The model can be extended by standard techniques for use in the area of viscoplasticity [15].

The calibration of the cyclic plasticity model was described briefly in this paper and was used with experimental data available for 08Ch18N10T. In future work, an automated process for identifying material parameters could be prepared in a similar way as in [16]. Some authors of this paper also work on the material parameters identification using results from DIC measurements in order to reduce number of necessary specimens for recently expensive technologies of 3D printing of metals [17].

Author Contributions: Conceptualization, J.F., R.H. and M.Š.; Data curation, R.P.; Investigation, J.F., R.P. and M.Š.; Methodology, J.F. and R.H.; Software, J.F.; Supervision, R.H. and M.Š.; Validation, J.F.; Visualization, J.F. and R.P.; Writing—original draft, J.F., R.H. and P.G.; Writing—review & editing, J.F., R.H., P.G. and M.Š.

Funding: This research was funded by the Czech Science Foundation (GACR), grant No. 19-03282S (Influence of Complex and Cyclic Loading Modes on Lifetime of Machine Parts Made by Additive Manufacturing) and by The Technology Agency of the Czech Republic in the frame of the project TN01000024 National Competence Center-Cybernetics and Artificial Intelligence. The paper has been done in connection with project Innovative and additive manufacturing technology—new technological solutions for 3D printing of metals and composite materials, reg. no. CZ.02.1.01/0.0/17_049/0008407 financed by Structural Founds of Europe Union.

Conflicts of Interest: The authors declare no conflict of interest.

Abbreviations

The following abbreviations are used in this manuscript:

DIC	digital image correlation
IDF	identification specimen series
FE	finite element
FEM	finite element method
UG	uniform-gage

Appendix A. Boundary Conditions of Simulations

Table A1. Boundary conditions of IDF specimens.

Specimen Name	Geometry Type	ΔL_{ext} [mm]	N_d
IDF-1	UG	0.030	37509
IDF-2	UG	0.050	4285
IDF-3	UG	0.075	916
IDF-4	UG	0.100	580
IDF-5	UG	0.125	254
IDF-6	E9	0.132	159
IDF-7	E9	0.154	381
IDF-8	E9	0.176	370
IDF-9	E9	0.198	161
IDF-10	E9	0.245	156
IDF-11	E9	0.264	124
IDF-12	E9	0.353	93

Table A2. Boundary conditions of E9 specimens.

Specimen Name	Geometry Type	ΔL_{ext} [mm]	N_d
E9-1	E9	0.0447	13382
E9-2	E9	0.0446	15104
E9-3	E9	0.0662	4053
E9-4	E9	0.0662	3887
E9-5	E9	0.0881	1529
E9-6	E9	0.0880	1853
E9-7	E9	0.1100	1158
E9-8	E9	0.1100	631
E9-9	E9	0.1320	748
E9-10	E9	0.1540	546
E9-11	E9	0.1770	406
E9-12	E9	0.1980	332
E9-13	E9	0.2200	253
E9-14	E9	0.2420	181
E9-15	E9	0.2420	195
E9-16	E9	0.2640	220
E9-17	E9	0.3520	128

Table A3. Boundary conditions of NT specimens.

Specimen Name	Geometry Type	$\Delta\phi_{ext}$ [°]	N_d
NT-1	NT	0.8703	5006
NT-2	NT	0.8694	6894
NT-3	NT	1.1423	2222
NT-4	NT	1.1414	2289
NT-5	NT	1.4031	2045
NT-6	NT	1.3772	1532
NT-7	NT	1.6554	1170
NT-8	NT	2.1492	925

Table A4. Boundary conditions of R1.2 specimens.

Specimen Name	Geometry Type	ΔL_{ext} [mm]	N_d
R1.2-1	R1.2	0.0245	1429
R1.2-2	R1.2	0.0246	946
R1.2-3	R1.2	0.0326	715
R1.2-4	R1.2	0.0406	523
R1.2-5	R1.2	0.0407	490
R1.2-6	R1.2	0.0489	290
R1.2-7	R1.2	0.0485	356
R1.2-8	R1.2	0.0560	241
R1.2-9	R1.2	0.0563	256
R1.2-10	R1.2	0.0639	134
R1.2-11	R1.2	0.0642	202
R1.2-12	R1.2	0.0721	171
R1.2-13	R1.2	0.0718	164
R1.2-14	R1.2	0.0794	112
R1.2-15	R1.2	0.0868	145
R1.2-16	R1.2	0.0869	114
R1.2-17	R1.2	0.0945	96
R1.2-18	R1.2	0.0944	105

Table A5. Boundary conditions of R2.5 specimens.

Specimen Name	Geometry Type	ΔL_{ext} [mm]	N_d
R2.5-1	R2.5	0.0228	5875
R2.5-2	R2.5	0.0341	1245
R2.5-3	R2.5	0.0340	1041
R2.5-4	R2.5	0.0454	607
R2.5-5	R2.5	0.0454	761
R2.5-6	R2.5	0.0568	378
R2.5-7	R2.5	0.0567	429
R2.5-8	R2.5	0.0718	242
R2.5-9	R2.5	0.0679	346
R2.5-10	R2.5	0.0794	265
R2.5-11	R2.5	0.0791	212
R2.5-12	R2.5	0.0904	210
R2.5-13	R2.5	0.0903	221
R2.5-14	R2.5	0.1015	205
R2.5-15	R2.5	0.1015	163
R2.5-16	R2.5	0.1126	189
R2.5-17	R2.5	0.1126	156
R2.5-18	R2.5	0.1237	132
R2.5-19	R2.5	0.1237	129
R2.5-20	R2.5	0.1419	106
R2.5-21	R2.5	0.1346	114

Table A6. Boundary conditions of R5 specimens.

Specimen Name	Geometry Type	ΔL_{ext} [mm]	N_d
R5-1	R5	0.0308	4427
R5-2	R5	0.0461	1700
R5-3	R5	0.0457	1072
R5-4	R5	0.0603	733
R5-5	R5	0.0589	953
R5-6	R5	0.0727	623
R5-7	R5	0.0747	527
R5-8	R5	0.0893	342
R5-9	R5	0.0869	543
R5-10	R5	0.1050	297
R5-12	R5	0.1010	374
R5-13	R5	0.1154	264
R5-14	R5	0.1156	290
R5-15	R5	0.1146	228
R5-16	R5	0.1287	152
R5-17	R5	0.1276	272
R5-18	R5	0.1418	179
R5-19	R5	0.1467	155
R5-20	R5	0.1403	177
R5-21	R5	0.1540	163
R5-22	R5	0.1531	174
R5-23	R5	0.1663	144
R5-24	R5	0.1685	189
R5-25	R5	0.1652	163

Appendix B. Abaqus USDFLD Subroutine

Appendix B.1. Full Fortran Code of Abaqus USDFLD Subroutine

```
C Material model by Miro Fumfera C
C        version 2019-11-10          C
CCCCCCCCCCCCCCCCCCCCCCCCCCCCCCCCCCCC
C      USDFLD Subroutine for 08Ch18N10T Austenitic Stainless Steel
C      Original modely by Radim Halama
C      modified by Miro Fumfera for 08Ch18N10T
       SUBROUTINE USDFLD(FIELD,STATEV,PNEWDT,DIRECT,T,CELENT,
      1 TIME,DTIME,CMNAME,ORNAME,NFIELD,NSTATV,NOEL,NPT,LAYER,
      2 KSPT,KSTEP,KINC,NDI,NSHR,COORD,JMAC,JMATYP,MATLAYO,LACCFLA)
       INCLUDE 'ABA_PARAM.INC'
       CHARACTER*80 CMNAME,ORNAME
       CHARACTER*3  FLGRAY(15)
       DIMENSION FIELD(NFIELD),STATEV(NSTATV),DIRECT(3,3),T(3,3),TIME(2)
       DIMENSION ARRAY(15),JARRAY(15),JMAC(*),JMATYP(*),COORD(*)
       parameter ZERO=0D0,ONE=1D0,TWO=2D0,THREE=3D0,TOLER=1D-12,
      + NTENS=6 !NTENS=4 for Axisymetric, NTENS=6 for 3D
       real*8 RMused,RM,RMmax,RMmin,oRM,dRM, RMRused,RMR,oRMR,dRMR,
      + heavisideG,DDP,G,DirVec(NTENS),DirVecR(NTENS)
       real*8 ALPHAv(NTENS),dALPHA1v(NTENS),ALPHA1v(NTENS),
      + dALPHA2v(NTENS),ALPHA2v(NTENS),dALPHA3v(NTENS),ALPHA3v(NTENS),
      + dALPHAv(NTENS),oALPHAv(NTENS),magALPHAv
       real*8 ALPHAr(NTENS),dALPHA1r(NTENS),ALPHA1r(NTENS),
      + dALPHA2r(NTENS),ALPHA2r(NTENS),dALPHA3r(NTENS),ALPHA3r(NTENS),
      + dALPHAr(NTENS),oALPHAr(NTENS),magALPHAr
       real*8 EPLAS(NTENS),oEPLAS(NTENS),dEPLAS(NTENS),EQPLAS,oEQPLAS,
      + dEQPLAS,FLOW(NTENS)
       real*8 R,oR,dR,AR,BR,CR,ER
       real*8 PhiInfty,dPHIcyc,PHIcyc,oPHIcyc,PHI0,PHI
       real*8 AInfty,BInfty,CInfty,DInfty,EInfty
       real*8 AOmega,BOmega,COmega
       real*8 KShear
       real*8 C1,GAMMA1,C2,GAMMA2,C3,GAMMA3
       integer K1,iEPLAS,iALPHA1v,iALPHA2v,iALPHA3v,iEQPLAS,iRM,iPHI,
      + iPHIcyc,iALPHAv,iR,iFIELD1,iFIELD2,iALPHA1r,iALPHA2r,iALPHA3r,
      + iRMR,iALPHAr
       parameter(iEPLAS=7,iALPHA1v=31,iALPHA2v=37,iALPHA3v=43,iEQPLAS=49,
      + iR=50,iRM=51,iPHI=52,iPHIcyc=53,iPhiInfty=54,iRMR=61,iALPHA1r=71,
      + iALPHA2r=77,iALPHA3r=83,iRMRused=95,iRMused=96,iALPHAv=97,
      + iALPHAr=94,iFIELD1=98,iFIELD2=99)
C      Material parameters
       C1 = 6.339971e+04
       GAMMA1 = 1.485569e+02
       C2 = 9.999778e+03
       GAMMA2 = 9.113512e+02
       C3 = 2000
       GAMMA3 = 0
       SYIELD = 150
```

```
      PHI0 = 2.317802e+00
      AInfty = -1.312737e-09
      BInfty = 1.798138e-06
      CInfty = -8.670490e-04
      DInfty = 1.667770e-01
      EInfty = -1.060028e+01
      RMmin = 1.305410e+02
      RMmax = 5.065918e+02
      BR = 3.011316e-01
      CR = 1.486489e-01
      ER = 1.181843e-02
      AOmega = 0
      BOmega = 2.002387e-13
      COmega = -4.859126e+00
      KShear = 1.50
C     get PE components
      call GETVRM('PE',ARRAY,JARRAY,FLGRAY,JRCD,JMAC,JMATYP,
     +  MATLAYO,LACCFLA)
C     EQPLAS
      EQPLAS = ARRAY(7)
      oEQPLAS = STATEV(iEQPLAS)
      dEQPLAS = EQPLAS - oEQPLAS
C     get PE
      do K1=1,NTENS
        oEPLAS(K1) = STATEV(iEPLAS-1+K1)
        EPLAS(K1) = ARRAY(K1)
        dEPLAS(K1) = EPLAS(K1) - oEPLAS(K1)
      enddo
C     get ALPHAv
      do K1=1,NTENS
        ALPHA1v(K1) = STATEV(iALPHA1v-1+K1)
        ALPHA2v(K1) = STATEV(iALPHA2v-1+K1)
        ALPHA3v(K1) = STATEV(iALPHA3v-1+K1)
        oALPHAv(K1) = STATEV(iALPHAv-1+K1)

        ALPHA1r(K1) = STATEV(iALPHA1r-1+K1)
        ALPHA2r(K1) = STATEV(iALPHA2r-1+K1)
        ALPHA3r(K1) = STATEV(iALPHA3r-1+K1)
        oALPHAr(K1) = STATEV(iALPHAr-1+K1)
      enddo
C     get FLOW vector
      if(dEQPLAS.gt.ZERO) then
        do K1=1,NDI
          FLOW(K1) = dEPLAS(K1)/dEQPLAS
        enddo
        do K1=NDI+1,NTENS
          FLOW(K1) = dEPLAS(K1)/TWO/dEQPLAS
        enddo
      else
        do K1=1,NTENS
```

```
        FLOW(K1) = ZERO
      enddo
    endif
C   RM
    RM = STATEV(iRM)
    oRM = RM
C   dALPHAv
    do K1=1, NDI
      dALPHA1v(K1) = (TWO/THREE*C1*dEQPLAS*FLOW(K1) -
   +    GAMMA1*ALPHA1v(K1)*dEQPLAS)/(ONE+GAMMA1*dEQPLAS)
      dALPHA2v(K1) = (TWO/THREE*C2*dEQPLAS*FLOW(K1) -
   +    GAMMA2*ALPHA2v(K1)*dEQPLAS)/(ONE+GAMMA2*dEQPLAS)
      dALPHA3v(K1) = (TWO/THREE*C3*dEQPLAS*FLOW(K1) -
   +    GAMMA3*ALPHA3v(K1)*dEQPLAS)/(ONE+GAMMA3*dEQPLAS)
      ALPHAv(K1) = (ALPHA1v(K1)+dALPHA1v(K1)) +
   +     (ALPHA2v(K1)+dALPHA2v(K1)) + (ALPHA3v(K1)+dALPHA3v(K1))
      !dALPHAv(K1) = ALPHAv(K1)-oALPHAv(K1)
      dALPHAv(K1) = dALPHA1v(K1) + dALPHA2v(K1) + dALPHA3v(K1)
    enddo
    do K1=NDI+1, NTENS
      dALPHA1v(K1) = (TWO/THREE*C1*dEQPLAS*FLOW(K1) -
   +    GAMMA1*KShear*ALPHA1v(K1)*dEQPLAS)/(ONE+GAMMA1*dEQPLAS)
      dALPHA2v(K1) = (TWO/THREE*C2*dEQPLAS*FLOW(K1) -
   +    GAMMA2*KShear*ALPHA2v(K1)*dEQPLAS)/(ONE+GAMMA2*dEQPLAS)
      dALPHA3v(K1) = (TWO/THREE*C3*dEQPLAS*FLOW(K1) -
   +    GAMMA3*KShear*ALPHA3v(K1)*dEQPLAS)/(ONE+GAMMA3*dEQPLAS)
      ALPHAv(K1) = (ALPHA1v(K1)+dALPHA1v(K1)) +
   +     (ALPHA2v(K1)+dALPHA2v(K1)) + (ALPHA3v(K1)+dALPHA3v(K1))
      !dALPHAv(K1) = ALPHAv(K1)-oALPHAv(K1)
      dALPHAv(K1) = dALPHA1v(K1) + dALPHA2v(K1) + dALPHA3v(K1)
    enddo
    do K1=1, NTENS
      ALPHA1v(K1) = ALPHA1v(K1) + dALPHA1v(K1)
      ALPHA2v(K1) = ALPHA2v(K1) + dALPHA2v(K1)
      ALPHA3v(K1) = ALPHA3v(K1) + dALPHA3v(K1)
      ALPHAv(K1) = ALPHA1v(K1) + ALPHA2v(K1) + ALPHA3v(K1)
    enddo
C   magALPHAv
    magALPHAv = ZERO
    do K1=1, NDI
      magALPHAv = magALPHAv + ALPHAv(K1)**2
    enddo
    do K1=NDI+1, NTENS
      magALPHAv = magALPHAv + TWO*ALPHAv(K1)**2
    enddo
    magALPHAv = sqrt(THREE/TWO*magALPHAv)
C   G function
    G = magALPHAv - RM
    if(magALPHAv.gt.ZERO) then
      do K1 = 1, NTENS
```

```
          DirVec(K1)=ALPHAv(K1)/magALPHAv
        enddo
      else
        do K1 = 1, NTENS
          DirVec(K1) = ZERO
        enddo
      endif
C     double dot product DDP
      DDP = ZERO
      do K1 = 1, NDI
        DDP = DDP+DirVec(K1)*dALPHAv(K1)
      enddo
      do K1 = NDI+1, NTENS
        DDP = DDP+TWO*DirVec(K1)*dALPHAv(K1)
      enddo
C     heaviside function of G
      if (G.gt.ZERO) then
        heavisideG = ONE
      elseif (abs(G).lt.TOLER) then
        heavisideG = ONE/TWO
      else
        heavisideG = ZERO
      endif
C     memory surface RM
      dRM = heavisideG*abs(DDP)
      RM = oRM + dRM
      if (RM.lt.RMmin) then
        RMused = RMmin
      elseif (RM.gt.RMmax) then
        RMused = RMmax
      else
        RMused = RM
      endif
C     RMR
      RMR = STATEV(iRMR)
      oRMR = RMR
      do K1=1, NDI
        dALPHA1r(K1) = (TWO/THREE*C1*dEQPLAS*FLOW(K1) -
     +     GAMMA1*ALPHA1r(K1)*dEQPLAS)/(ONE+GAMMA1*dEQPLAS)
        dALPHA2r(K1) = (TWO/THREE*C2*dEQPLAS*FLOW(K1) -
     +     GAMMA2*ALPHA2r(K1)*dEQPLAS)/(ONE+GAMMA2*dEQPLAS)
        dALPHA3r(K1) = (TWO/THREE*C3*dEQPLAS*FLOW(K1) -
     +     GAMMA3*ALPHA3r(K1)*dEQPLAS)/(ONE+GAMMA3*dEQPLAS)
        ALPHAr(K1) = (ALPHA1r(K1)+dALPHA1r(K1)) +
     +      (ALPHA2r(K1)+dALPHA2r(K1)) + (ALPHA3r(K1)+dALPHA3r(K1))
        !dALPHAr(K1) = ALPHAr(K1)-oALPHAr(K1)
        dALPHAr(K1) = dALPHA1r(K1) + dALPHA2r(K1) + dALPHA3r(K1)
      enddo
      do K1=NDI+1, NTENS
        dALPHA1r(K1) = (TWO/THREE*C1*dEQPLAS*FLOW(K1) -
```

```
     +      GAMMA1*KShear*ALPHA1r(K1)*dEQPLAS)/(ONE+GAMMA1*dEQPLAS)
              dALPHA2r(K1) = (TWO/THREE*C2*dEQPLAS*FLOW(K1) -
     +      GAMMA2*KShear*ALPHA2r(K1)*dEQPLAS)/(ONE+GAMMA2*dEQPLAS)
              dALPHA3r(K1) = (TWO/THREE*C3*dEQPLAS*FLOW(K1) -
     +      GAMMA3*KShear*ALPHA3r(K1)*dEQPLAS)/(ONE+GAMMA3*dEQPLAS)
              ALPHAr(K1) = (ALPHA1r(K1)+dALPHA1r(K1)) +
     +       (ALPHA2r(K1)+dALPHA2r(K1)) + (ALPHA3r(K1)+dALPHA3r(K1))
              !dALPHAr(K1) = ALPHAr(K1)-oALPHAr(K1)
              dALPHAr(K1) = dALPHA1r(K1) + dALPHA2r(K1) + dALPHA3r(K1)
            enddo
            do K1=1, NTENS
              ALPHA1r(K1) = ALPHA1r(K1) + dALPHA1r(K1)
              ALPHA2r(K1) = ALPHA2r(K1) + dALPHA2r(K1)
              ALPHA3r(K1) = ALPHA3r(K1) + dALPHA3r(K1)
              ALPHAr(K1) = ALPHA1r(K1) + ALPHA2r(K1) + ALPHA3r(K1)
            enddo
C           magALPHAr
            magALPHAr = ZERO
            do K1=1, NDI
              magALPHAr = magALPHAr + ALPHAr(K1)**2
            enddo
            do K1=NDI+1, NTENS
              magALPHAr = magALPHAr + TWO*ALPHAr(K1)**2
            enddo
            magALPHAr = sqrt(THREE/TWO*magALPHAr)
C           G function
            G = magALPHAr - RMR
            if(magALPHAr.gt.ZERO) then
              do K1 = 1, NTENS
                DirVecR(K1)=ALPHAr(K1)/magALPHAr
              enddo
            else
              do K1 = 1, NTENS
                DirVecR(K1) = ZERO
              enddo
            endif
C           double dot product DDP
            DDP = ZERO
            do K1 = 1, NDI
              DDP = DDP+DirVecR(K1)*dALPHAr(K1)
            enddo
            do K1 = NDI+1, NTENS
              DDP = DDP+TWO*DirVecR(K1)*dALPHAr(K1)
            enddo
C           heaviside function of G
            if (G.gt.ZERO) then
              heavisideG = ONE
            elseif (abs(G).lt.TOLER) then
              heavisideG = ONE/TWO
            else
```

```
        heavisideG = ZERO
      endif
C     memory surface RMR
      dRMR = heavisideG*abs(DDP)
      RMR = oRMR + dRMR
      if (RMR.lt.RMmin) then
        RMRused = RMmin
      elseif (RMR.gt.RMmax) then
        RMRused = RMmax
      else
        RMRused = RMR
      endif
C     R
      oR = STATEV(iR)
      AR = CR*exp(ER*RMRused)
      dR = AR*((EQPLAS+dEQPLAS)**BR-EQPLAS**BR)
      R = oR + dR;
C     PHIinfty
      PhiInfty = AInfty*RMused**4 + BInfty*RMused**3 +
     + CInfty*RMused**2 + DInfty*RMused + EInfty
C     Omega
      OMEGA~= AOmega+BOmega*(RMused)**-COmega
C     PHIcyc
      oPHIcyc = STATEV(iPHIcyc)
      dPHIcyc = OMEGA*(PhiInfty-oPHIcyc)*DEQPLAS
      PHIcyc = oPHIcyc + dPHIcyc
C     PHI
      PHI = PHI0 + PHIcyc
C     save STATEV
      STATEV(iEQPLAS) = EQPLAS
      do K1=1,NTENS
        STATEV(iEPLAS-1+K1) = EPLAS(K1)
        STATEV(iALPHA1v-1+K1) = ALPHA1v(K1)
        STATEV(iALPHA2v-1+K1) = ALPHA2v(K1)
        STATEV(iALPHA3v-1+K1) = ALPHA3v(K1)
        STATEV(iALPHAv-1+K1) = ALPHAv(K1)
        STATEV(iALPHA1r-1+K1) = ALPHA1r(K1)
        STATEV(iALPHA2r-1+K1) = ALPHA2r(K1)
        STATEV(iALPHA3r-1+K1) = ALPHA3r(K1)
        STATEV(iALPHAr-1+K1) = ALPHAr(K1)
        STATEV(120+K1) = dALPHAv(K1)
      enddo
      STATEV(iR) = R
      STATEV(iRM) = RM
      STATEV(iRMR) = RMR
      STATEV(iRMused) = RMused
      STATEV(iRMRused) = RMRused
      STATEV(iPHIcyc) = PHIcyc
      STATEV(iPHI) = PHI
      STATEV(iPhiInfty) = PhiInfty
```

```
      STATEV(127) = SYIELD+R
      STATEV(128) = DDP
C     FIELD(1)
      FIELD(1) = SYIELD+R
      STATEV(iFIELD1) = FIELD(1)
C     FIELD(2)
      FIELD(2) = PHI
      STATEV(iFIELD2) = FIELD(2)
      RETURN
      END
```

Appendix B.2. Material Parameters Definition in the Abaqus Input File

The example of material parameters definition in Abaqus input file:

```
*Material, name=Material-1
*Depvar
    128
*Elastic
210000.0,0.3
*Plastic, dependencies=2, hardening=COMBINED, datatype=PARAMETERS,
number backstresses=3
** Material data as~a~function of FIELD1 and~FIELD2 follows:
SYIELD,C1,GAMMA1,C2,GAMMA2,C3,GAMMA3,FIDEL1,FIELD2
%%
```

In the last material data line, the numeric values of material parameters are written. The material data line repeats for different values of variables $FIELD1$ and $FIELD2$. Variables definitions are: $SYIELD = Y$, $C1 = C_1$, $GAMMA1 = \phi \cdot \gamma_1$, $C2 = C_2$, $GAMMA2 = \phi \cdot \gamma_2$, $C3 = C_3$, $GAMMA3 = \phi \cdot \gamma_3$, $FIELD1 = Y$, $FIELD2 = \phi$. So, for presented material model, few material data lines can look like this:

```
** Material data as~a~function of FIELD1 and~FIELD2 follows:
** ...
250.0,63399.70889,222.83539,9999.77788,1367.02686,2000.0,0.0,250.0,1.5
150.0,63399.70889,237.69108,9999.77788,1458.16199,2000.0,0.0,150.0,1.6
151.0,63399.70889,237.69108,9999.77788,1458.16199,2000.0,0.0,151.0,1.6
** ...
```

References

1. Halama, R.; Sedlák, J.; Šofer, M. Phenomenological Modelling of Cyclic Plasticity. In *Numerical Modelling*; Miidla, P., Ed.; IntechOpen: Rijeka, Croatia, 2012; pp. 329–354.
2. Halama, R.; Fumfera, J.; Gál, P.; Kumar, T.; Makropulos, A. Modeling the Strain-Range Dependent Cyclic Hardening of SS304 and 08Ch18N10T Stainless Steel with a Memory Surface. *Metals* **2019**, *9*, 832. [CrossRef]
3. Jiang, Y.; Zhang, J. Benchmark experiments and characteristic cyclic plastic deformation behavior. *Int. J. Plast.* **2008**, *24*, 1481–1515. [CrossRef]
4. Facheris, G.; Janssens, K.G.F.; Foletti, S. Multiaxial fatigue behavior of AISI 316L subjected to strain-controlled and ratcheting paths. *Int. J. Fatigue* **2014**, *68*, 195–208. [CrossRef]
5. Kim, C. Nondestructive Evaluation of Strain-Induced Phase Transformation and Damage Accumulation in Austenitic Stainless Steel Subjected to Cyclic Loading. *Metals* **2018**, *8*, 14. [CrossRef]
6. Borodii, M.; Shukayev, S. Additional cyclic strain hardening and its relation to material structure, mechanical characteristics, and lifetime. *Int. J. Fatigue* **2007**, *29*, 1184–1191. [CrossRef]

7. Jin, D.; Tian, D.J.; Li, J.H.; Sakane, M. Low-cycle fatigue of 316L stainless steel under proportional and nonproportional loadings. *Fatigue Fract. Eng. Mater. Struct.* **2016**, *39*, 850–858. [CrossRef]
8. Xing, R.; Dunji, Y.; Shouwen, S.; Xu, C. Cyclic deformation of 316L stainless steel and constitutive modeling under non-proportional variable loading path. *Int. J. Plast.* **2019**, *120*, 127–146. [CrossRef]
9. Srnec Novak, J.; Benasciutti, D.; De Bona, F.; Stanojevic, A.S.; De Luca, A.; Raffaglio, Y. Estimation of Material Parameters in Nonlinear Hardening Plasticity Models and Strain Life Curves for CuAg Alloy. *IOP Conf. Ser. Mater. Sci. Eng.* **2016**, *119*, 012020. [CrossRef]
10. Benallal, A.; Marquis, D. Constitutive Equations for Nonproportional Cyclic Elasto-Viscoplasticity. *J. Eng. Mater. Technol.* **1987**, *109*, 326–336. [CrossRef]
11. Tanaka, E. A nonproportionality parameter and a cyclic viscoplastic constitutive model taking into account amplitude dependences and memory effects of isotropic hardening. *Eur. J. Mech. A/Solids* **1994**, *13*, 155–173.
12. Jiang, Y.; Sehitoglu, H. Modeling of cyclic ratchetting plasticity, part I: Development of constitutive relations. *J. Appl. Mech.* **1996**, *63*, 720–725. [CrossRef]
13. ASTM Standard E606-92. *Standard Practise for Strain-Controlled Fatigue Testing*; ASTM International: West Conshohocken, PA, USA, 1998.
14. Fumfera, J.; Halama, R.; Kuželka, J.; Španiel, M. Strain-Range Dependent Cyclic Plasticity Material Model Calibration for the 08Ch18N10T Steel. In Proceedings of the 33rd Conference with International Participation on Computational Mechanics, Špičák, Czech Republic, 6–8 November 2017.
15. Abdel-Karim, M.; Ohno, N. Kinematic hardening model suitable for ratchetting with steady-state. *Int. J. Plast.* **2000**, *16*, 225–240. [CrossRef]
16. Peč, M.; Šebek, F.; Zapletal, J.; Petruška, J.; Hassan, T. Automated calibration of advanced cyclic plasticity model parameters with sensitivity analysis for aluminium alloy 2024-T351. *Adv. Mech. Eng.* **2019**, *11*, 1–14. [CrossRef]
17. Halama, R.; Pagáč, M.; Paška, Z.; Pavlíček, P.; Chen, X. Ratcheting Behaviour of 3D Printed and Conventionally Produced SS316L Material. In Proceedings of the ASME 2019 Pressure Vessels & Piping Conference PVP2019, San Antonio, TX, USA, 14–19 July 2019; Paper Number PVP2019-93384.

Review

Decade of Twist Channel Angular Pressing: A Review

Adéla Macháčková

Faculty of Materials Science and Technology, VŠB–Technical University of Ostrava, 17. Listopadu 15, 708 00 Ostrava, Czech Republic; adela.machackova@vsb.cz; Tel.: +420-597-324-344

Received: 16 March 2020; Accepted: 2 April 2020; Published: 7 April 2020

Abstract: The methods of severe plastic deformation (SPD) have gained attention within the last decades primarily owing to their ability to substantially refine the grains within metallic materials and, therefore, significantly enhance the properties. Among one of the most efficient SPD methods is the equal channel angular pressing (ECAP)-based twist channel angular pressing (TCAP) method, combining channel twist and channel bending within a single die. This unique die affects the processed material with three independent strain paths during a single pass, which supports the development of substructure and efficiently refines the grains. This review is intended to summarize the characteristics of the TCAP method and its main features documented within the last decade, since its development in 2010. The article is supplemented with a brief characterization of other known SPD methods based on the combination of ECAP and twist extrusion (TE) within a single die.

Keywords: twist channel angular pressing; severe plastic deformation; microstructure; mechanical properties

---

1. The Principles of Severe Plastic Deformation

Rapid development of novel construction components, innovative tools, and modern features in the industry and commerce goes hand in hand with the need to research materials with enhanced performance and increased longevity. Generally, the mechanical, physical, and utility properties of metallic materials are strongly affected by their structures—grain size in particular. By this reason, the ultra-fine grained (UFG) and nanomaterials featuring grain sizes between 100 nm and 1 μm and 1 nm and 100 nm, respectively, have been developed and intensively researched.

The Hall–Petch relation, although expressed in various ways according to the considered structural features and deformation history of the examined material [1], generally expresses the reciprocal proportion of the strength of the material and grain size (Equation (1)),

$$\sigma = \sigma_0 + k \cdot d^{-\frac{1}{2}} \tag{1}$$

where σ_0 and k are constants depending on the material chemical composition and deformation history, and d is the grain size.

In other words, the smaller the grains, the higher the strength. Despite the fact that this relation only applies for grain sizes above the critical value, the ratio of grain boundaries to grain interiors below which increases rapidly (which leads to the grain boundary sliding phenomenon), the validity of the Hall–Petch relation has been proven even for metals with grain sizes between 10 and 100 nm [1]. Nevertheless, such a fine microstructure is hardly achievable by any known method of plastic deformation. Therefore, having in mind the Hall–Petch relation, methods of severe plastic deformation (SPD) have been developed.

The principle of SPD methods lies in modifying microstructures of bulk samples without significantly changing their external shapes. SPD methods can thus generally be considered to positively affect the strength of processed metallic materials via grain refinement. During SPD processing, free material flow is restricted as the stress state aimed to be achieved and maintained during processing is the hydrostatic pressure; such a stress state is necessary for a high density of defects to be introduced to the crystal lattice. This phenomenon is required to achieve a significant grain refinement without reducing the cohesion of the processed material. However, the resulting material's performance not only depends on the achieved grain size, but also on substructure development, as it is a function of the density of dislocations and their (re)arrangement into specific shapes—networks.

The efficiency of SPD processes is ensured by two main parameters, which generally characterize them, as follows:

(1) Severe shear strain—regardless of the selected processing method, the shape of the die, and the number and character of the strain paths affecting the processed sample, all SPD methods are based on imposing severe shear strain into the material;

(2) Processing temperature—the temperature should be kept under the recrystallization temperature, that is, lower than 0.4–0.5 of the melting temperature of the particular metal, to aggravate recrystallization-induced grain growth and support grain refinement via accumulation of dislocations, formation of dislocation cells and walls, and subsequent polygonization, that is, formation of subgrains, which finally transform to fine new grains featuring high angle grain boundaries (HSGBs) [2]. Nevertheless, the substantial amount of energy introduced via the severe shear strain can impart recrystallization, resulting in grain fragmentation even at low temperatures [3]; the grain refining mechanisms that can occur in severely deformed metallic materials were thoroughly characterized by Glezer and Metlov [4]. However, the occurring grain refinement mechanisms also depend on the character of the processed metal (e.g., its stacking fault energy value), and the original material state, grain size, and deformation history in particular [5]. For example, metals with FCC lattices were documented to primarily deform via dislocation glide and twinning, but their ratio primarily depends on the applied strain rate and processing temperature; decreasing temperature and/or increasing strain rate support the dominance of twinning.

2. Severe Plastic Deformation Processes

The first ever attempt to perform severe plastic deformation via combinations of high pressure and severe shear strain was by professor P.W. Bridgman, who received the Nobel Prize in Physics in 1946 [6] (this idea later resulted in the invention of the high pressure torsion (HPT) process [7–9]). The breakthrough in SPD processing came later, in the 1980s, when Dr. V. M. Segal introduced the method of equal channel angular pressing (ECAP) [10–12] to the world.

In the early beginnings, a few processes were developed, some of which were also adjusted to enable processing of hollow billets (tubes), as well as powders and composite materials. However, the human ambition to improve the existing led to a massive development of SPD techniques. The everlasting research has brought about not only advancements of the already introduced methods, but also the design and subsequent verification of new methods, all of which have led to gradual amelioration of the original processes and tools and/or equipment geometry, allowing processing of not only typical bulk metallic materials, but also materials with low formability or hard-to-deform metals [13]. Although numerous SPD processes have been introduced, not all the existing methods can be listed here owing to the continuing research in this field. However, the examples are high pressure sliding [14], high pressure tube twisting [15], friction stir processing [16], equal channel angular drawing [17], ECAP–partial back pressure [18], accumulative back extrusion [19], accumulative roll bonding [20], accumulative spin bonding [21], constrained groove pressing [22], constrained studded pressing [23], parallel tubular channel angular pressing [24], tube channel pressing [25], elliptical cross-section spiral equal-channel extrusion [26], C-shape equal channel reciprocating extrusion [27], half channel angular extrusion [28], cyclic extrusion compression [29], cyclic closed die forging [30], cyclic expansion extrusion [31], cyclic extrusion compression angular pressing [32], (variable lead) axi-symmetric forward spiral extrusion [33,34], forward shear normal extrusion [35], rotary extrusion [36], torsion extrusion (TE) [37], twist extrusion (TE) [38], simple shear extrusion [39], continuous shearing [40], continuous confined strip shearing [41], continuous frictional angular extrusion [42], repetitive corrugation straightening [43], severe torsion straining [44], and so on.

Each of the above-mentioned SPD methods have pros and cons. Nevertheless, all of them aim to impart substantial structure refinement and the best possible structure homogeneity via imposing severe shear strain. Certain methods, such as HPT, are even able to introduce nano-scale features and provoke the formation of nanostructure. Nevertheless, the ECAP method still remains one of the most researched SPD technologies and keeps inspiring researchers worldwide.

It is quite complicated to characterize SPD methods according to strict criteria. Among the main criteria characterizing the SPD methods are primarily whether they are continuous or discontinuous, and the maximum size of the sample that can be processed by the method [45]—for example, HPT is only suitable for discontinuous processing of thin, coin-like samples, whereas ECAP–conform is suitable for processing of bulk rods and can also be applied industrially [46]. From the viewpoint of process efficiency, the maximum possible imposed strain is crucial. Imposing the maximum shear strain via a single pass while keeping the sample consistent and defect-free is favourable. However, not only the amount of the imposed strain, but also its homogeneity across the cross section of the sample is an important factor that consequently affects the homogeneity of properties within the processed material, as well as the possible development of residual stress.

Similar to virtually all the SPD methods, the maximum amount of the shear strain that can be introduced during a single pass while maintaining reasonable cross-sectional homogeneity and consistency of the processed material is limited for the basic and most researched method of angular pressing—ECAP. Despite the fact that the final material performance can be affected via changing processing parameters (friction, temperature, die geometry, and so on), the required modifications of particular processing steps necessarily bring about demands on the used equipment and reduce process effortlessness. The process of twist extrusion (TE) advantageously introduces high shear strain (which can be modified by changing the twist angle γ_{max}), but, on the other hand, also features high strain inhomogeneity across the cross section. Under optimized conditions, that is, having reduced/eliminated their negative aspects, both the ECAP and TE methods are very convenient for possible application in the industry. Considering these factors, combining the ECAP and TE methods is a favourable way to take advantage of the positive aspects of both and create an advantageous processing methodology for the preparation of ultra-fine-grained (UFG) materials. The combination of both can either be performed by consequent processing of the billet (which would require equipment for both the methods to be in hand, as well as optimization of the process so that the processed material is not affected by the dwells and varying processing conditions), or by introducing a novel single process. This solution preferably offers a wide possibility of variations while keeping the production process quick and easy. In accordance with the aim to improve the efficiency of SPD processing, the method of twist channel angular pressing was developed.

3. Twist Channel Angular Pressing

The method of twist channel angular pressing (TCAP) was firstly introduced by Kocich et al., in the 2010 [47]. TCAP is inspired by two individual SPD techniques, equal channel angular pressing (ECAP) and twist extrusion (TE), combining the positive aspects of both into a unique novel method. As mentioned above, ECAP and TE can be combined in several ways, as shown in Figure 1a to Figure 1c. Nevertheless, as experimentally proven [48], only one of the variants introduces reasonable processing parameters (e.g., punch load, temperature distribution) and preserves the constant shape of the sample, thus ensuring the repeatability of the entire process; Variant *I* (Figure 1a).

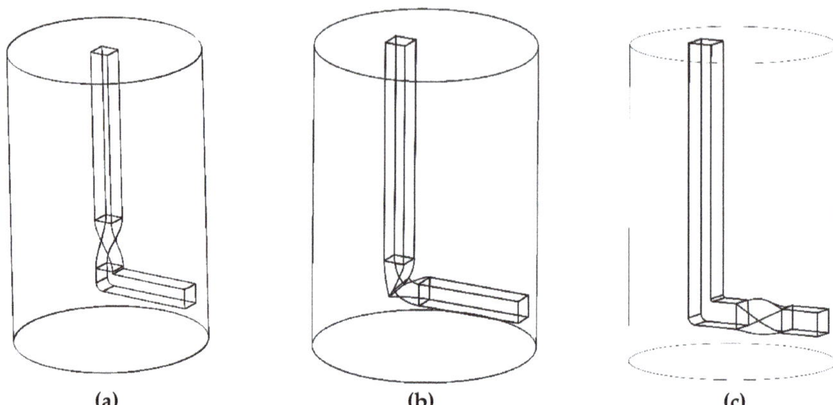

Figure 1. Experimentally investigated variants of twist channel angular pressing (TCAP): (**a**) Variant *I*, (**b**) Variant *II*, and (**c**) Variant *III*.

Although Variant *II* (Figure 1b) also combines both the ECAP and TE processes, this variant brings about several issues, probably the most severe of which is that positioning the twist directly to the bending introduces substantial local reduction of the channel cross section. In other words, the process then locally involves forward extrusion, which necessarily changes the processed sample cross section and introduces changes in its dimensions, by which the repeatability of the process, the crucial factor of all SPD processes, is disabled. This phenomenon can only be compensated by the application of back pressure. The second significant issue for this variant is a substantial increase in the punch load, which becomes even more substantial by the effect of applied back pressure. Among others, numerous cross-sectional reductions typical for this variant introduce a complex stress state, which can be unfavourable for the processed material.

The number of issues is reduced for Variant *III* (Figure 1c), although they are not eliminated. The main issue is related to the location of the twist being positioned behind the bending, which leads to a partially twisted final shape of the processed sample. Besides, by locating the twist behind the bending, the issues of the TE method, that is, partial cross-sectional deformation of the processed sample, as well as cross-sectional inhomogeneity of the imposed strain, are not eliminated; unlike Variant *I*, Variant *III* does not feature any final deformation zone that could suppress/reduce the strain inhomogeneity imparted by the twist.

Finite element analyses (FEAs) were further performed to evaluate the individual variants [45]. All of the modelled TCAP variants featured identical processing conditions, that is, room temperature processing with identical extrusion velocity, friction, and boundary conditions. The graphical dependencies depicted in Figure 2a,b show the differences between Variant *I* and Variant *II*. As shown by the temperature dependencies in Figure 2a, the increase in temperature for Variant *II* is more than twice as high as for Variant *I*, which can negatively affect the structures of processed materials featuring low melting temperatures. Moreover, the increase in temperature introduced by the development of deformation heat does not correspond to the increase in the imposed strain, which is only by approximately 20% (Figure 2b). By these reasons, only Variant *I* is further characterized (referred to just as TCAP).

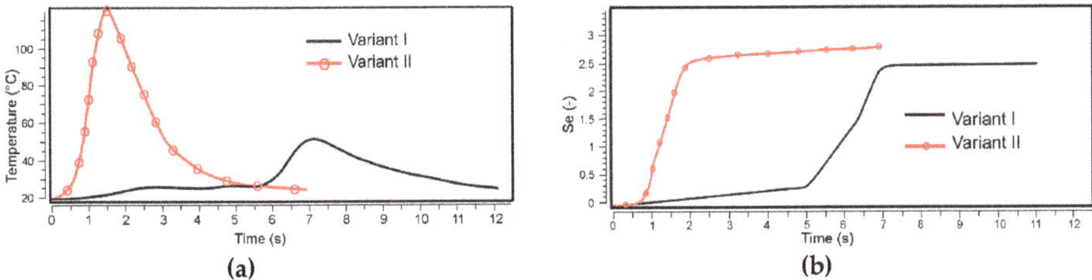

Figure 2. Comparison of deformation parameters: (**a**) temperature, (**b**) effective imposed strain.

The TCAP die (Figure 3) consists of a single channel bent under the desired channel bending angle φ, preferably between 90° and 120° (depending on the processed material and its formability), and the first channel part contains a twist defined with multiple parameters, such as the twist slope angle β, twist rotation angle ω, and distance between the end of twist part within the channel and the channel bending L, all of which affect the final properties of the material and the amount and distribution of the imposed strain, as documented by the preliminary TCAP studies characterizing the method and describing its influence on the material behaviour, with the help of numerical modelling and subsequent practical experiments [47,49].

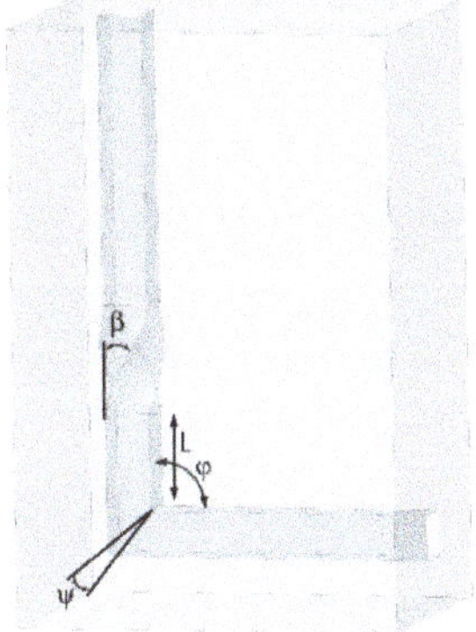

Figure 3. Schematic depiction of the TCAP die.

The most recent FEA study, which was applied to analyse mutual correlations of strain distribution and punch load, focused on the optimization of the TCAP die geometry from the viewpoint of minimizing the punch load while imposing a high shear strain [50]. Nine TCAP dies with varied twist slope angle β of 35°, 45°, and 55°, and channel bending angle ψ of 90°, 100°, and 110°, were designed. The results showed that, among all the combinations, the die with the twist slope angle of 45° and channel bending angle of 110° features a reasonable strain rate and relatively low punch load. The results were experimentally validated with sufficient correlation by performing TCAP of AA6061-T6 alloy. Moreover, to supplement the study with data on the effects of the processing temperature, the samples were processed at room temperature, 150 °C, and 250 °C; the sample processed at room temperature exhibited the best mechanical properties achieved.

3.1. Imposed Strain

The fact that the TCAP process consists of two individual methods has to be taken into account when calculating the mean value of the total imposed strain. In other words, the strain imposed in both the deformation zones, the twist and bending channel parts, has to be calculated separately.

3.1.1. Channel Bending Part

The used symbols and the particular geometrical features are summarized in Figure 4a–c.

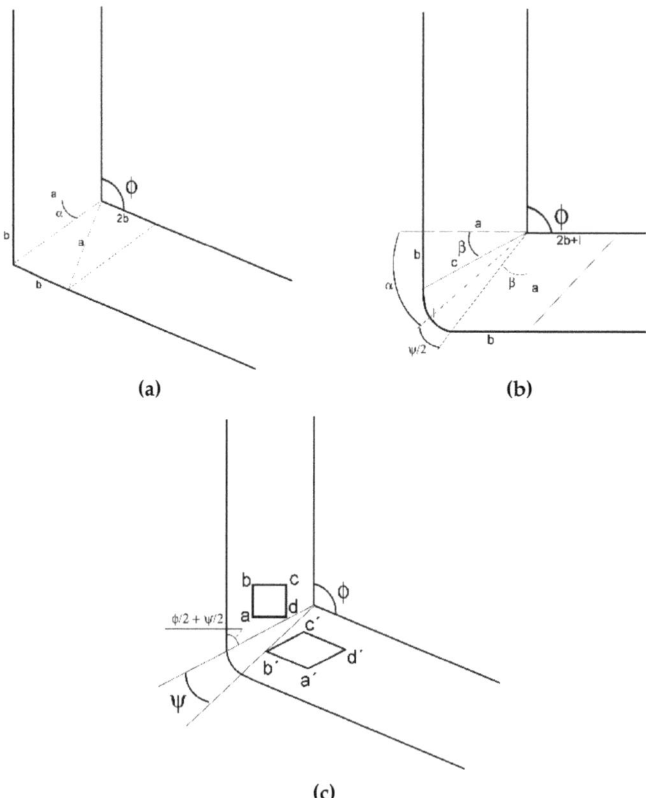

Figure 4. Schematic depiction of individual geometrical features for the following cases: (**a**) $\psi = 0$, (**b**) $\psi \neq 0$; (**c**) effect of varying angle ψ on the individual deformation zone geometrical parameters.

The following equations (Equations (2)–(4)) lead to the mathematical expression of the imposed shear strain shown in Equation (5) for the case when $\psi = 0$, that is, the outer channel bending radius is neglected. The used symbols are the die geometrical features, as referred to in Figure 4a.

$$\alpha = \frac{\pi - \varphi}{2} = \frac{\pi}{2} - \frac{\varphi}{2} \tag{2}$$

$$\frac{b}{a} = tg\alpha = \cot g\frac{\varphi}{2} \tag{3}$$

$$\frac{b}{a} = \cot g\frac{\varphi}{2} \Rightarrow b = a \cdot \cot g\frac{\varphi}{2} \tag{4}$$

$$\gamma = \frac{2b}{a} = \frac{2 \cdot a \cdot \cot g\left(\frac{\varphi}{2}\right)}{a} = 2 \cdot \cot g\left(\frac{\varphi}{2}\right) \tag{5}$$

Considering the imposed strain to be $\varepsilon = \frac{\gamma}{\sqrt{3}}$, the final simplified equation for the calculation of the imposed shear strain is then expressed as Equation (6).

$$\varepsilon = \frac{2}{\sqrt{3}} \cdot \cot g\left(\frac{\varphi}{2}\right) \tag{6}$$

For the case of $\psi \neq 0$, that is, when the outer channel bending radius is not neglected, the following equations, Equations (7)–(10), ensue from Equation (2), and from angle β, which is expressed as follows:

$$\beta = \alpha - \frac{\psi}{2} = \frac{\pi}{2} - \frac{\varphi}{2} - \frac{\psi}{2} \tag{7}$$

$$\frac{b}{a} = tg\beta = \cot g\left(\frac{\varphi}{2} + \frac{\psi}{2}\right) \tag{8}$$

$$\frac{a}{c} = \cos\beta = \cos\left(\frac{\pi}{2} - \frac{\varphi}{2} - \frac{\psi}{2}\right) = \sin\left(\frac{\varphi}{2} + \frac{\psi}{2}\right) \Rightarrow c = \frac{a}{\sin\left(\frac{\varphi}{2} + \frac{\psi}{2}\right)} = a \cdot \cos ec\left(\frac{\varphi}{2} + \frac{\psi}{2}\right) \tag{9}$$

$$l = \psi \cdot c = \psi \cdot a \cdot \cos ec\left(\frac{\varphi}{2} + \frac{\psi}{2}\right) \tag{10}$$

The final equation for the calculation of the imposed shear strain is then expressed as Equation (11), resp. Equation (12).

$$\gamma = \frac{2b + l}{a} = \frac{2b}{a} + \frac{l}{a} = 2 \cdot \cot g\left(\frac{\varphi}{2} + \frac{\psi}{2}\right) + \frac{\psi \cdot a \cdot \cos ec\left(\frac{\varphi}{2} + \frac{\psi}{2}\right)}{a} \tag{11}$$

$$\gamma = 2 \cdot \cot g\left(\frac{\varphi}{2} + \frac{\psi}{2}\right) + \psi \cdot \cos ec\left(\frac{\varphi}{2} + \frac{\psi}{2}\right) \tag{12}$$

Considering the imposed strain to be $\varepsilon = \frac{\gamma}{\sqrt{3}}$, the final simplified equation for the calculation of the imposed shear strain is then expressed as Equation (13).

$$\varepsilon = \frac{1}{\sqrt{3}} \cdot \left[2 \cdot \cot g\left(\frac{\varphi}{2} + \frac{\psi}{2}\right) + \psi \cdot \cos ec\left(\frac{\varphi}{2} + \frac{\psi}{2}\right)\right] \tag{13}$$

All of the above mentioned equations calculate the assumption of simplified boundary conditions and constant processing parameters, such as homogeneous plastic flow, ideally plastic material behaviour, absence of friction between the sample and die, complete filling of the channel with extruded material, and so on. In other words, certain parameters having complex effects, such as widening of the deformation zone, are usually neglected. Implementing the Segal theory of slip lines [51] to predict widening of the deformation zone, the angle characterizing the boundaries of the deformation zone can be depicted as β for dies with no outer channel bending radius ($\Psi = 0$); angle $\beta = 0$ when neglecting friction and increases up to $\beta = \pi/2$ with increasing friction. Assuming that the material exhibits ideal plastic flow, the deformation zone boundaries are given by angle α, that is, $\beta = \pi/2 - 2\alpha$ for angle $\Phi = 90°$. On the basis of these facts, the imposed strain for dies featuring angle $\Phi = 90°$ can be expressed via Equation (14):

$$\varepsilon = \sqrt{\frac{1}{3}}\sqrt{1 + 2a(1 - b + a) + b^2} \tag{14}$$

where $a = 2\alpha$, $b = (1 + \beta)tg\alpha$.

The ε value increases from 0.5774 for $\beta = \pi/2$, up to 1.1547 for $\beta = 0$. Simple shear occurring in the plane of intersection of both channel parts ($\beta = 0$), as well as shear occurring behind the deformation zone for dies featuring the outer channel bending radius ($\Psi > 0$), should, theoretically, impart homogenous shear strain distribution [52]. Nevertheless, deformation hardening occurring during ECAP processing leads to asymmetrical distribution of the strain rate in the deformation zone region, which results in the development of a dead zone and occurrence of strain heterogeneity. However, friction can affect the size of the dead zone; increasing friction between the die and sample surfaces attributes to decreasing the size of the dead zone region. In case that the die is filled with the processed material completely, the shape and dimensions of the deformation zone region are primarily given by the die geometry, that is, Ψ angle; friction has a neglectable effect. Assuming elimination of the dead zone, friction affects the imposed strain locally at the surface of the processed sample, which increases strain heterogeneity across the sample cross section. On the contrary, if a dead zone is present, friction affects the imposed strain throughout the entire sample volume. Increasing friction generally increases the imposed strain heterogeneity, decreases angle β, and decreases size of the dead zone region. Increasing the strain hardening rate results in increasing angle Ψ and increasing angle Ψ, and increasing the size of the dead zone region increases angle β. Generally, however, $\Psi \neq \beta$.

3.1.2. Channel Twist Part

For the channel part featuring the twist, calculation of the imposed strain has to be performed considering, among others, different flow velocities in the particular material regions. The individual v_1 and v_2 components of plastic flow velocity in the twist channel part are schematically depicted in Figure 5a,b, respectively, and mathematically expressed via the following equations.

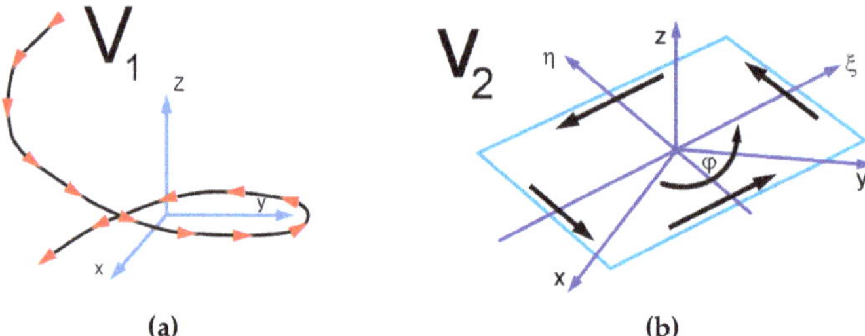

Figure 5. Schematic depiction of individual material flow velocities: (**a**) v_1, (**b**) v_2.

The individual plastic flow velocity components in the individual directions can be expressed as follows (Equations (15)–(20)) [53]:

$$v_{1x} = -\frac{y v_0 tg\gamma}{R} \tag{15}$$

$$v_{1y} = \frac{x v_0 tg\gamma}{R} \tag{16}$$

$$v_{1z} = v_0 \tag{17}$$

$$v_{2x} = \frac{\partial(\Omega P)}{\partial y} \tag{18}$$

$$v_{2y} = -\frac{\partial(\Omega P)}{\partial x} \tag{19}$$

$$v_{2z} \equiv 0 \tag{20}$$

where Ω is a function defining the cross-sectional shape, $\Omega = 0$ at the periphery of the sample cross section, $\Omega > 0$ within the sample cross section, and $\Omega < 0$ outside the cross section. P is a parameter characterizing the variable part of the velocity field; P is defined by variation methods and its value at the periphery of the sample cross section is $/P/ = /v_2/$. R is half of the length of the sample's edge.

The final velocity of material flow when passing through the twist channel part is then expressed via Equation (21):

$$v = v_1 + v_2 \tag{21}$$

where v_1 is the velocity field component expressing flow of the cross section as a whole, whereas v_2 is the velocity field component expressing transversal material flow across the cross section.

The shear strain is subsequently expressed in dependence on the particular cross-sectional location within the processed material, regarding axis—cross-section periphery of the sample. The maximum component of the shear strain expressed via the v_1 velocity component, that is, neglecting the v_2 velocity component, is formulated as Equation (22):

$$\gamma_{\max} = \sqrt{3}\cdot tg\beta_{\max} \tag{22}$$

where $\gamma_{\max}$ is the shear strain and $\beta_{\max}$ is the twist slope angle.

Consequently, the shear strain can in this case be expressed as Equation (23).

$$\varepsilon_{\max} = tg\beta_{\max} \tag{23}$$

When the v_2 velocity component is not neglected, the shear strain is expressed as Equation (24):

$$\gamma_{\min} = \sqrt{3}\cdot(0.4 + 0.1\cdot tg\beta_{\max}) \tag{24}$$

where $\gamma_{\min}$ is the shear strain and $\beta_{\max}$ is the twist slope angle.

The imposed shear strain can then be expressed as Equation (25).

$$\varepsilon_{\min} = 0.4 + 0.1\cdot tg\beta_{\max} \tag{25}$$

For the twist part of the TCAP channel, the value of the imposed shear strain can thus be calculated using both the Equations (23) and (25). However, the mentioned equations are only applicable for the axial and peripheral cross-sectional regions of the processed sample, respectively. Relations characterizing the shear strain imposed during TCAP processing in the near-axial and near-peripheral cross-sectional regions of the processed sample would involve combinations of both Equations (23) and (25), as the final equation would take into account increments from both locations, depending on the particular monitored region. The simplified assumption of simply adding up the individual increments from both locations can, however, only be made if no vortex-like material flow is presupposed. Nevertheless, the results of numerical predictions showed that the vortex-like flow is most probably the key factor contributing to final homogenization of the imposed strain across the sample cross section. The imposed strain is not homogeneous throughout the sample cross section after ECAP, especially for dies with corner radii, or dead zone regions. The smallest values of the imposed shear strain are recorded at the peripheral regions of the processed sample, that is, in regions adjacent to the outer channel radius. Similar phenomena, that is, the occurrence of a dead zone and shift of the deformation zone from the central die region, are also imparted via deformation hardening during ECAP, as proven by Kim et al. [52], who documented plastic flow velocity gradients to be present in both mentioned regions.

The mean value of the imposed shear strain is finally derived from relations for strain rates in the individual directions (Equations (26)–(30)):

$$\dot{e}_{xx} = P\frac{\partial^2\omega}{\partial x\partial y} \tag{26}$$

$$\dot{e}_{yy} = -P\frac{\partial^2\omega}{\partial x\partial y} \tag{27}$$

$$\dot{e}_{xy} = \frac{1}{2}P\left(\frac{\partial^2\omega}{\partial y^2} - \frac{\partial^2\omega}{\partial x^2}\right) \tag{28}$$

$$\dot{e}_{xz} = \frac{1}{2}\frac{\partial^2(\omega P)}{\partial y \partial z} - \frac{yV_0}{2R\cos^2\beta}\frac{d\beta}{dz} \tag{29}$$

$$\dot{e}_{yz} = -\frac{1}{2}\frac{\partial^2(\omega P)}{\partial x \partial z} + \frac{xV_0}{2R\cos^2\beta}\frac{d\beta}{dz} \tag{30}$$

Strain rate intensity from the above equations can further be expressed (Equation (31)).

$$S\dot{e} = \frac{\sqrt{2}}{3}\sqrt{\left(\dot{e}_{xx} - \dot{e}_{yy}\right)^2 + \left(\dot{e}_{xx} - \dot{e}_{zz}\right)^2 + \left(\dot{e}_{zz} - \dot{e}_{yy}\right)^2 + 6\left(\dot{e}_{xy}{}^2 + \dot{e}_{xz}{}^2 + \dot{e}_{yz}{}^2\right)} \tag{31}$$

Integration of Equation (31) consequently results in the mean value of the imposed strain for the twist part of the channel (Equation (32)):

$$\varepsilon = \int S\dot{e}\, dt \tag{32}$$

The final relation defining the mean value of the imposed strain during TCAP process can eventually be expressed via Equation (33).

$$\varepsilon = \frac{1}{\sqrt{3}} \cdot \left[2 \cdot \cot g\left(\frac{\varphi}{2} + \frac{\psi}{2}\right) + \psi \cdot \cos ec\left(\frac{\varphi}{2} + \frac{\psi}{2}\right)\right] + \int S\dot{e}\, dt \tag{33}$$

3.2. Die Geometry

On the basis of not only the mentioned studies, TCAP enables to impose a substantially high effective strain into the material during a single pass (up to ~2.3), which significantly reduces the number of passes necessary to acquire a material with an ultra-fine (nano) structure, and thus makes the entire processing less time-consuming and more efficient [47]. As reported by Bagherpour et al. [5], the favourable combination of both the ECAP and TE methods introduces relatively large plastic strains into the processed material by combining different deformation modes, which eventually leads to very fine grains and a high fraction of high angle grain boundaries (HAGBs). In addition, the advantageous combination leads to homogenization of final properties throughout the sample cross section and length, while keeping the shape unchanged [47]. Sole application of TE leads to a certain permanent deformation of the processed sample (primarily imparted by the friction during processing). This phenomenon can be compensated by the application of back pressure, however, back pressure is not necessary to be applied during ECAP, that is, the use of back pressure can be eliminated by selecting the favourable die design, that is, TCAP Variant *I*. However, the overall efficiency of Variant *I* varies depending on the die geometry (see Figure 3). The distance between the end of twist part of the channel and the channel bending L was shown to have a neglectable influence on the amount of the imposed strain, but to substantially affect the strain homogeneity; increasing the distance between the twist and bending channel parts suppresses the vortex-like flow and improves strain homogeneity across the sample cross section. Regarding twist slope angle β and twist rotation angle ω, both parameters significantly affect both the amount and homogeneity of the imposed strain; higher values of β and ω angles result in higher imposed strain, but also in higher inhomogeneity. The channel bending angle φ especially affects the amount of the imposed strain. However, its negative effect on strain homogeneity via the formation of dead zones for large bending angles should be mentioned, too. In summary, the most significant effect on increasing the amount of the imposed strain was increasing the twist rotation angle ω, and mutually increasing the twist slope angle β and decreasing the channel bending angle φ. The fact that the dead zone corner gap is substantially smaller for the TCAP die than for ECAP die with identical channel bending angles φ needs to be stressed, too.

3.3. Strain Path

The description of the strain path affecting the material when passing through the TCAP die ensues from the simple shear model developed by Latypov et al. [54]. According to the model, the material processed through the ECAP die is subjected to simple shear in a limited intersection plane located in the channel bending; this model can be applied to describe the bending part of TCAP. The twist part of the die is then considered to be comparable to twist extrusion (TE), during which the material is subjected to simple shear in two intersecting planes. The TE process was already characterized to be more efficient than ECAP from the viewpoint of the occurring grain refinement at identical imposed strains owing to the presence of two independent shear strain paths [55]. However, the TCAP process imposes shear strain into the processed material along three independent shear planes, which ensures its higher efficiency compared with ECAP featuring a single shear plane, as well as TE having two shear planes.

Given by the three independent shear strain paths, the material flow during TCAP is rather complex [47]. Moreover, its intensity depends on the particular monitored location and selected cutting plane. Numerical prediction of the plastic flow during TCAP processing was performed using three superimposed meshes, as depicted in Figure 6a. As can be seen in the figure, all of the selected cutting planes with the superimposed meshes *1*, *2*, and *3* are situated to the sample extrusion axis and rotated by 45° from one another; this particular rotation is selected in accordance with the rotation imparted to the material by the twist channel part. Superimposed meshes *1* and *3*, depicted in Figure 6b,d, respectively, reveal the twist channel part to impart the most substantial plastic flow to the axial sample region (the plastic flow intensity decreases across the sample cross section towards its periphery). This flow inhomogeneity is then reduced in the channel bending part, the flow velocities of the axial and peripheral material regions in which equalize. Therefore, the imposed shear strain exhibits high homogeneity along the cutting planes depicted via superimposed meshes *1* and *3*. On the other hand, superimposed mesh *2*, the detail of which is depicted in Figure 6c, features differences when compared with meshes *1* and *3*. In this particular cutting plane, the plastic flow imparted by the twist channel part is more intense in the axial sample region than in its peripheral regions and, as a result of the vortex-like flow, this character of plastic flow remains visible even after passing through the channel bending. Among the significant factors affecting the plastic flow homogeneity is also the friction between the channel and extruded material; increasing friction not only increases the amount of the imposed shear strain, but also imparts flow inhomogeneities across the sample cross section [56].

Similar to ECAP, TCAP can also be performed in multiple passes. However, given the unique strain path and complex material flow, the shear systems activated during ECAP and TCAP when applying the conventional ECAP routes are not identical. This was demonstrated by Kocich et al. [49], by performing individual experiments in which they applied the known ECAP deformation routes, that is, route *A* (no rotation between passes), route *Ba* (±90° rotation between passes), route *Bc* (+90° rotation between passes, clockwise, and counter-clockwise), and route *C* (180° rotation between passes), for both ECAP and TCAP. For example, processing of the sample via ECAP, route *A*, results in repetitive activation of identical shear systems. On the other hand, processing of the sample via TCAP, route *A*, activates multiple shear systems owing to the implemented twist channel part. Activation of different slip systems will also lead to proportional changes in the energy dissipation modes characteristic for the SPD processing, especially regarding the ratio of dynamic recrystallization and dislocation-disclination accommodation [57]. A comparison of the shear systems activated during processing via the ECAP and TCAP deformation routes in multiple passes is shown in Figure 7.

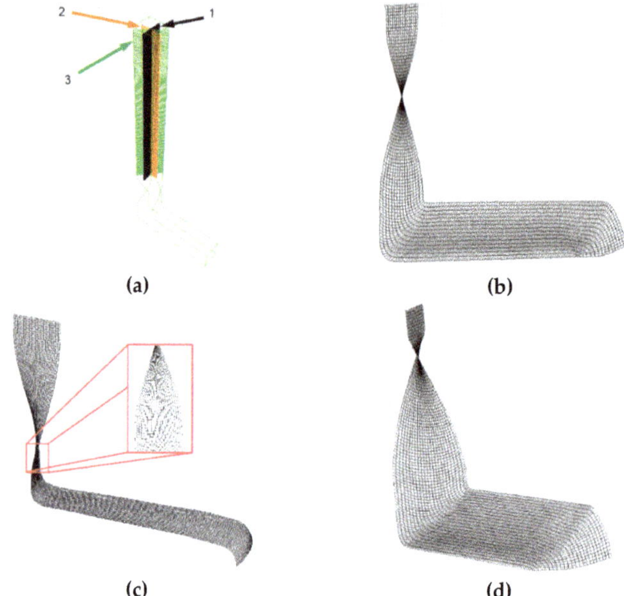

Figure 6. Predicted material flow through the TCAP die, three superimposed meshes: (**a**) detail of superimposed mesh *1* (**b**); superimposed mesh *2* (**c**); and superimposed mesh *3* (**d**).

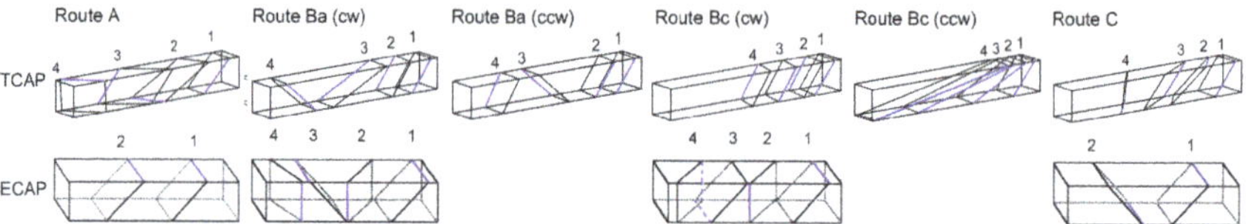

Figure 7. Comparison of activated shear systems during processing via multiple equal channel angular pressing (ECAP) and TCAP methods.

3.4. Structure and Properties

The general comparison of TCAP with the TE and ECAP technologies by which TCAP was inspired shows that TCAP ensures higher homogeneity as well as higher efficiency of the imposed strain across the cross section of the processed material.

The numerically predicted high homogeneity of the imposed strain throughout the sample when extruded with optimized processing parameters was validated by experimental investigations of mechanical properties. The stress–strain curves recorded during the tensile testing of specimens taken from the axial and peripheral regions of the commercial purity (CP) Cu TCAP-processed sample both reached the maximum value of 440 MPa, and their courses featured only minor deviations [47]. Similar trends were documented for the CP Cu microhardness, which increased by 80% after TCAP processing, to the average value of 108 HV, and exhibited only minor deviations along the measured diagonals across the square TCAP sample cross section [56].

The high efficiency of TCAP can primarily be attributed to the three independent shear strain paths, supporting development of lattice distortions acting as obstacles for dislocations movement, and nucleation sites for recovery. The comparison of stress–strain curves for CP Al processed via ECAP and TCAP showed the strength to significantly increase after a single ECAP pass, and then increase slightly again after the second pass via the *A* and *Bc* routes up to 300 MPa for the *Bc* route processed sample. Nevertheless, the sample processed via a single TCAP pass exhibited the strength of approximately 340 MPa, which can be attributed to two main phenomena—the presence of strengthening secondary particles, and substantial grain refinement. The efficiency of multiple TCAP on grain refinement was thoroughly documented for CP Cu [49], whereas the effects of the strain paths acting during a single pass through the TCAP die were characterized via studies on samples of CP Al [58,59]. The analyses showed that the original unprocessed Al sample exhibited relatively coarse grains with bimodal grain size distribution (the average grain diameter was 39 μm) and more or less random grains orientations. After a single ECAP pass, the grains refined significantly and featured the average diameter of slightly more than 10 μm. Processing by two ECAP passes via route *A* led to grain refinement to the average diameter of ~8 μm, while processing via two ECAP passes via route *Bc*, which is considered to be the most efficient ECAP route from the viewpoint of grain refinement [60], resulted in the average grain size of 7.4 μm. The most intensive grain refinement was recorded after TCAP, the average grain size after a single pass of which was ~5.8 μm (68% of grains were smaller than 5 μm). The grains also featured a high dislocations density and well developed substructure, as documented by Figure 8a [61]. The sample processed by two ECAP passes via route *Bc* also featured refined grains and a deformed substructure, but exhibited no highly developed subgrains [58].

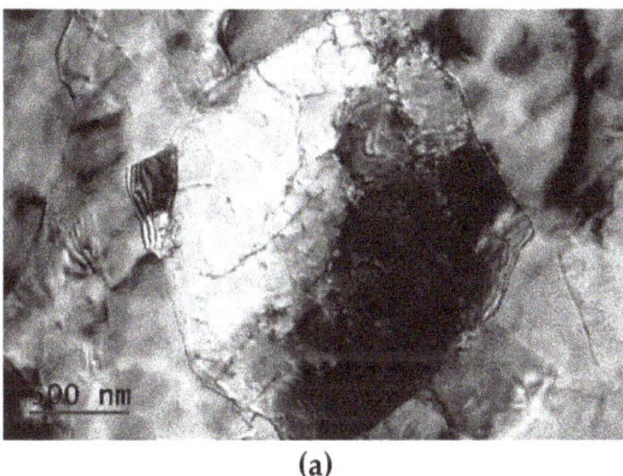

(a)

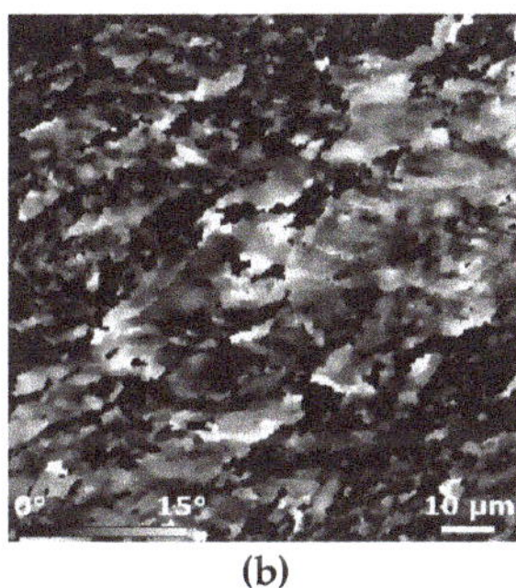

(b)

Figure 8. Transmission electron microscopy image showing the detail of CP Al substructure after single pass TCAP (**a**); residual stress depicted via internal grains misorientations in scale from 0° (black) to 15° (white) for the two-year-old CP Al sample (**b**).

Taking into consideration the high imposed energy and substructure development occurring during a single pass, the structure of TCAP-processed material can be considered to be metastable, and further structure changes can be supposed to occur over time. This presupposition was confirmed in a study dealing with two-year-old CP Al samples [62]. After natural ageing, the restoration processes provoked the grains to further refine down to the average grain size of ~1 μm. Natural ageing also supported precipitation of strengthening particles; small angle neutron scattering (SANS) analyses showed the particles to preferentially precipitate in orientations corresponding to the active shear strain paths. The aged material also exhibited relaxation of residual stress; the study depicts the distribution of residual stress (via internal grains misorientations in scale from 0° to 15°) within the structure of the CP Al TCAP sample right after processing, and after the two years of natural ageing. As evident in Fig. 8b, the misorientations from 0° to 15° in which are depicted in the black to white scale, the aged structure has a tendency to relax and homogenize the level of residual stress throughout the structure. Stress-free locations start to form within the aged sample, too. Evaluation of mechanical properties performed via mapping of microhardness throughout the naturally aged the CP Al sample revealed that the homogeneity of mechanical properties remains stable over time. The map showed a decrease in the average HV value resulting from the occurrence of restoration processes and residual stress relaxation, although it exhibited high homogeneity.

3.5. Texture

The comparison of textures within CP Al after TCAP and ECAP was also performed [58,59]; however, comparing textures of materials processed by different SPD processes can be ambiguous, as texture is affected by various factors and, above all, the strain paths along which the shear strain is imposed (ideal texture components are characterized according to the description performed by Beyerlein and Tóth [63]). The original weak cubic texture present in the unprocessed Al transformed into *A* ideal texture component, and *B* texture fibre parallel to the shear direction (<110>||SD) after a single pass ECAP; similar texture formation tendencies were also observed after two ECAP *Bc* route passes, whereas strong *A* ideal texture component {111} <011> formed in the structure after two ECAP *A* route passes. Processing via TCAP led to the formation of ideal <110>||SD and <100>||SD *B* texture fibres, as well as the *A* ideal texture component and *C* ideal texture component. Nevertheless, given the substantial substructure development and grain fragmentation, the maximum texture intensities were lower than two times random for all of the components, as also confirmed by a 3D EBSD (electron back-scattered diffraction) study [59]. Figure 9a shows inverse pole figures (IPFs) for the CP Al sample processed via a single pass TCAP, while Figure 9b depicts the intensities of the characteristic ideal texture orientations for this sample.

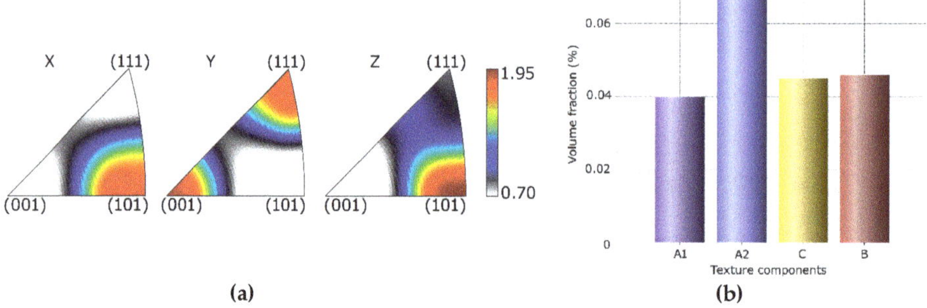

Figure 9. Inverse pole figures (**a**) and ideal texture components (**b**) for CP Al after single pass TCAP.

4. Twist Channel Multi Angular Pressing

The twist channel multi angular pressing (TCMAP) method was invented in order to explore the efficiency limits of TCAP. The TCMAP die basically consists of a TCAP die modified by implementation of another deformation section, that is, it contains multiple bendings. Kocich et al. [64] studied two different variants of the TCMAP die; the first variant consisted of the TCAP die supplemented with an additional bending situated between the original TCAP twist and bending channel parts, whereas the second variant consisted of the TCAP die having the additional channel bending located in front of the conventional TCAP twist and bending parts.

The results of numerical simulations showed the imposed effective strain to increase when compared with the conventional TCAP. The maximum effective strain was higher for the second TCMAP variant having the additional bending implemented as the first die feature (maximum strain of ~3.2 compared with ~2.8 for the first variant having the additional bending between the TCAP twist and bending channel parts). However, the strain homogeneity was higher for the first variant. Among the positive aspects of TCMAP is the elimination of dead corner zones. On the other hand, the full contact of the extruded material with the channel, together with its complicated construction, increases the load on the punch, and consequently the power demands of the entire extrusion process. Measurements of microhardness values for the TCMAP-processed CP Al sample showed a substantial increase in the average HV value (increase by 97% compared with the original CP Al, to ~90 HV), as well as high microhardness homogeneity along the diagonals across the square TCMAP sample cross section [65].

5. Twisted Multi Channel Angular Pressing

The twisted multi channel angular pressing (TMCAP) method based on the combination of ECAP and TE was very recently presented by S. M. Alavizadeh et al. [66]. The TMCAP die consists of three independent parts, the first one of which is a rectangular extrusion channel with the length of 55 mm. The second part, bending under the angle of 156°, is followed by the third part being a twist with the rotation angle of 60° and total length of 16 mm. The last deformation feature within the extrusion channel is two bendings of 132° and 113°. The die is designed to enable the repeatability of the entire process.

Numerical predictions showed that the process imposes a substantially high effective strain onto the processed material, although the inhomogeneity of the imposed strain throughout the sample cross section is notable; the average effective imposed strain was predicted to be approximately 2.4, but the local strain maximum reached the value of 5. The inhomogeneity was also documented by the experimental microhardness testing performed on extruded CP Al. Nevertheless, selecting an optimized strain path while performing multiple TMCAP contributed to homogenization of the imposed strain and, correspondingly, the mechanical properties throughout the sample cross section. The yield strength of the processed AL1050 alloy increased by 164 % after four TMCAP passes.

6. Planar Twist Channel Angular Extrusion

Last, but not least, the planar twist channel angular extrusion (PTCAE) method is mentioned [67]. This process combines the ECAP technique with the planar twist extrusion (PTE) method within a single die, and thus simultaneously imposes severe shear strain to the processed material in three perpendicular planes within a single deformation zone. In the deformation zone, the processed sample is subjected to severe deformation from its original square cross-sectional shape at the channel inlet, to the parallelogram shape with the maximum distortion angle α in middle part of the channel, and finally returns to the square shape at the channel outlet [68]. Further, the processed material is subjected to shear strain, similar to ECAP, in the plane parallel to the two intersected channels. The die is designed to enable processing of the sample via multiple passes.

As proven by experiments with CP Al, the PTCAE process features two main advantages over the ECAP process; not only is it more efficient, but it also features high homogeneity of the imposed shear strain, and consequent uniformity of the mechanical properties; microhardness increased from the original 29 HV to 41 HV after processing via ECAP, and to 49 HV after one pass of PTCAE [69]. The efficiency of the entire process can primarily be optimized via changing the planar twist angle α, as shown by numerical simulations; decreasing the planar twist angle decreases the imposed effective strain and increases the cross-sectional area of the deformed sample. Further, increasing the planar twist angle results in increasing the deformation load owing to the higher strain imposed along the planar twist path in the deformation zone, and higher friction increased by the increased contact surface of the die and processed sample.

7. Conclusions

The review was focused on characterization of the main features of the equal channel angular pressing (ECAP)-based twist channel angular pressing (TCAP) method of severe plastic deformation (SPD). The unique construction of the die combining twist and bending in a single channel enables to impart severe shear strain into the processed material via three independent shear strain paths, which makes the process more efficient from the viewpoint of grain refinement and improvement of mechanical properties than ECAP. These conclusions were supported by numerous experimental studies, all documenting the efficiency of the TCAP process via substantial grain refinement and enhancement of mechanical properties. Numerical predictions of plastic flow via three independent superimposed meshes, together with characterization of the effects of the individual TCAP die features on the material behaviour, as well as calculations of the imposed shear strain, were performed as well. The study was supplemented with data characterizing the behaviour of CP Al samples processed via TCAP and subjected to two years of natural ageing, after which structure restoration, relaxation of residual stress, and grain refinement down to the ultra-fine scale that were observed. Having characterized the TCAP method, the twist channel multi angular pressing (TCMAP) and twisted multi channel angular pressing (TMCAP) methods featuring the combinations of TE and ECAP with the implementation of additional deformation features were presented. Last, but not least, the planar twist channel angular extrusion (PTCAE) method combining ECAP with planar twist extrusion (PTE) was presented.

Author Contributions: The contributions of the individual authors were as follows: methodology, writing—original draft preparation, review and editing, project administration, funding acquisition, A.M. The author has read and agreed to the published version of the manuscript.

Funding: This research was supported by the 19-15479S PROJECT OF THE GRANT AGENCY OF THE CZECH REPUBLIC.

Conflicts of Interest: The authors declare no conflict of interest. The funders had no role in the design of the study; in the collection, analyses, or interpretation of data; in the writing of the manuscript; or in the decision to publish the results. All figures used within this manuscript are original unpublished data acquired directly from the archive of the inventor, Assoc. Prof. Radim Kocich, with his kind permission.

References

1. Hansen, N. Hall–Petch relation and boundary strengthening. *Scr. Mater.* **2004**, *51*, 801–806. [CrossRef]
2. Verlinden, B.; Driver, J.; Samajdar, I.; Doherty, R.D. *Thermo-Mechanical Processing of Metallic Materials*; Elsevier: Amsterdam, The Netherlands, 2007; ISBN 9780080444970.
3. Fang, Y.; Chen, X.; Madigan, B.; Cao, H.; Konovalov, S. Effects of strain rate on the hot deformation behavior and dynamic recrystallization in China low activation martensitic steel. *Fusion Eng. Des.* **2016**, *103*, 21–30. [CrossRef]
4. Glezer, A.M.; Metlov, L.S. Physics of megaplastic (Severe) deformation in solids. *Phys. Solid State* **2010**, *52*, 1162–1169. [CrossRef]
5. Bagherpour, E.; Pardis, N.; Reihanian, M.; Ebrahimi, R. An overview on severe plastic deformation: Research status, techniques classification, microstructure evolution, and applications. *Int. J. Adv. Manuf. Technol.* **2019**, *100*, 1647–1694. [CrossRef]
6. Langdon, T.G. Twenty-five years of ultrafine-grained materials: Achieving exceptional properties through grain refinement. *Acta Mater.* **2013**, *61*, 7035–7059. [CrossRef]
7. Qian, C.; He, Z.; Liang, C.; Cha, Y.; Ji, W. Superior intracluster conductivity of metallic lithium-ion battery anode achieved by high-pressure torsion. *Mater. Lett.* **2020**, *260*, 126933. [CrossRef]
8. Voronova, L.M.; Chashchukhina, T.I.; Gapontseva, T.M.; Krasnoperova, Y.G.; Degtyarev, M.V.; Pilyugin, V.P. Effect of the deformation temperature on the structural refinement of BCC metals with a high stacking fault energy during high pressure torsion. *Russ. Metall.* **2016**, *2016*, 960–965. [CrossRef]
9. Kunčická, L.; Lowe, T.C.; Davis, C.F.; Kocich, R.; Pohludka, M. Synthesis of an Al/Al_2O_3 composite by severe plastic deformation. *Mater. Sci. Eng. A* **2015**, *646*, 234–241. [CrossRef]
10. Kocich, R.; Greger, M.; Macháčková, A. Finite element investigation of influence of selected factors on ECAP process. In Proceedings of the METAL 2010: 19th International Metallurgical and Materials Conference, Roznov pod Radhostem, Czech Republic, 20–22 May 2010; Tanger Ltd.: Ostrava, Czech Republic, 2010; pp. 166–171.
11. Elhefnawey, M.; Shuai, G.L.; Li, Z.; Nemat-Alla, M.; Zhang, D.T.; Li, L. On achieving superior strength for Al–Mg–Zn alloy adopting cold ECAP. *Vacuum* **2020**, *174*, 109191. [CrossRef]
12. Hlaváč, L.M.; Kocich, R.; Gembalová, L.; Jonšta, P.; Hlaváčová, I.M. AWJ cutting of copper processed by ECAP. *Int. J. Adv. Manuf. Technol.* **2016**, *86*, 885–894. [CrossRef]
13. Valiev, R.Z.; Langdon, T.G. Principles of equal-channel angular pressing as a processing tool for grain refinement. *Prog. Mater. Sci.* **2006**, *51*, 881–981. [CrossRef]

14. Tang, Y.; Sumikawa, K.; Takizawa, Y.; Yumoto, M.; Otagiri, Y.; Horita, Z. Multi-pass high-pressure sliding (MP-HPS) for grain refinement and superplasticity in metallic round rods. *Mater. Sci. Eng. A* **2019**, *748*, 108–118. [CrossRef]
15. Tóth, L.S.; Arzaghi, M.; Fundenberger, J.J.; Beausir, B.; Bouaziz, O.; Arruffat-Massion, R. Severe plastic deformation of metals by high-pressure tube twisting. *Scr. Mater.* **2009**, *60*, 175–177. [CrossRef]
16. Kunčická, L.; Král, P.; Dvořák, J.; Kocich, R. Texture evolution in biocompatible mg-y-re alloy after friction stir processing. *Metals* **2019**, *9*, 1181. [CrossRef]
17. Zisman, A.A.; Rybin, V.V.; Van Boxel, S.; Seefeldt, M.; Verlinden, B. Equal channel angular drawing of aluminium sheet. *Mater. Sci. Eng. A* **2006**, *427*, 123–129. [CrossRef]
18. Kocich, R.; Macháčková, A.; Andreyachshenko, V.A. A study of plastic deformation behaviour of Ti alloy during equal channel angular pressing with partial back pressure. *Comput. Mater. Sci.* **2015**, *101*, 233–241. [CrossRef]
19. Fatemi-Varzaneh, S.M.; Zarei-Hanzaki, A.; Izadi, S. Shear deformation and grain refinement during accumulative back extrusion of AZ31 magnesium alloy. *J. Mater. Sci.* **2010**, *46*, 1937–1944. [CrossRef]
20. Kocich, R.; Macháčková, A.; Fojtík, F. Comparison of strain and stress conditions in conventional and ARB rolling processes. *Int. J. Mech. Sci.* **2012**, *64*, 54–61. [CrossRef]
21. Mohebbi, M.S.; Akbarzadeh, A. Accumulative spin-bonding (ASB) as a novel SPD process for fabrication of nanostructured tubes. *Mater. Sci. Eng. A* **2010**, *528*, 180–188. [CrossRef]
22. Gupta, A.K.; Maddukuri, T.S.; Singh, S.K. Constrained groove pressing for sheet metal processing. *Prog. Mater. Sci.* **2016**, *84*, 403–462. [CrossRef]
23. Torkestani, A.; Dashtbayazi, M.R. A new method for severe plastic deformation of the copper sheets. *Mater. Sci. Eng. A* **2018**, *737*, 236–244. [CrossRef]
24. Abdolvand, H.; Sohrabi, H.; Faraji, G.; Yusof, F. A novel combined severe plastic deformation method for producing thin-walled ultrafine grained cylindrical tubes. *Mater. Lett.* **2015**, *143*, 167–171. [CrossRef]
25. Zangiabadi, A.; Kazeminezhad, M. Development of a novel severe plastic deformation method for tubular materials: Tube Channel Pressing (TCP). *Mater. Sci. Eng. A* **2011**, *528*, 5066–5072. [CrossRef]
26. Wang, C.; Li, F.; Li, Q.; Li, J.; Wang, L.; Dong, J. A novel severe plastic deformation method for fabricating ultrafine grained pure copper. *Mater. Des.* **2013**, *43*, 492–498. [CrossRef]
27. Wang, Q.D.; Chen, Y.J.; Lin, J.B.; Zhang, L.J.; Zhai, C.Q. Microstructure and properties of magnesium alloy processed by a new severe plastic deformation method. *Mater. Lett.* **2007**, *61*, 4599–4602. [CrossRef]
28. Kim, K.; Yoon, J. Evolution of the microstructure and mechanical properties of AZ61 alloy processed by half channel angular extrusion (HCAE), a novel severe plastic deformation process. *Mater. Sci. Eng. A* **2013**, *578*, 160–166. [CrossRef]
29. Richert, M.; Liu, Q.; Hansen, N. Microstructural evolution over a large strain range in aluminium deformed by cyclic-extrusion–compression. *Mater. Sci. Eng. A* **1999**, *260*, 275–283. [CrossRef]
30. Ghosh, A.K.; Huang, W. Severe Deformation Based Process for Grain Subdivision and Resulting Microstructures. In *Investigations and Applications of Severe Plastic Deformation*; Springer: Dordrecht, The Netherlands, 2000; pp. 29–36.
31. Pardis, N.; Talebanpour, B.; Ebrahimi, R.; Zomorodian, S. Cyclic expansion-extrusion (CEE): A modified counterpart of cyclic extrusion-compression (CEC). *Mater. Sci. Eng. A* **2011**, *528*, 7537–7540. [CrossRef]
32. Ensafi, M.; Faraji, G.; Abdolvand, H. Cyclic extrusion compression angular pressing (CECAP) as a novel severe plastic deformation method for producing bulk ultrafine grained metals. *Mater. Lett.* **2017**, *197*, 12–16. [CrossRef]
33. Khoddam, S.; Farhoumand, A.; Hodgson, P.D. Axi-symmetric forward spiral extrusion, a kinematic and experimental study. *Mater. Sci. Eng. A* **2011**, *528*, 1023–1029. [CrossRef]
34. Farhoumand, A.; Hodgson, P.D.; Khoddam, S. Finite element analysis of plastic deformation in variable lead axisymmetric forward spiral extrusion. *J. Mater. Sci.* **2013**, *48*, 2454–2461. [CrossRef]
35. Ebrahimi, G.R.; Barghamadi, A.; Ezatpour, H.R.; Amiri, A. A novel single pass severe plastic deformation method using combination of planar twist extrusion and conventional extrusion. *J. Manuf. Process.* **2019**, *47*, 427–436. [CrossRef]
36. Yu, J.; Zhang, Z.; Wang, Q.; Hao, H.; Cui, J.; Li, L. Rotary extrusion as a novel severe plastic deformation method for cylindrical tubes. *Mater. Lett.* **2018**, *215*, 195–199. [CrossRef]

37. Mizunuma, S. Large Straining Behavior and Microstructure Refinement of Several Metals by Torsion Extrusion Process. *Mater. Sci. Forum* **2006**, *503–504*, 185–192. [CrossRef]
38. Noor, S.V.; Eivani, A.R.; Jafarian, H.R.; Mirzaei, M. Inhomogeneity in microstructure and mechanical properties during twist extrusion. *Mater. Sci. Eng. A* **2016**, *652*, 186–191. [CrossRef]
39. Pardis, N.; Ebrahimi, R. Deformation behavior in Simple Shear Extrusion (SSE) as a new severe plastic deformation technique. *Mater. Sci. Eng. A* **2009**, *527*, 355–360. [CrossRef]
40. Utsunomiya, H.; Hatsuda, K.; Sakai, T.; Saito, Y. Continuous grain refinement of aluminum strip by conshearing. *Mater. Sci. Eng. A* **2004**, *372*, 199–206. [CrossRef]
41. Lee, J.-C.; Seok, H.-K.; Suh, J.-Y. Microstructural evolutions of the Al strip prepared by cold rolling and continuous equal channel angular pressing. *Acta Mater.* **2002**, *50*, 4005–4019. [CrossRef]
42. Huang, Y.; Prangnell, P.B. Continuous frictional angular extrusion and its application in the production of ultrafine-grained sheet metals. *Scr. Mater.* **2007**, *56*, 333–336. [CrossRef]
43. Huang, J.Y.; Zhu, Y.T.; Jiang, H.; Lowe, T.C. Microstructures and dislocation configurations in nanostructured Cu processed by repetitive corrugation and straightening. *Acta Mater.* **2001**, *49*, 1497–1505. [CrossRef]
44. Nakamura, K.; Neishi, K.; Kaneko, K.; Nakagaki, M.; Horita, Z. Development of Severe Torsion Straining Process for Rapid Continuous Grain Refinement. *Mater. Trans.* **2004**, *45*, 3338–3342. [CrossRef]
45. Kocich, R.; Lukáč, P. SPD Processes-Methods for Mechanical Nanostructuring. In *Handbook of Mechanical Nanostructuring*; Wiley-VCH Verlag GmbH & Co. KGaA: Weinheim, Germany, 2015; pp. 235–262.
46. Semenova, I.P.; Polyakov, A.V.; Raab, G.I.; Lowe, T.C.; Valiev, R.Z. Enhanced fatigue properties of ultrafine-grained Ti rods processed by ECAP-Conform. *J. Mater. Sci.* **2012**, *47*, 7777–7781. [CrossRef]
47. Kocich, R.; Greger, M.; Kursa, M.; Szurman, I.; Macháčková, A. Twist channel angular pressing (TCAP) as a method for increasing the efficiency of SPD. *Mater. Sci. Eng. A* **2010**, *527*, 6386–6392. [CrossRef]
48. Jamili, A.M.; Zarei-Hanzaki, A.; Abedi, H.R.; Mosayebi, M.; Kocich, R.; Kunčická, L. Development of fresh and fully recrystallized microstructures through friction stir processing of a rare earth bearing magnesium alloy. *Mater. Sci. Eng. A* **2019**, *775*, 138837. [CrossRef]
49. Kocich, R.; Fiala, J.; Szurman, I.; Macháčková, A.; Mihola, M. Twist-channel angular pressing: Effect of the strain path on grain refinement and mechanical properties of copper. *J. Mater. Sci.* **2011**, *46*, 7865–7876. [CrossRef]
50. Iqbal, U.M.; Muralidharan, S. Optimization of die design parameters and experimental validation on twist channel angular pressing process of AA6061-T6 aluminium alloy. *Mater. Res. Express* **2019**, *6*, 0865f2. [CrossRef]
51. Segal, V.M. Slip line solutions, deformation mode and loading history during equal channel angular extrusion. *Mater. Sci. Eng. A* **2003**, *345*, 36–46. [CrossRef]
52. Kim, H.S.; Seo, M.H.; Hong, S.I. On the die corner gap formation in equal channel angular pressing. *Mater. Sci. Eng. A* **2000**, *291*, 86–90. [CrossRef]
53. Orlov, D.; Beygelzimer, Y.; Synkov, S.; Varyukhin, V.; Horita, Z. Evolution of Microstructure and Hardness in Pure Al by Twist Extrusion. *Mater. Trans.* **2008**, *49*, 2–6. [CrossRef]
54. Latypov, M.I.; Lee, M.G.; Beygelzimer, Y.; Kulagin, R.; Kim, H.S. On the simple shear model of twist extrusion and its deviations. *Met. Mater. Int.* **2015**, *21*, 569–579. [CrossRef]
55. Bahadori, S.R.; Dehghani, K.; Akbari Mousavi, S.A.A. Comparison of microstructure and mechanical properties of pure copper processed by twist extrusion and equal channel angular pressing. *Mater. Lett.* **2015**, *152*, 48–52. [CrossRef]
56. Kocich, R.; Kunčická, L.; Mihola, M.; Skotnicová, K. Numerical and experimental analysis of twist channel angular pressing (TCAP) as a SPD process. *Mater. Sci. Eng. A* **2013**, *563*, 86–94. [CrossRef]
57. Glezer, A.M.; Sundeev, R.V. General view of severe plastic deformation in solid state. *Mater. Lett.* **2015**, *139*, 455–457. [CrossRef]
58. Kunčická, L.; Kocich, R.; Král, P.; Pohludka, M.; Marek, M. Effect of strain path on severely deformed aluminium. *Mater. Lett.* **2016**, *180*, 280–283. [CrossRef]
59. Kocich, R.; Kunčická, L.; Král, P.; Macháčková, A. Sub-structure and mechanical properties of twist channel angular pressed aluminium. *Mater. Charact.* **2016**, *119*, 75–83. [CrossRef]
60. Stolyarov, V.V.; Zhu, Y.T.; Alexandrov, I.V.; Lowe, T.C.; Valiev, R.Z. Influence of ECAP routes on the microstructure and properties of pure Ti. *Mater. Sci. Eng. A* **2001**, *299*, 59–67. [CrossRef]

61. Kunčická, L.; Kocich, R. Structure Development after Twist Channel Angular Pressing. *Acta Phys. Pol. A* **2017**, *134*, 681–685. [CrossRef]
62. Kunčická, L.; Kocich, R.; Ryukhtin, V.; Cullen, J.C.T.; Lavery, N.P. Study of structure of naturally aged aluminium after twist channel angular pressing. *Mater. Charact.* **2019**, *152*, 94–100. [CrossRef]
63. Beyerlein, I.J.; Tóth, L.S. Texture evolution in equal-channel angular extrusion. *Prog. Mater. Sci.* **2009**, *54*, 427–510. [CrossRef]
64. Kocich, R.; Macháčková, A.; Kunčická, L. Twist channel multi-angular pressing (TCMAP) as a new SPD process: Numerical and experimental study. *Mater. Sci. Eng. A* **2014**, *612*, 445–455. [CrossRef]
65. Kocich, R.; Kunčická, L.; Macháčková, A. Twist Channel Multi-Angular Pressing (TCMAP) as a method for increasing the efficiency of SPD. *IOP Conf. Ser. Mater. Sci. Eng.* **2014**, *63*, 012006. [CrossRef]
66. Alavizadeh, S.M.; Abrinia, K.; Parvizi, A. Twisted Multi Channel Angular Pressing (TMCAP) as a Novel Severe Plastic Deformation Method. *Met. Mater. Int.* **2020**, *26*, 260–271. [CrossRef]
67. Shamsborhan, M.; Ebrahimi, M. Production of nanostructure copper by planar twist channel angular extrusion process. *J. Alloys Compd.* **2016**, *682*, 552–556. [CrossRef]
68. Shokuhfar, A.; Shamsborhan, M. Finite element analysis of planar twist channel angular extrusion (PTCAE) as a novel severe plastic deformation method. *J. Mech. Sci. Technol.* **2014**, *28*, 1753–1757. [CrossRef]
69. Shamsborhan, M.; Shokuhfar, A.; Nejadseyfi, O.; Kakemam, J.; Moradi, M. Experimental and numerical comparison of equal channel angular extrusion (ECAE) with planar twist channel angular extrusion (PTCAE). *Proc. Inst. Mech. Eng. Part C* **2015**, *229*, 3059–3067. [CrossRef]

Article

Effect of Imposed Shear Strain on Steel Ring Surfaces during Milling in High-Speed Disintegrator

Karel Dvořák [1,*], Adéla Macháčková [2], Simona Ravaszová [1] and Dominik Gazdič [1]

1 Faculty of Civil Engineering, Brno University of Technology, Veveří 331/95, 602 00 Brno, Czech Republic; ravaszova.s@fce.vutbr.cz (S.R.); gazdic.d@fce.vutbr.cz (D.G.)

2 Faculty of Materials Science and Technology, VŠB–Technical University of Ostrava, 17. listopadu 15, 708 00 Ostrava, Czech Republic; adela.machackova@vsb.cz

* Correspondence: dvorak.k@fce.vutbr.cz

Received: 13 April 2020; Accepted: 11 May 2020; Published: 13 May 2020

Abstract: This contribution characterizes the performance of a DESI 11 high-speed disintegrator working on the principle of a pin mill with two opposite counter-rotating rotors. As the ground material, batches of Portland cement featuring 6–7 Mohs scale hardness and containing relatively hard and abrasive compounds with the specific surface areas ranging from 200 to 500 m^2/kg, with the step of 50 m^2/kg, were used. The character of the ground particles was assessed via scanning electron microscopy and measurement of the absolute/relative increase in their specific surface areas. Detailed characterization of the rotors was performed via recording the thermal imprints, evaluating their wear by 3D optical microscopy, and measuring rotor weight loss after the grinding of constant amounts of cement. The results showed that coarse particles are ground by impacting the front faces of the pins, while finer particles are primarily milled via mutual collisions. Therefore, the coarse particles cause higher abrasion and wear on the rotor pins; after the milling of 20 kg of the 200 m^2/kg cement sample, the wear of the rotor reached up to 5% of its original mass and the pins were severely damaged.

Keywords: disintegrator; microscopy; wear; high energy milling; cement

---

1. Introduction

The necessity to enhance the utility properties of materials and increase the longevity of construction components has led research and development in various industrial branches, such as materials processing [1] or construction [2,3], towards the design of new materials and improving the performance of contemporary ones. Generally, enhancement of the mechanical and utility properties can be achieved via variations in the chemical composition of the particular material, as documented, for example, by Zhang et al. [4], and structural modifications (grain refinement), which was demonstrated also for pure and commercially pure metals [5,6]. Verlinden et al. claimed that grain refinement introduces lattice defects, such as dislocations and grain boundaries, which increase the materials' intrinsic energy and act as strengthening agents [7]. Kunčická et al. studied syntheses of various kinds of alloys and reported that the grain size can typically be decreased in two ways, the first of which is application of the the methods of intensive/severe plastic deformation (SPD) decreasing the grain size down to sub-micron scale via imparting severe shear strain [8] (e.g., equal channel angular pressing [9] and high-pressure torsion [10]). The second is the implementation of powder metallurgy, i.e., the production of materials from original powders, which is typically used for challenging materials, such as Ti composites [11] or tungsten heavy alloys [12]. Combinations consisting of the processing of challenging powder-based materials, such as Ti-based [13], and W-based [14] alloys, via SPD methods are also advantageous.

The process of production of a component from powders involves final shape and surface treatments, which typically follow deformation processing and/or variations of sintering and heat treatments. The first production step is, however, preparation of the original powders. These are usually manufactured by milling. The grain size distribution and morphology of the grains have fundamental effects not only on the sintering process itself, as mentioned by Wang et al. [15], but also on the final structure of the alloys as described by Macháčková et al. [16]. Surzhenkov et al. studied the wear resistance and mechanisms of abrasion during milling. Based on the results of their research, it can be stated that the grain refinement in mills is typically ensured via several fundamental mechanisms, among which are compression, shear (attrition), compression pulse, impact (stroke or collision), impact and shear, tension/bending, splitting, and cutting. Milling can be performed by collisions of particles with the working tools, via mutual collisions between particles, or between particles and their environment [17].

Among the recent trends in milling is the high-energy milling (HEM) method. The possibilities of application of HEM are in various industrial fields, from milling of demanding composites in metallurgy, as documented, for example, by Muroi et al. [18] or Sazavi et al. [19], through fabrication of various intermetallic and ceramic materials, such as those mentioned by Serena et al. [20], to the preparation of nanomaterials (e.g., Rojac et al. [21]). High-speed grinding (HSG) is a particular type of HEM carried out by applying high amounts of energy using very short and intense power pulses. The amount of energy that is effectively transferred to the material is higher in HSG than in the case of conventional grinding in mills with identical power inputs. Among the types of mills suitable for HSG is a high-speed pin mill with two counter-rotating rotors, known as the disintegrator. This type of mill, the material in which is refined by high-frequency changes in mechanical strength, was extensively studied by Hint [22]. The disintegrator is particularly suitable for the grinding and activation of fine powder materials, as documented by Bumanis [23], and Bumanis and Bajare [24]. The principle of a disintegrator lies in accelerating the material to a high speed by means of the pins on the grinding rotor. The particles than collide with other particles or with the pins on the rotor which rotates in opposite direction. Disintegrators only use the impact, impact and shear, shear (attrition), and tension/bending mechanisms. However, other processes similar to those in attritors or jet mills occur as well. These are introduced by turbulent flow and rapid compression and expansion between the rotors, as described by Kovalev [25]. Baláž wrote in his study [26] that the main advantages of a high-speed disintegrator are continuity of the grinding process, and variety of working tools that can be employed. On the other hand, Surzhenkov et al. stated the main disadvantage to be that the grinding elements are prone to abrasion/wear [17].

Collisions of particles and hard surfaces can be advantageous since the shear strain introduced to the hard surface via the impacting particles typically induces the formation of hard structure phases at the surface of the base material (such as martensite for stainless steels as reported by Staman et al. [27]), by the effect of which the surface hardness increases, and the surface hardening consequently reduces the tendency of the base material (i.e., rotors) to abrasion/wear, as claimed by Silva et al. [28] and Han et al. [29]. On the other hand, continuous high-speed shelling of the base material (rotors) with hard and/or sharp-edged particles inevitably leads to abrasion/wear in time. Gåhlin and Jacobson [30] mentioned that the intensity of wear depends on the mineralogy, morphology, and granulometry of the ground material. However, the effect of particle size of the milled material on wear is not yet fully understood. Misra and Finnie [31] suggested that increasing the particle size increases the wear rate. Small particles lead to penetrations that do not pass through the surface layers of the grinding elements; according to the theory introduced by Larsen-Bads [32], sufficiently small particles are in elastic contact with the grinding tools and do not contribute to abrasion. The presupposed effect of particles' shapes on the degree of abrasion is based on a theory assuming that blunt shapes exert less pressure on the grinding tools and consequently act more like particles providing surface hardening via imposing shear strain and produce less wear than sharp-edged particles. For milled material featuring wider size fractions, the largest grains exert strong point loads on the surfaces of the grinding elements; i.e., such material is more abrasive [17,30].

If the pins are damaged due to abrasion, the efficiency of grinding is reduced. By this reason, the geometry of the milling elements is favourable to be kept constant for as long as possible during milling. The presented study is focused on characterization of the behaviour of rotors of a high-speed disintegrator during milling of a Portland cement, because the HSG technology seems to be a very effective milling technology, moreover, with the benefit of mechanical activation. However, certain issues in this field still remain uncovered. The standard Portland cement is relatively fragile and contains four main minerals with the average hardness of 6–7 (according to the Mohs scale). It features sharp-edged particles, which makes it a very abrasive material. Grinding by a disintegrator is also supposed to be very effective for this particular material since it responds very well in milling by impact or compression pulse [33]. The study is supplemented with characterization of correlations between entry granulometry and abrasion of the rotors during milling.

2. Materials and Methods

The equipment investigated is the DESI 11 disintegrator (Desintegraator Tootmise OÜ, Tallinn, Estonian Republic), which is a laboratory version of a high-speed pin mill with two counter-rotating rotors. The total installed output of the mill is 4.1 kW. The rotor rotation frequency is up to 12,000 RPM, and the maximum circumferential speed of each rotor is 92.4 m/s. The material is fed into the device by a continuous feeder and enters the grinding chamber through the middle of the left rotor; a schematic depiction of the principle of the disintegrator is shown in Figure 1a, and also described in [17]. The real machine is presented in Figure 1b. The construction of the mill allows for a choice of working tools. For this experiment, the CR type rotors designed and manufactured by the FF servis s.r.o. company (Prag, Czech Republic) were used. The rotors were manufactured from C45 steel, sometimes also characterized as SI 1045 steel. This medium-carbon steel was selected for its high quality, relatively high strength, and easy machinability; the rotors were machined from normalized hot-rolled bars. The external diameter of both the rotors was 147 mm. The left rotor has two rows, while the right rotor has three rows of $3 \times 3 \times 3$ mm cubic pins. The designs of both the rotors are demonstrated in Figure 1c,d.

The Portland cement used for this experiment was prepared in a ball mill by collective milling of cement clinker (from Hranice cement plant) and chemical gypsum (Pregips), the chemical compositions of which are depicted in Table 1. The chemical composition of the selected clinker is typical for Portland clinkers. The gypsum (highly pure) contained relatively high humidity and it was further dried before milling to decrease the moisture to under 5%. The ratio of clinker to gypsum was 95:5.

Table 1. Partial chemical compositions of the ground material.

Material	Component								
	SiO_2	CaO	Al_2O_3	Fe_2O_3	SO_3	$CaSO_4{\cdot}2H_2O$	H_2O	$CaSO_4$	Others
Clinker	20.29	65.33	5.21	5.04	0.79	-	-	-	3.34
Gypsum	-	-	-	-	-	84.0	11.0	2.4	2.6

For the subsequent milling process, a set of seven individual batches with specific surface areas ranging from 200 to 500 m^2/kg was prepared. The batches of cement with the selected granulometries were fed into the disintegrator continuously. The dosing rate was 5.5 $g{\cdot}s^{-1}$ and the cement particles are supposed to spend approximately 1 second in the milling chamber due to the pin mill principle. All the samples were milled at the maximum speed of 12,000 RPM.

Grinding of all the samples was carried out under standard laboratory conditions, at 22 °C and relative humidity of 56%. After grinding of each 1 kg batch, thermal imprint on the rotor was acquired using a Flir E4 thermal camera (FLIR Systems Inc., Wilsonville, OR, USA). Between milling of the individual batches, the disintegrator was cooled by an air stream to the initial temperature of 22 °C and the rotors were replaced with a new set.

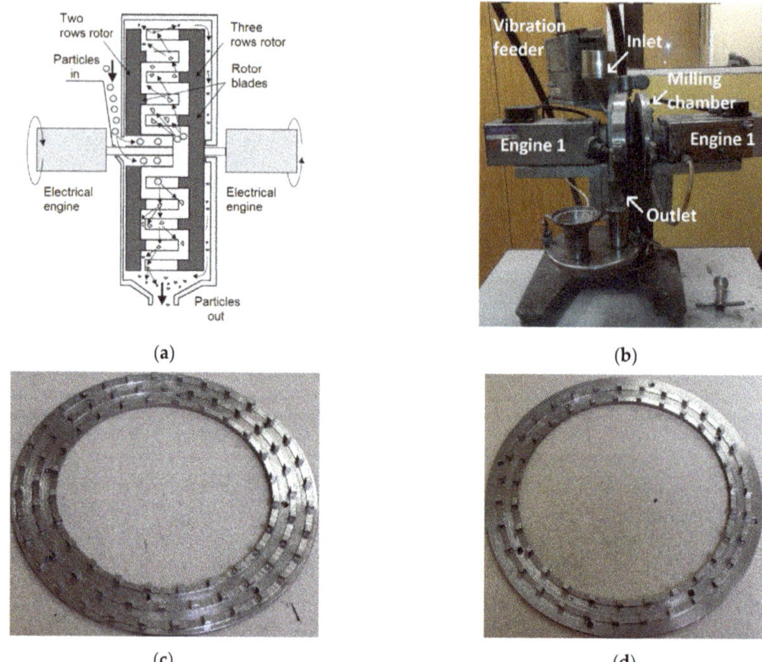

Figure 1. (**a**) Schematic depiction of the investigated disintegrator; (**b**) laboratory DESI 11 HSG mill (**c**) rotor with three rows of pins; (**d**) rotor with two rows of pins.

The increase in the Blaine specific surface area and the morphology of grains were determined for all the samples. The Blaine specific surface area was measured using a PC-Blaine-Star automatic device (Zünderwerke Ernst Brün GmbH, Haltern am See, Germany) with a measuring cell with the volume of 7.95 cm^3. Measurement was repeated three times and averaged to minimize errors. To evaluate the impact of the input granulometry on the abrasion of working elements, two 20 kg batches of cement were prepared in a ball mill. The first sample featured the specific surface area of 200 m^2/kg, while for the second one it was 450 m^2/kg. The milling parameters were identical as for the previous case; rotor speed of 12,000 RPM and dosing rate of 5.5 $g \cdot s^{-1}$.

The particle size and particle size distribution were determined for both the input samples using a Malvern Mastersizer 2000 (Malvern Panalytical B.V., Almelo, The Netherlands) with a Hydro 2000 G fluid dispersing unit, 2-isopropanol was used as a dispersing agent. The morphology and grain shapes were observed and assessed by scanning electron microscopy (SEM) using a Tescan MIRA 3XMU (Tescan Brno s.r.o., Brno, Czech Republic) equipment.

Evaluation of the abrasion-milling elements was performed via calculating the ratio between the weight of the eroded rotors after milling of 1, 10 and 20 kg of abradant, and the initial weight of the rotors. The accuracy of the weight loss measurements was 0.01 mg.

Also, the impact of rotor wear on grinding efficiency was measured by monitoring the increase in the specific surface area of the cement. The abrasion of the rotors was also determined by 3D scanning. Detailed 3D scans of the rotor pins were performed and evaluated using an Olympus DSX1000 digital microscope (Olympus Czech Group, s.r.o., Prague, Czech Republic).

3. Results

Abrasion speed is not constant and depends on the input granulometry and on the shape of the ground particles. Morphology of the cement particles was investigated via SEM-BSE; images of the cement particles within the individual samples are depicted in Figure 2a–f within which individual pairs of figures represent morphology of the samples before and after milling in the HSG mill. Figure 2a,d depict the samples with the initial specific surface area of 200 m^2/kg, whereas Figure 2b,e show the samples with the initial specific surface area of 300 m^2/kg, and Figure 2c,f depict the samples with the initial specific surface area of 450 m^2/kg.

Figure 2. Comparison of the original and post-milling shapes of cement particles for: 200 m^2/kg sample, (**a**) original particles; (**d**) milled particles; 300 m^2/kg sample, (**b**) original particles; (**e**) milled particles; 450 m^2/kg sample, (**c**) original particles; (**f**) milled particles.

The particles from the sample with the specific surface area of 200 m^2/kg exhibited sharp edges, however, they were slightly abraded. Similar results were acquired for the sample with the specific surface area of 300 m^2/kg, but these particles exhibited a more substantial abrasion. The particles from the finest fractions then exhibited more or less spherical shapes and the tendency to agglomerate. Agglomerated clusters were also present in the sample with the initial specific surface area of 450 m^2/kg.

Laser granulometry revealed that the batch with the input-specific surface area of 200 m^2/kg contained 20.8% of particles larger than 100 μm, while the batch with the input specific surface area of 450 m^2/kg contained only 10.7% of particles larger than 100 μm. Aggregates were most probably not present because the measurements were performed under wet conditions.

The resulting specific surface areas for all the seven ground samples, together with the absolute and relative values of their increases, are depicted in Figure 3. The results show that the absolute specific surface area values increased significantly when grinding coarse fractions (specific surface areas of 200 and 250 m^2/kg). In contrast, both the absolute and relative specific surface area values were lower for the input fractions of 300 and 350 m^2/kg. The absolute increase in the specific surface area then increases gradually with continuing refinement of the input material, whereas the relative increase in the specific surface area remains more or less constant from the 300 m^2/kg input fraction.

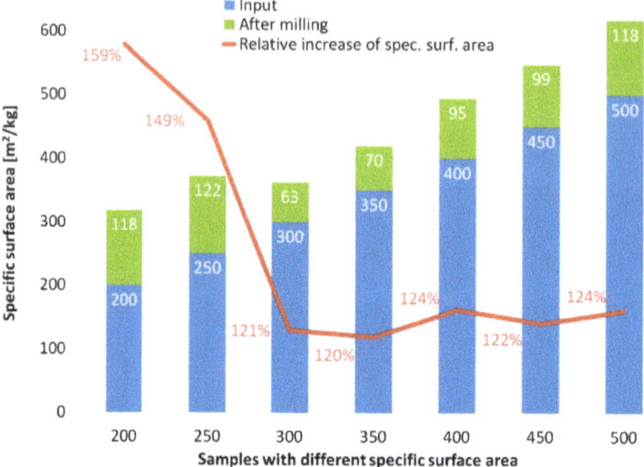

Figure 3. Increase in specific surface area for all ground samples.

Thermal imprints on the rotors before and after milling of 1 kg batches of 200 m^2/kg to 500 m^2/kg samples are shown in Figure 4a–i, respectively. The imprints clearly show that grinding of the coarse batches (200 and 250 m^2/kg) was primarily performed by the front faces of the rotor pins, the temperature of which increased (Figure 4c,d, respectively). As the specific surface area of the input cement increased, the areas in the vicinity of the pins started to exhibit increased temperature, too. The 300, 350, and 400 m^2/kg batches contained plenty of coarse particles, the grinding of which was performed directly on the rotor pins. However, mutual grinding of fine particles against each other in the spaces between the pins occurred as well. This phenomenon was then dominant for the batches with the specific surface areas of 450, and 500 m^2/kg, during the milling of which the pins acted more like breaking wedges directing the milled material into the spaces between them. For these fine batches, the particles were primarily milled by their mutual collisions, and milling by direct impacts with the pins' front faces was of low significance (Figure 4h,i clearly show the pins to be cooler than the surrounding area).

The impact of input granulometry on wear of the grinding rotors is demonstrated by Figure 5 showing wear of the rotors (expressed as loss of rotor mass in relation to the original rotor mass) is dependent on the amount of the ground material. The figure indicates that a coarser material damages the grinding rotors of the disintegrator more significantly than a finer one.

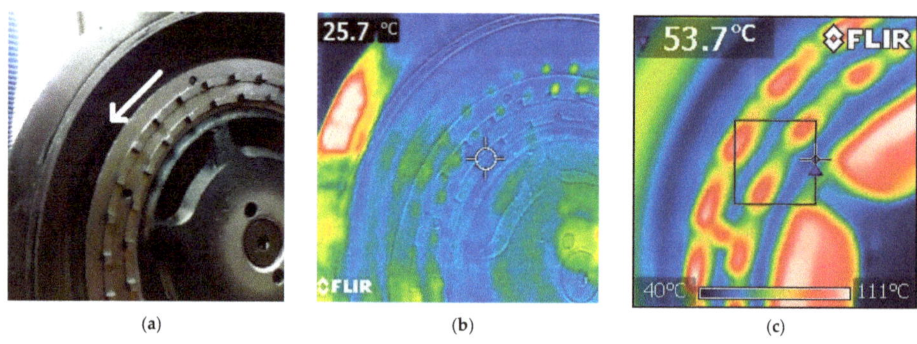

(a) (b) (c)

Figure 4. *Cont.*

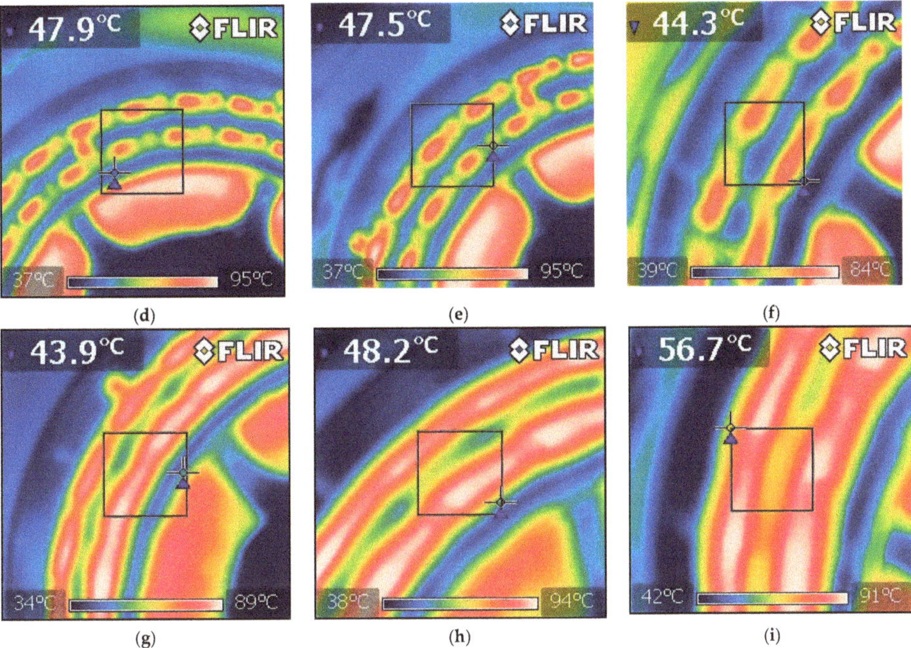

(d) (e) (f) (g) (h) (i)

Figure 4. Thermal imprints on rotors before and after milling of 1 kg of individual batches: (**a**) new rotor with two rows of pins (rotation direction is marked by white arrow); (**b**) new rotor; (**c**) 200 m^2/kg; (**d**) 250 m^2/kg; (**e**) 300 m^2/kg; (**f**) 350 m^2/kg; (**g**) 400 m^2/kg; (**h**) 450 m^2/kg; (**i**) 500 m^2/kg.

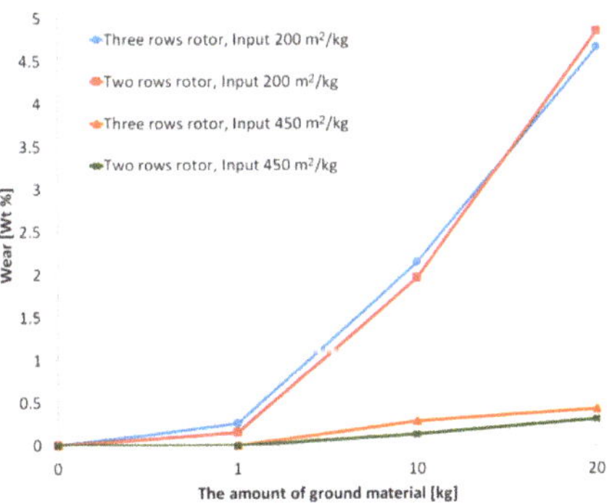

Figure 5. Effect of the amount of ground material on rotor wear for individual samples.

After milling of only a 10 kg batch of the testing cement with the specific surface area of 200 m^2/kg, the wear of the grinding rotor with three rows of pins was 2.14%, and the grinding rotor with two rows of pins exhibited wear of 1.96%. The wear directly influenced the increase in the specific surface area of particles, as the grinding was about 1.61% less effective. Table 2 summarizes the results of measurements of the specific surface area for the cement powder at the beginning of the experiment, and after milling of 10 and 20 kg batches. Table 3 then presents the reductions of grinding efficiency, characterized as reduction of the specific surface area. When grinding the fine-grained input material with the specific surface area of 450 m^2/kg, the abrasion occurred to a lesser extent. The weight loss measured for the rotor with three rows of pins after grinding of a 10 kg batch was 0.28%, while for the rotor with two rows of pins it was 0.13%; the reduction of the efficiency of grinding by the worn rotors was negligible. Milling of a 20 kg batch of the testing cement resulted in further rotor weight loss; after grinding an additional 10 kg batch, the weight loss of the three-row rotor was 0.42%, while for the two-row rotor it was only 0.30%. The efficiency of the rotors was only 1.28% lower when compared to new rotors.

Table 2. Specific surface area of the cement powder at the beginning and after 10 and 20 kg.

Input Specific Surface Area (m^2/kg)	Specific Surface Area of Resulting Powders (m^2/kg)		
	1 kg	**10 kg**	**20 kg**
200	310	305	273
450	548	543	541

Table 3. Reduction of grinding efficiency.

Input Specific Surface Area (m^2/kg)	Grinding Efficiency Reduction (%)		
	1 kg	**10 kg**	**20 kg**
200	-	1.61	11.94
450	-	0.91	1.28

Figure 6a–h show 3D optical microscopy images of the new and worn rotor pins. All the scans were performed on the rotors with three rows of pins. Figure 6a,b depict the 3D scan and height profile, respectively, of the pins of a new rotor, while Figure 6c,d depict the 3D scan and height profile, respectively, of the pins of the rotor used for milling of a 20 kg batch of the 450 m^2/kg sample. Figure 6e,f then depict the 3D scan and height profile, respectively, of the pins of the rotor used for the milling of a 20 kg batch of the 200 m^2/kg sample. The figures confirm the wear analyses, as they clearly show that the wear was the most significant for the rotor used for milling of the 200 m^2/kg batch; scans of a pin from the most worn rotor and its coloured height profile are shown in Figure 6g,h. The figures also show that the wear of the pins was the most significant at their front faces, and also at the face on the side from which the cement particles were injected into the disintegrator. The figures also show evident "shade" behind the pins, which corresponds to the above described occurring milling mechanisms for the individual sample batches.

(**a**)

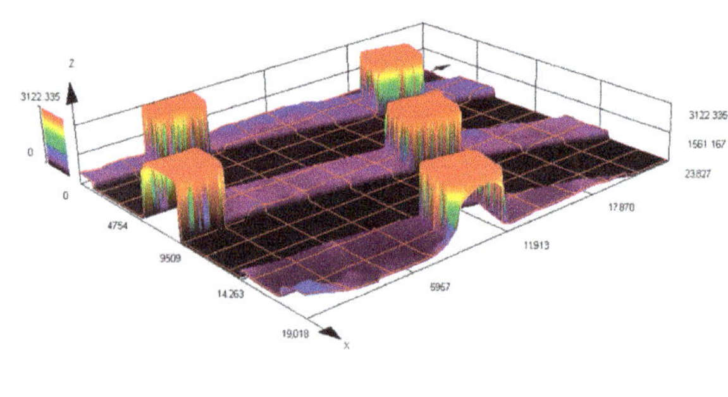

(**b**)

Figure 6. *Cont.*

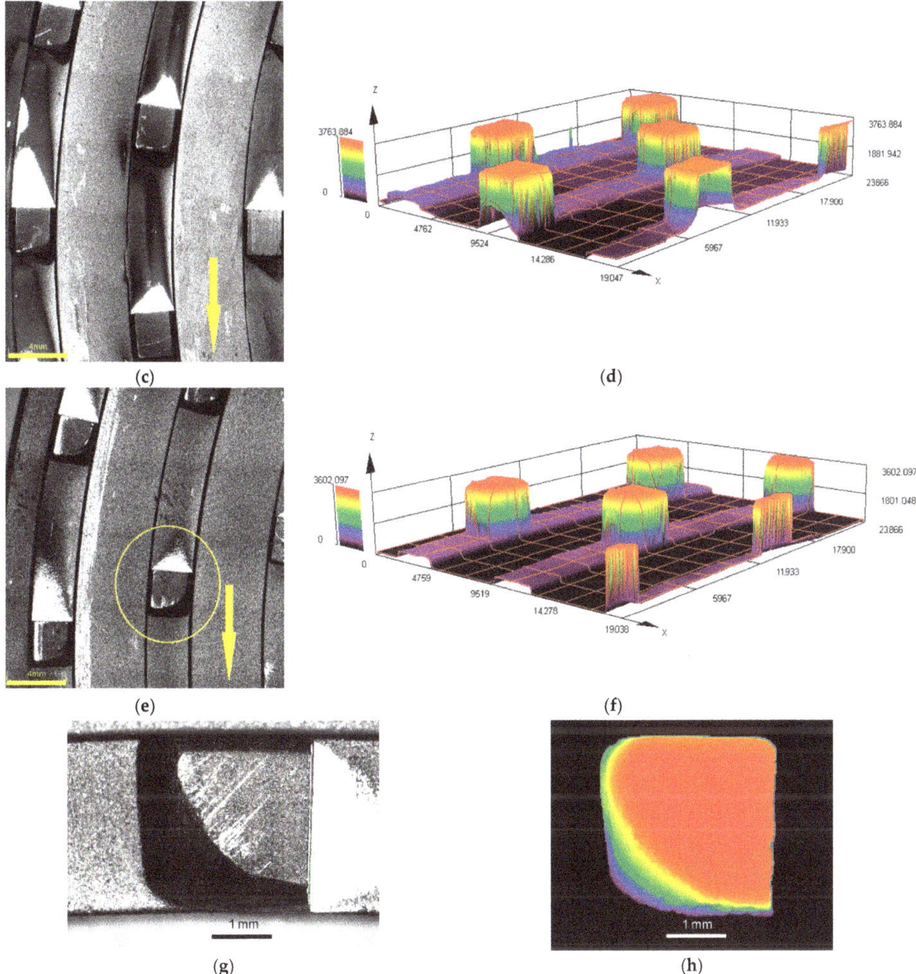

Figure 6. Optical images of new rotor: (**a**) 3D scan and (**b**) height profile of pins; optical images of rotor used for milling of 20 kg batch of 450 m^2/kg sample: (**c**) 3D scan and (**d**) height profile of pins; optical images of rotor used for milling of 20 kg batch of 200 m^2/kg sample: (**e**) 3D scan and (**f**) height profile of pins. Detailed images of pin marked by circle in (**f**) from rotor used for milling of 20 kg batch of 200 m^2/kg sample: (**g**) detailed scan and (**h**) coloured profile. (Rotation directions marked by arrows).

4. Discussion

The different acting particle grinding principles imparted by different input granulometries have considerable effects on the abrasion of the grinding rotors, especially the pins. When milling the samples with the input-specific surface areas of 200 and 250 m^2/kg, the pins of the grinding rotors are strongly heated due to the direct contact with the material being ground, as confirmed by the thermal imprints shown in Figure 4a,b. The direct crushing and friction of the coarse particles on the front faces of the rotor pins also resulted in a substantial increase in the specific surface areas of the particles. The negative aerodynamic pressure occurring right behind the rotor pins is not sufficient to impart swirling motion to the solid particles. For this reason, the particles are not milled by mutual collisions and thus maintain their sharp-edged morphology during passing through the mill; only a slight abrasion of their edges occurs. The wear of the rotors was the highest for these batches and expressed as mass % of the original rotor mass was almost 5% for both the rotors with two and three rows of pins. The shapes of the rotor pins were also severely damaged after milling of the batches with coarse particles, as demonstrated in Figure 6e–h. These findings are in accordance with the theory of Misra and Finnie [31]; large particles impact the front faces of the pins, easily penetrating the surface layers, and cause rapid erosion.

The results acquired during grinding of the samples with very fine particles having the specific surface areas of over 400 m^2/kg were different; the cement particles exhibited intensive refinement and gained spherical shapes, however, they also exhibited a tendency to agglomerate. The momentum of the particles is sufficiently low for them to be drawn into the area of lower pressure in the space behind the rotor pins; the pins act more like breaking wedges and regulate the material flow. The milling is primarily performed by mutual particle collisions in the turbulent area right behind the rotor pins, as demonstrated by the thermal imprints on the rotors in Figure 4e–g, the locations with the highest temperatures in which were situated behind the rotor pins. In contrast to the coarse samples, the finer material with the specific surface area of 450 m^2/kg flows around the pins. Thus small particles do not penetrate the surface layer of the grinding pins and the wear on the rotors is significantly lower for the fine samples. The 3D profiles of the rotor pins depicted in Figure 6c,d exhibited only a slight abrasion, which is consistent with the measured wear being lower than 0.5% of the original rotor mass.

Characterization of the grinding process for the batches featuring the input specific surface areas of 300 to 350 m^2/kg is quite complex. Increases in the specific surface areas for these samples were lower than for the other, coarser as well as finer, batches. The thermal imprints depicted in Figure 4c,d show that a portion of the ground material is in direct contact with the front faces of the grinding pins. Nevertheless, an increase in the temperature in the spaces between and behind the rotor pins is observed as well. This phenomenon implies that during milling of these samples, reduction of particles' sizes proceeds not only via direct contact with the rotor pins, but also, to some extent, via mutual collisions of the particles behind the grinding pins. The amount of coarse grains that are ground on the front surfaces of the rotor pins is lower than for coarser samples. At the same time, the amount of fine grains that have lower momentum and are milled by mutual collisions is not sufficient. The total increase in the specific surface area is thus lower than for the finer and coarser samples. The sharp edges of the particles are abraded after passing through the mill, however, the shape of the coarse particles does not change significantly. For these reasons, assessment of the surface hardening effect for these particular batches would require a deeper study.

5. Conclusions

A variety of modern materials with enhanced properties is nowadays fabricated via methods of powder metallurgy, the initial powders for which are typically manufactured using milling machines. This study focused on the characterization of the performance of a DESI 11 high-speed disintegrator. For the purpose of the study, a variety of samples of Portland cement, containing relatively hard and abrasive particles with specific surface areas ranging from 200 to 500 m^2/kg, were used. The results showed that:

- Grinding of coarse angular-shaped particles proceeds primarily via impacts of particles with the front faces of rotor pins, and such particles cause their significant damage and abrasion.
- Fine cement particles are refined in the area between and behind the pins. This grinding principal supports wear reduction and the grinding efficiency is not substantially affected when milling larger batches of finer cement samples.
- Without sustaining significant damage, this type of high-speed mill is suitable for final grinding of fine particles with the Blaine specific surface area of at least 400 m^2/kg.
- Coarser materials can be ground more efficiently in other types of mill.

- A reasonable specific surface increase with minimum wear of the grinding tools can be achieved in a relatively short time for materials with suitable input granulometries.

The high-speed disintegrator is, thus, an advantageous, versatile piece of equipment for the preparation of powders subsequently used to prepare modern materials.

Author Contributions: Conceptualization, K.D.; methodology, K.D.; software, A.M.; validation, D.G.; formal analysis K.D.; investigation, K.D. and S.R.; resources K.D.; data curation, K.D., A.M., D.G. and S.R.; writing—original draft preparation, K.D.; writing—review and editing, A.M.; visualization, D.G. and S.R.; supervision, K.D.; project administration, K.D. and A.M.; funding acquisition, K.D. and A.M. All authors have read and agreed to the published version of the manuscript.

Funding: We acknowledge the financial support of Czech Science Foundation under the project "Effects of mechano-chemical activation on the process of formation, structure and stability of selected clinker minerals" (20-00676S), and FV40286 Ministry of Industry and Trade TRIO programme.

Acknowledgments: The authors would like to thank Lenka Kunčická for her help with analysing the rotors.

Conflicts of Interest: The authors declare no conflict of interest.

References

1. Thomasová, M.; Seiner, H.; Sedlák, P.; Frost, M.; Ševčík, M.; Szurman, I.; Kocich, R.; Drahokoupil, J.; Šittner, P.; Landa, M. Evolution of macroscopic elastic moduli of martensitic polycrystalline NiTi and NiTiCu shape memory alloys with pseudoplastic straining. *Acta Mater.* **2017**, *123*, 146–156. [CrossRef]
2. Leo Samuel, D.G.; Dharmasastha, K.; Shiva Nagendra, S.M.; Maiya, M.P. Thermal comfort in traditional buildings composed of local and modern construction materials. *Int. J. Sustain. Built Environ.* **2017**, *6*, 463–475. [CrossRef]
3. Naser, M.Z. Properties and material models for modern construction materials at elevated temperatures. *Comput. Mater. Sci.* **2019**, *160*, 16–29. [CrossRef]
4. Kunčická, L.; Kocich, R. Comprehensive characterisation of a newly developed Mg-Dy-Al-Zn-Zr alloy structure. *Metals* **2018**, *8*, 73. [CrossRef]
5. Zhang, P.; Li, Z.; Liu, B.; Ding, W. Effect of chemical compositions on tensile behaviors of high pressure die-casting alloys Al-10Si-yCu-xMn-zFe. *Mater. Sci. Eng. A* **2016**, *661*, 198–210. [CrossRef]
6. Edalati, K.; Cubero-Sesin, J.M.; Alhamidi, A.; Mohamed, I.F.; Horita, Z. Influence of severe plastic deformation at cryogenic temperature on grain refinement and softening of pure metals: Investigation using high-pressure torsion. *Mater. Sci. Eng. A* **2014**, *613*, 103–110. [CrossRef]
7. Kunčická, L.; Kocich, R.; Král, P.; Pohludka, M.; Marek, M. Effect of strain path on severely deformed aluminium. *Mater. Lett.* **2016**, *180*, 280–283. [CrossRef]
8. Verlinden, B.; Driver, J.; Samajdar, I.; Doherty, R.D. *Thermo-Mechanical Processing of Metallic Materials*; Elsevier: Amsterdam, The Netherlands, 2007; ISBN 9780080444970.
9. Kunčická, L.; Kocich, R.; Drápala, J.; Andreyachshenko, V.A. FEM simulations and comparison of the ecap and ECAP-PBP influence on Ti6Al4V alloy's deformation behaviour. In Proceedings of the Metal 2013: 22nd International Conference on Metallurgy and Materials, Brno, Czech Republic, 15–17 May 2013; pp. 391–396.
10. Kunčická, L.; Lowe, T.C.; Davis, C.F.; Kocich, R.; Pohludka, M. Synthesis of an Al/Al2O3composite by severe plastic deformation. *Mater. Sci. Eng. A* **2015**, *646*, 234–241. [CrossRef]
11. Joshi, P.B.; Marathe, G.R.; Murti, N.S.S.; Kaushik, V.K.; Ramakrishnan, P. Reactive synthesis of titanium matrix composite powders. *Mater. Lett.* **2002**, *56*, 322–328. [CrossRef]
12. Kunčická, L.; Macháčková, A.; Lavery, N.P.; Kocich, R.; Cullen, J.C.T.; Hlaváč, L.M. Effect of thermomechanical processing via rotary swaging on properties and residual stress within tungsten heavy alloy. *Int. J. Refract. Met. Hard Mater.* **2020**, *87*, 105120. [CrossRef]
13. Stolyarov, V.V.; Zhu, Y.T.; Lowe, T.C.; Islamgaliev, R.K.; Valiev, R.Z.R. Processing nanocrystalline Ti and its nanocomposites from micrometer-sized Ti powder using high pressure torsion. *Mater. Sci. Eng. A* **2000**, *282*, 78–85. [CrossRef]
14. Kunčická, L.; Kocich, R.; Hervoches, C.; Macháčková, A. Study of structure and residual stresses in cold rotary swaged tungsten heavy alloy. *Mater. Sci. Eng. A* **2017**, *704*, 25–31. [CrossRef]

15. Wang, X.; Chen, Y.; Xu, L.; Xiao, S.; Kong, F.; Woo, K. Do Ti-Nb-Sn-hydroxyapatite composites synthesized by mechanical alloying and high frequency induction heated sintering. *J. Mech. Behav. Biomed. Mater.* **2011**, *4*, 2074–2080. [CrossRef]
16. Macháčková, A.; Krátká, L.; Petrmichl, R.; Kunčická, L.; Kocich, R. Affecting Structure Characteristics of Rotary Swaged Tungsten Heavy Alloy Via Variable deformation temperature. *Materials* **2019**, *12*, 4200. [CrossRef] [PubMed]
17. Surzhenkov, A.; Viljus, M.; Simson, T.; Tarbe, R.; Saarna, M.; Casesnoves, F. Wear resistance and mechanisms of composite hardfacings at abrasive impact erosion wear. *J. Phys. Conf. Ser.* **2017**, *843*, 012060. [CrossRef]
18. Muroi, M.; Street, R.; McCormick, P.G. Structural and Magnetic Properties of Ultrafine La0.7Ca0.3MnOz Powders Prepared by Mechanical Alloying. *J. Solid State Chem.* **2000**, *152*, 503–510.
19. Razavi Tousi, S.S.; Yazdani Rad, R.; Salahi, E.; Mobasherpour, I.; Razavi, M. Production of Al–20 wt.% Al2O3 composite powder using high energy milling. *Powder Technol.* **2009**, *192*, 346–351. [CrossRef]
20. Serena, S.; Caballero, A.; Turrillas, X.; Martin, D.; Sainz, M.A. Effect of ceramic nanoparticles on the solid-state reaction mechanism of dolomite–zirconium oxide followed by neutron thermodiffraction measurements. *J. Nanopart. Res.* **2009**, *11*, 869–878. [CrossRef]
21. Rojac, T.; Kosec, M.; Malič, B.; Holc, J. Mechanochemical synthesis of NaNbO3. *Mater. Res. Bull.* **2005**, *40*, 341–345. [CrossRef]
22. Hint, J. *Fundamentals of Silicalcite Production*; Strojizdat: Moscow, Russia, 1962. (In Russian)
23. Bumanis, G.; Bajare, D.; Goljandin, D. Performance evaluation of cement mortar and concrete with incorporated micro fillers obtained by collision milling in disintegrator. *Ceram. Silik.* **2017**, *61*, 231–243. [CrossRef]
24. Bumanis, G.; Bajare, D. Compressive Strength of Cement Mortar Affected by Sand Microfiller Obtained with Collision Milling in Disintegrator. *Procedia Eng.* **2017**, *172*, 149–156. [CrossRef]
25. Kovalev, P.I. Influence of shock destruction of solid and liquid particles in the supersonic flow around a solid two-phase flow. *J. Appl. Phys.* **2008**, *78*, 40–46.
26. Baláž, P. High-Energy Milling. In *Mechanochemistry in Nanoscience and Minerals Engineering*; Springer: Berlin/Heidelberg, Germany, 2008; pp. 103–132.
27. Starman, B.; Hallberg, H.; Wallin, M.; Ristinmaa, M.; Halilovič, M. Differences in phase transformation in laser peened and shot peened 304 austenitic steel. *Int. J. Mech. Sci.* **2020**, *176*, 105535. [CrossRef]
28. Silva, K.H.S.; Carneiro, J.R.; Coelho, R.S.; Pinto, H.; Brito, P. Influence of shot peening on residual stresses and tribological behavior of cast and austempered ductile iron. *Wear* **2019**, *440–441*, 203099. [CrossRef]
29. Han, X.; Zhang, Z.; Hou, J.; Barber, G.C.; Qiu, F. Tribological behavior of shot peened/austempered AISI 5160 steel. *Tribol. Int.* **2020**, *145*, 106197. [CrossRef]
30. Misra, A.; Finnie, I. On the size effect in abrasive and erosive wear. *Wear* **1981**, *65*, 359–373. [CrossRef]
31. Larsen-Badse, J. Influence of grit diameter and specimen size on wear during sliding abrasion. *Wear* **1968**, *12*, 35–53. [CrossRef]
32. Gåhlin, R.; Jacobson, S. The particle size effect in abrasion studied by controlled abrasive surfaces. *Wear* **1999**, *224*, 118–125. [CrossRef]
33. Hewlett, P. *Lea's Chemistry of Cement and Concrete*; Elsevier: Amsterdam, The Netherlands, 2004; ISBN 9780750662567.

Article

The Effect of Processing Route on Properties of HfNbTaTiZr High Entropy Alloy

Jaroslav Málek [1,2,*], Jiří Zýka [1], František Lukáč [3,4], Monika Vilémová [3], Tomáš Vlasák [4], Jakub Čížek [4], Oksana Melikhova [4], Adéla Macháčková [5] and Hyoung-Seop Kim [6]

1 UJP PRAHA a.s., Nad Kamínkou 1345, 156 10 Prague-Zbraslav, Czech Republic; zyka@ujp.cz
2 CTU in Prague-Faculty of Mechanical Engineering, Karlovo Náměstí 13, 121 35 Praha 2, Czech Republic
3 Institute of Plasma Physics CAS, 182 00 Praha 8, Czech Republic; lukac@ipp.cas.cz (F.L.); vilemova@ipp.cas.cz (M.V.)
4 Faculty of Mathematics and Physics, Charles University, 180 00 Praha 8, Czech Republic; tomas.vlasak@seznam.cz (T.V.); jakub.cizek@mff.cuni.cz (J.Č.); Oksana.Melikhova@mff.cuni.cz (O.M.)
5 Faculty of Materials Science and Technology, VŠB-Technical University of Ostrava, 17. Listopadu 15, 708 33 Ostrava 8, Cech Republic; adela.machackova@vsb.cz
6 Department of Materials Science and Engineering, POSTECH, Pohang 790-784, Korea; hyoungseopkim@gmail.com
* Correspondence: jardamalek@seznam.cz

Received: 5 November 2019; Accepted: 27 November 2019; Published: 3 December 2019

Abstract: High entropy alloys (HEA) have been one of the most attractive groups of materials for researchers in the last several years. Since HEAs are potential candidates for many (e.g., refractory, cryogenic, medical) applications, their properties are studied intensively. The most frequent method of HEA synthesis is arc or induction melting. Powder metallurgy is a perspective technique of alloy synthesis and therefore in this work the possibilities of synthesis of HfNbTaTiZr HEA from powders were studied. Blended elemental powders were sintered, hot isostatically pressed, and subsequently swaged using a special technique of swaging where the sample is enveloped by a titanium alloy. This method does not result in a full density alloy due to cracking during swaging. Spark plasma sintering (SPS) of mechanically alloyed powders resulted in a fully dense but brittle specimen. The most promising result was obtained by SPS treatment of gas atomized powder with low oxygen content. The microstructure of HfNbTaTiZr specimen prepared this way can be refined by high pressure torsion deformation resulting in a high hardness of 410 HV10 and very fine microstructure with grain size well below 500 nm.

Keywords: high-entropy alloy; powder metallurgy; plastic deformation; microstructure

---

1. Introduction

High entropy alloys (HEAs), complex concentrated alloys (CCAs), and multi-principal element alloys (MPEAs) are the most common names of a new group of materials [1–5] introduced by Yeh et al. [6] and Cantor et al. [7] in the beginning of this century. Many definitions were proposed for such materials, but in general they consist of multiple (usually at least five) elements in equiatomic or near-equiatomic composition [2]. This approach is different from the traditional alloy design using one (or maximum two) principal element and other minor elements. The presence of many elements in equiatomic (or near) composition leads to high mixing entropy and thus to many interesting properties [2,5]. One of such effects is the existence of a random solid solution and simple body centered cubic (bcc) or faced centered cubic (fcc) structure. Recently HEAs with hexagonal closed packed (hcp) structure have been reported as well [8,9]. A new approach to HEA design using the valence electron concentration (VEC) was recently proposed with the aim of increasing the plasticity of these alloys [10,11]. Other authors showed that decreasing the solid solution phase stability promotes the stress-induced transformation during loading and therefore may result in increased plasticity of the alloy [12–14].

The group of HEAs has the potential to overcome currently used materials in many applications. Some of them have excellent high temperature properties and are therefore candidates for replacement of Ni-superalloys, which are currently used for turbine blades, etc. [15]. Cryogenic applications are also potential fields of interest as some HEAs retain high plasticity even at cryogenic temperatures [5].

In this work the HfNbTaTiZr (equiatomic) alloy was studied. This alloy has the potential to be used in high temperature applications or as a biocompatible material due to its chemical composition consisting of refractory metals [16–18], which are biocompatible [19–21] elements. This alloy is usually prepared by an arc melting process [22–26]. Power metallurgy was studied as an alternative route of HEA synthesis [27–30]. Mechanical alloying (MA) followed by hot isostatic pressing (HIP) [30] or spark plasma sintering (SPS) [27,28] are the most common powder metallurgy techniques. HfNbTaTiZr alloy was recently produced from commercially available elemental powders that were cold isostatically pressed (CIP) and sintered. Commercially available elemental powders were used for $TiNbTa_{0.5}Zr$ and $TiNbTa_{0.5}Zr_{0.2}Al$ alloy synthesis by Cao et al. [29]. This way of preparation may overcome the main drawbacks of the arc melted alloys, which suffer from very coarse dendritic microstructures. The dendritic microstructure may lead to phase separation (a mixture of two phases with slightly different chemical composition) due to variation of chemical composition in dendrite and interdendritic regions [23,31]. The microstructure of specimens produced by powder metallurgy exhibits no dendrites and also the grain size is on average significantly finer than in the arc melted samples. On the other hand, residual porosity present after sintering deteriorates the mechanical properties of HEAs significantly [29,32,33]. This is probably the main drawback of powder metallurgy processed HEAs. Moreover the initial powders have large active surfaces and could therefore be vulnerable to oxidation. Hot forging deformation can be one of the ways to eliminate the porosity. Hence, forging may in general improve microstructural and mechanical properties of HEAs [34]. Our previous experiments showed that the residual porosity of as-sintered HfNbTaTiZr specimens was eliminated during hot compression tests. However, it should be pointed out that the hot compression tests and real forging processes may significantly differ from each other. Therefore in this work various techniques of removing residual porosity were examined. Microstructure and mechanical properties of HfNbTaTiZr alloy prepared by various methods of powder metallurgy are reported in this paper. The possibility of using such techniques for bigger specimens under real conditions was accentuated as well.

2. Materials and Methods

Several methods of powder metallurgy were used to prepare the equimolar HfNbTaTiZr HEA. Preparation routes of various samples are summarized in Table 1. Elemental powders (purity >99%) and granularity of −325 mesh (<44 μm) were supplied by Huarui Co. (Chengdu, China). Weighting and mixing of powders was performed under Ar protective atmosphere. Three different kinds of initial powders were used:

(i) elemental powders mixed in appropriate ratio in Turbula 2F device for 10 h at 45 rpm and subsequently pressed into green compacts (denoted here HEAP) using CIP under the pressure of 400 MPa;

(ii) MA powder prepared from the elemental powders (granularity −325 mesh) by high energy ball milling in an Ar atmosphere for 42 h using tungsten carbide balls. The mean particle size of the MA powder was ≈3 μm and MA particles consisted of nanocrystalline grains

(iii) atomized powder (AT) prepared from arc melted HfNbTaTiZr alloy by crucible-free electrode induction-melting gas atomization in a protective Ar atmosphere to suppress undesirable oxidation. The AT powder had a broad particle size distribution covering the range from 10 to 300 μm.

Green compacts (HEAP) were further processed via different routes. Sintering at 1400 °C for 16 h was carried out in a vacuum furnace and sintered samples were slowly cooled in the furnace (Sintered specimens are denoted "HEAP-S"). Some of the sintered specimens were subjected to HIP performed at 1400 °C for 2 h under the pressure of 190 MPa (these specimens are denoted "HEAP-S-HIP"). Another sintered specimens were hot swaged to remove residual porosity (these samples are denoted "HEAP-S-SW"). Specimens were swaged at 950 °C in several consecutive steps with small section reduction with reheating between each step. The same process was also performed using an envelope of beta-titanium alloy in which the specimens were inserted and swaged using the same procedure.

The effect of HIP on specimens directly after CIP was studied as well (the specimens are denoted "HEAP-HIP"). The specimen after HIP was subsequently sintered (1400 °C/14 h/furnace cooled) in order to ensure the chemical homogeneity (the specimen is denoted as "HEAP-HIP-S").

MA and AT powders were compacted by SPS in an SPS 10-4 device (Santa Rosa, California, CA, USA) using a graphite die (Ø 20 mm) and punches to compact the powder using pressure of 80 MPa. The SPS processing was performed in vacuum. Tungsten foil was used to separate the samples from the graphite tools. The sample was heated with the heating rate of 700 °C/min up to the sintering temperature of 1300 °C at which it was kept for 2 min.

Contrary to CIP followed by HIP, sintering (or swaging), which is suitable for fabrication of large samples, produces relatively small disc shape samples with diameter of 20 mm and thickness of a few mm. Such samples can be strained by high pressure torsion (HPT) [35] in order to refine their structure. Severe plastic deformation applied during HPT processing results in many materials into an extreme grain refinement down to the nanoscale. In the present work HPT straining was performed at room temperature using pressure of 2.5 GPa and 15 revolutions. The samples produced using this route are denoted MA-SPS-HPT and AT-SPS-HPT, respectively.

Table 1. Specimens and their processing routes.

Specimen	Process Route
HEAP	CIP
HEAP-S	CIP + sintering
HEAP-S-HIP	CIP + sintering + HIP
HEAP-HIP	CIP + HIP
HEAP-HIP-S	CIP + HIP + sintering
HEAP-S-SW	CIP + sintering + hot swaging
MA-SPS	mechanical alloying + SPS
AT-SPS	atomized powder + SPS
MA-SPS-HPT	mechanical alloying + SPS + HPT
AT-SPS-HPT	atomized powder + SPS + HPT

The microstructure of specimens was studied by light microscopy (LM), using a Nikon EPIPHOT 3000 (Nikon, Melville, New York, USA) microscope. Scanning electron microscopes (SEM) JEOL 7650F (JEOL, Akishima, Tokyo, Japan) and FEI Quanta 200F (FEI, Hillsboro, OR, USA) were used for microstructural observations as well. The SEM observations were done in back scattered electron mode (BSE) unless otherwise noted. The mean particles size was determined by the analysis of SEM micrographs. NIS Elements software was used for image analysis. Energy dispersive X-ray spectroscopy (EDS) was employed for used analysis of chemical composition. Specimens for LM and SEM observations were prepared by a standard metallographic process (ground up to #4000 with SiC papers and polished with Struers OP-S emulsion with the addition of H_2O_2). For etching, 3 mL of HF + 8 mL of HNO_3 + 100 mL of H_2O etchants were used. The phase identification was also carried out using the X-ray diffraction analysis (XRD), on a Bruker D8 Discover diffractometer (Bruker, Karlsruhe, Germany) using Cu Kα radiation. XRD investigations were performed in the symmetrical Bragg–Brentano geometry. Quantitative Rietveld refinement analysis was performed by TOPAS V5 software (Bruker, Karlsruhe, Germany).

Vickers hardness was determined using a Zwick/Roell ZHU 250 top hardness tester (Zwick/Roell, Ulm, germany) with the load of 98.1 N (according to the ISO 6507 standard). At least seven values were determined for each measurement. The oxygen content in samples studied was determined by a Bruker Galileo G8 gas fusion analyzer (Bruker, Karlsruhe, Germany). At least three measurements for each specimen were performed. Residual porosity in the samples was determined by the image analysis of SEM micrographs.

Positron lifetime (LT) spectroscopy was employed for characterization of lattice defects in the samples. A ^{22}Na radioisotope with activity of 1 MBq deposited on a 2 μm thick mylar foil was used as a positron source. LT measurements were carried out on a digital spectrometer (Hamamatsu Photonics, Hamamatsu, Japan) [36] with time resolution of 145 ps (FWHM of the resolution function). At least 10^7 positron annihilation events were collected in each LT spectrum. The source contribution to the LT spectra consisted of two components with lifetimes of 368 ps and 1.5 ns and corresponding relative intensities of 11% and 1%, representing contributions of positrons annihilating in the ^{22}Na source spot and the covering mylar foil, respectively.

3. Results

3.1. Initial Powders

The initial powders used for further processing were characterized as their properties may influence the final product.

3.1.1. HEAP Green Compact

Figure 1a shows the microstructure of the initial green compact (HEAP). The green compact obviously consisted of multiple phases. The XRD diffraction pattern for the green compact is plotted in Figure 2. The sample contained the bcc (Nb and Ta powders have both the bcc lattice with similar parameters) and hcp phases (Hf, Ti, and Zr elements). The oxygen content determined in the samples studied is listed in Table 2. The HEAP green compacts exhibited an oxygen concentration of ≈0.55 wt %.

Table 2. Properties (porosity, grain size, hardness, and oxygen content) of studied specimens.

Specimen	Porosity (%)	Grain Size (μm)	Hardness (HV10)	Oxygen (wt %)
HEAP	N/A	N/A	N/A	0.55 ± 0.05
HEAP-S	6.5 ± 1	~35	330 ± 13	0.65 ± 0.07
HEAP-S-HIP	4.5 ± 0.5	~45	325 ± 13	0.83 ± 0.02
HEAP-HIP	5 ± 0.5	~20	151 ± 7	0.80 ± 0.05
HEAP-HIP-S	5 ± 1	~25	225 ± 10	0.84 ± 0.06
HEAP-S-SW	N/A [1]	N/A	405 ± 15 [2]	0.75 ± 0.03
MA-SPS	0.9 ± 0.3	~10	584 ± 10	1.07 ± 0.10
AT-SPS	0	~50	350 ± 5	0.12 ± 0.02
MA-SPS-HPT	N/A [1]	~4	680 ± 5 [2]	1.07 ± 0.03
AT-SPS-HPT	0	~0.5	410 ± 5	0.12 ± 0.02

[1] Porosity was not measured due to damaged specimens. [2] Hardness of HEAP-S-SW and MA-SPS-HPT is disputable due to defects in the microstructure (e.g., cracks).

Results of LT measurements are listed in Table 2. The HEAP sample exhibited a two-component LT spectrum. The short component with lifetime τ_1 represented a contribution of free positrons (not trapped at defects) while the longer component with lifetime τ_2 could be attributed to positrons trapped at vacancy-like misfit defects at interfaces between precipitates and the matrix and at dislocations introduced into the sample by CIP. Assuming that the component τ_2 arose exclusively from positrons trapped at dislocations, the mean dislocation density $\rho_D \approx 1.48 \times 10^{14}$ m^{-2} in the sample could be calculated from the two-state trapping model [37]. This value should be considered as an upper level of the actual dislocation density in the sample since some fraction of positrons contributing to the component τ_2 was trapped at misfit defects. This was indicated by the lifetime of trapped positrons value of $\tau_2 \approx 165$ ps, which was slightly lower than the value determined for dislocations in the HfNbTaTiZr alloy [38].

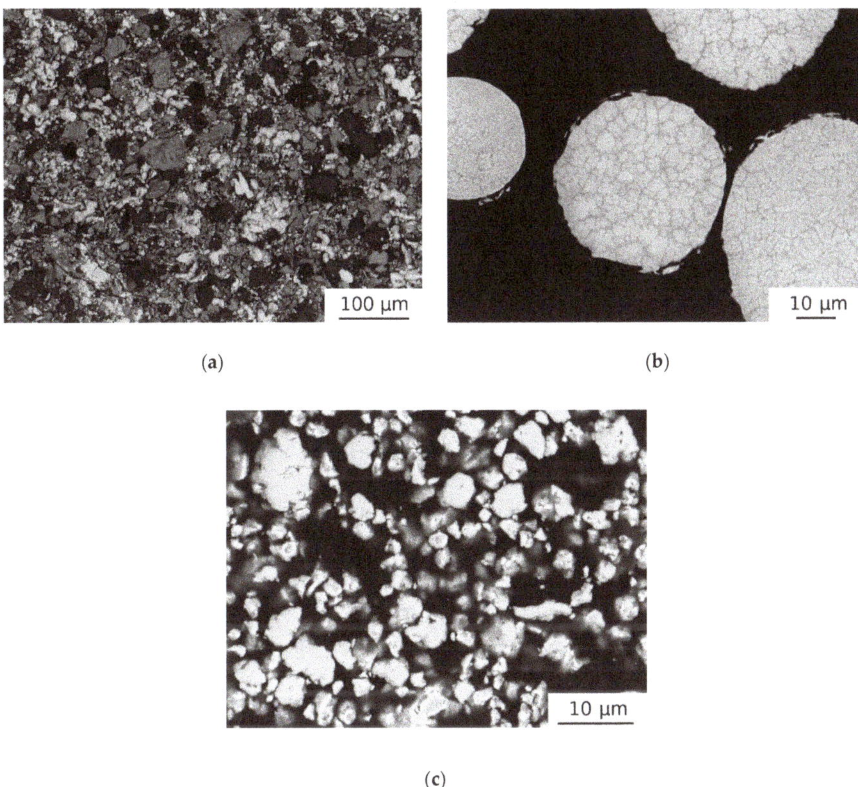

(a) (b) (c)

Figure 1. SEM micrographs showing microstructure of (**a**) HEAP green compact, (**b**) AT initial powder, and (**c**) MA initial powder.

3.1.2. AT Powder

The microstructure of gas atomized powder (AT) is shown in Figure 1b. The particle sizes of the powder fell into a broad range from 10 to 300 μm. One can see in Figure 1b that the AT particles exhibited a dendritic structure typical for conventionally cast HfNbTaTiZr alloy [39] and consisted of two chemically different phases with bcc structures: one bcc phase, denoted bcc1, was slightly rich in Nb and Ta, and the second bcc phase, denoted bcc2, was rich in Hf, Zr. The XRD pattern for the AT powder is plotted in Figure 3. The two bcc phases had very similar lattice parameters so that their XRD reflections overlapped each other, making the reflections asymmetrical. The lattice parameter a = 3.4024(5) Å was determined by Rietveld refinement. The AT powder exhibited a single component LT spectrum with a lifetime of ≈165 ps. It represented a contribution of positrons trapped at misfit defects at the interfaces between bcc1 and bcc2 phases. Since bcc1 and bcc2 phases had slightly different lattice parameters, the discontinuity of lattice planes resulted in the formation of misfit defects at interfaces between these two phases [27].

AT powder exhibited low oxygen content (ten times lower than MA) since crucible-free gas atomization was performed in a protective Ar atmosphere and particles of AT powder were rather coarse, i.e., the surface-to-volume ration was relatively low.

3.1.3. MA Powder

The XRD diffraction pattern for MA powder is shown in Figure 3 as well. The sample exhibited very broad XRD reflections at positions corresponding to the bcc phase with lattice parameter a = 3.409(1) Å, which was slightly higher than that for the AT powder. Broad XRD reflections testified that MA powder had a nanocrystalline structure. The average size of coherently scattering domains determined by Rietveld refinement was 22.7(6) nm.

The MA powder exhibited rather high oxygen content of ≈1.07 wt %, which was ten times higher than that in the AT powder. Oxygen was incorporated into the MA powder obviously during high energy ball milling due to high surface area of nanocrystalline powder particles. Note that LT investigations of nanocrystalline MA powder were not performed because of its high reactivity with oxygen.

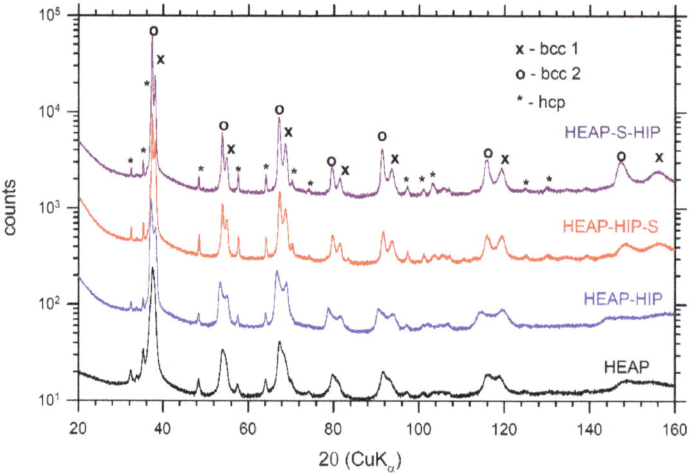

Figure 2. XRD patterns for HEAP, HEAP-HIP, HEAP-HIP-S, and HEAP-S-HIP specimens. The samples contain two bcc phases (bcc1, bcc2) and a hcp phase. The XRD patterns were shifted vertically for better visibility.

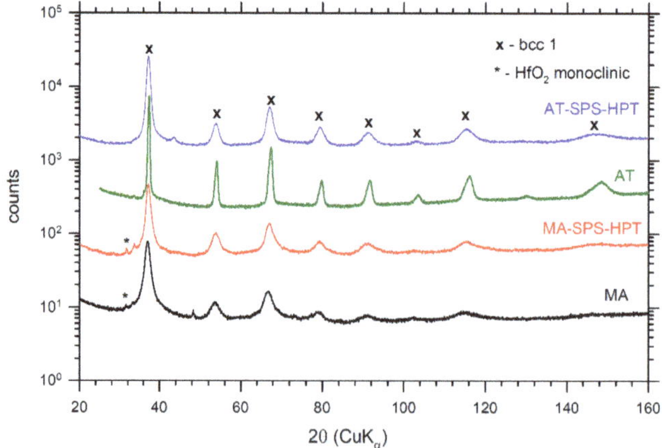

Figure 3. XRD patterns for samples AT, MA, AT-SPS, MA-SPS, AT-SPS-HPT, and MA-SPS-HPT. The samples contain a bcc phase and a small fraction of a monoclinic HfO_2 phase.

3.2. Effect of HIP and Sintering

Figure 4a,b show the microstructure of the HEAP-S specimen. Sintering resulted in equiaxed and relatively fine grains. The chemical homogeneity was good after 16 h sintering as determined by SEM and EDS microstructure analysis. Secondary phases (needle-like precipitates) were observed in the microstructure along with irregular shaped particles of the hcp phase. HIP processing of sintered specimens (HEAP-S-HIP) introduced no significant changes. The microstructure of green compact subjected to HIP treatment (HEAP-HIP) is shown in Figure 4c,d. The chemical homogeneity of HEAP-S was better than that of HEAP-HIP (cf. Figure 4a,c). Several chemical inhomogeneities (i.e., areas with higher Ta, or Nb concentration-probably at the positions of original Ta particles) were observed in the HEAP-HIP sample, as seen in Figure 4c (see EDS maps of typical region in HEAP-HIP specimen in Supplementary Materials). Therefore additional sintering (1400 °C/14 h) was performed resulting in a HEAP-HIP-S specimen with good chemical homogeneity. However, porosity in the HEAP-HIP-S sample remained present in the specimen, and it still consisted of bcc1, bcc2, and hcp phases.

A significant number of pores was observed in the microstructure. The approximate porosity values in the samples studied are listed in Table 2. The as-sintered specimen (HEAP-S) exhibited the highest porosity. The HIP process caused a slight decrease in porosity, but the specimens still remained rather porous, as seen in Figure 4c.

The oxygen content in the initial powder used for sintering (HEAP sample) was relatively high (~0.55 wt %), as seen in Table 2. However, it could be seen that it increased during further processing. The oxygen content increased from 0.55 wt % to 0.65 wt % during sintering. HIP caused even more significant increase in the oxygen content, up to ~0.80 wt %.

The XRD diffraction patterns of HEAP specimens subjected to HIP and sintering are plotted in Figure 2. From inspection of the Figure 2 it becomes clear that HIP led only to slight broadening of XRD reflections and peak separation (c.f. samples HEAP and HEAP-HIP). This was caused by the fact that the constituting elements were not fully dissolved during the HIP process of green powders (see Figure 4c,d). On the other hand the sintering led to a quite homogeneous microstructure, which resulted in an increase of the concentration of the bcc2 phase at the expense of bcc1. The sintering caused narrowing of XRD profiles, which indicated grain growth occurring during sintering. LT investigations revealed that sintering reduced dislocation density and also the concentration of misfit defects, which was indicated by increases of τ_2 towards the value of ≈180 ps, typical for dislocations in HfNbTaTiZr alloy [38]. The latter effect was obviously caused by grain growth during sintering. The HIP treatment of sintered sample led to a slight increase of dislocation density, as seen in Table 3, due to dislocations introduced by HIP processing.

Table 3. Results of LT spectroscopy: lifetimes τ_i and relative intensities I_i of the components resolved in LT spectra. The mean density of dislocations ρ_D calculated using the two-state simple trapping model is shown in the Table as well.

Specimen	τ_1 (ps)	I_1 (%)	τ_2 (ps)	I_2 (%)	ρ_D (10^{14} m^{-2})
HEAP	69(2)	12(1)	165(2)	88(1)	1.48(8)
HEAP-S	85(2)	22(2)	175(4)	78(2)	0.94(9)
HEAP-S-HIP	81(4)	20(2)	178(3)	80(2)	1.1(1)
HEAP-HIP	81(3)	17(3)	168(3)	83(3)	1.1(1)
HEAP-HIP-S	86(4)	23(2)	176(4)	77(2)	0.9(1)
HEAP-S-SW	45(4)	9(1)	181(2)	91(2)	3.0(1)
AT	-	-	165(2)	100	<0.1
MA-SPS	80(6)	8(1)	150(2)	92(1)	1.1(1)
AT-SPS	148(1)	100	-	-	<0.01
MA-SPS-HPT	-	-	180(1)	100	>5
AT-SPS-HPT	-	-	183(1)	100	>5

Mechanical properties of samples studied were represented by hardness values, which are listed in Table 2. One can see in the table that hardness of HEAP-S and HEAP-S-HIP specimens were similar (both around 320 HV10). Interestingly the hardness of the HEAP-HIP specimen was around 150 HV10, which was significantly lower value than that of HEAP-S. The hardness of the HEAP-HIP specimen significantly increased during sintering (1400 °C/14 h) to 225 HV10, but did not reach the hardness of sintered specimens. On the other hand the HEAP-S-HIP specimen had a hardness comparable to that of the sintered specimen.

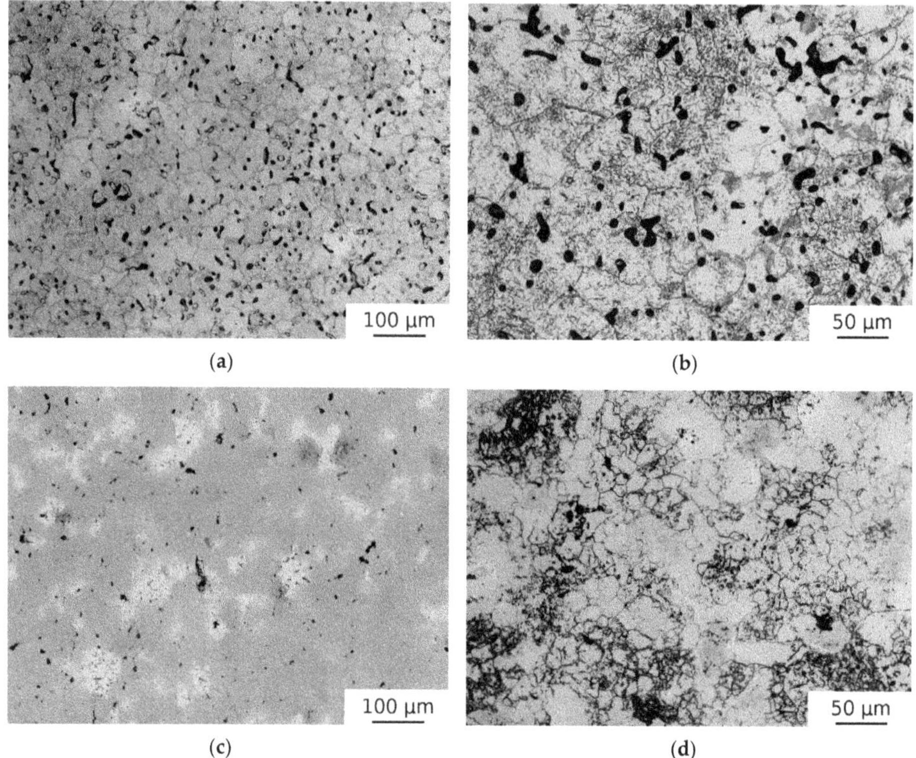

Figure 4. Micrographs of (**a**) HEAP-S, (**b**) HEAP-S—detail of etched specimen, (**c**) HEAP-HIP (SEM-BSE), and (**d**) detail of HEAP-HIP (LM-etched).

3.3. *Effect of Swaging*

The as-sintered specimen (HEAP-S) was hot swaged to remove residual porosity. However, the swaging process was not completed due to cracking of the specimens, which can be clearly seen in Figure 5. The microstructure of the swaged sample was similar to that of the HEAP-S specimen. The cracking of specimens during hot swaging was at first attributed to a decrease of temperature during contact of the die with the specimen. Due to this reason the specimens were enveloped in a titanium alloy and subsequently swaged in order to ensure sufficiently high temperature of the whole specimen during the process. Unfortunately, the same result was obtained and the specimens were damaged, i.e., cracking occurred as well. It should be pointed out that this specimen contained relatively high amount of oxygen (~0.75 wt %), which may have caused the brittleness of this alloy at the given deformation conditions (i.e., temperature and strain rate). It is also possible that the deformation temperature was too low and the strain rate was too high for the given specimens.

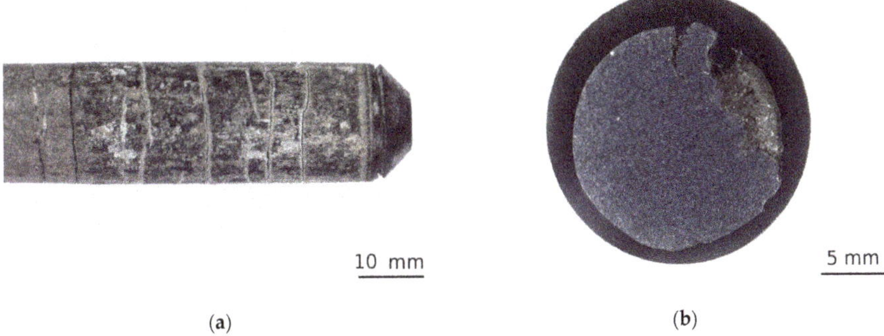

Figure 5. The microstructure of the hot swaged (HEAP-S-SW) specimen: (**a**) overview with radial cracks; (**b**) after forging in Ti-envelope.

Porosity was not measured in the HEAP-S-SW specimen because the swaging process was interrupted due to specimen cracking, and the distribution of porosity was inhomogeneous. Nevertheless it seems that some regions in the swaged samples had significantly lower porosity and therefore swaging had some potential to eliminate the porosity. Swaging caused a slight increase of the oxygen concentration as well, as seen in Table 2.

Table 2 shows that hardness increased during hot swaging to more than 400 HV10, indicating strengthening of the material by dislocations introduced by swaging. A significant increase of dislocation density up to $\approx 3 \times 10^{14}$ m^{-2} in the swaged sample was confirmed by LT spectroscopy, as seen in Table 3. However, it should be accentuated that the specimen was damaged during swaging and therefore this hardness value may not be representative. If the indent was made in the region containing a crack, then it became larger due to a lateral shift of the material caused by opening of the crack. As a consequence the determined hardness value was lower than in the region without the crack.

3.4. Effect of SPS and HPT Processing

The microstructure of the MA-SPS-HPT specimen consisted of an equiaxed very fine (grain size ~4 μm) microstructure with numerous precipitates and some amount of porosity (see Figure 6). It should be accentuated that the MA-SPS specimen was damaged during the HPT process, as many cracks emerged, and the specimen was separated into several smaller pieces. The microstructure of this specimen shown in Figure 6 was obtained from one of these pieces. Figure 6a presents an overview of the microstructure while Figure 6b shows detail of the microstructure with higher magnification. Pores are marked as "1" in the Figure. The particle labelled "2" is the hcp phase rich in Hf and Zr. The bcc1 phase (Nb and Ta rich) distributed mainly on grain boundaries is denoted as "3". The particles labeled "4" are HfO_2 oxides (with some minor other elements concentrations). The XRD pattern of the MA-SPS-HPT sample is plotted in Figure 3. The sample contained a bcc phase with lattice parameter a = 3.404(5) Å and a small concentration of HfO_2 oxide.

The MA-SPS-HPT sample had the highest oxygen concentration among the samples studied, as seen in Table 2, which indicates that a significant concentration of oxygen was introduced into the alloy during the MA processing.

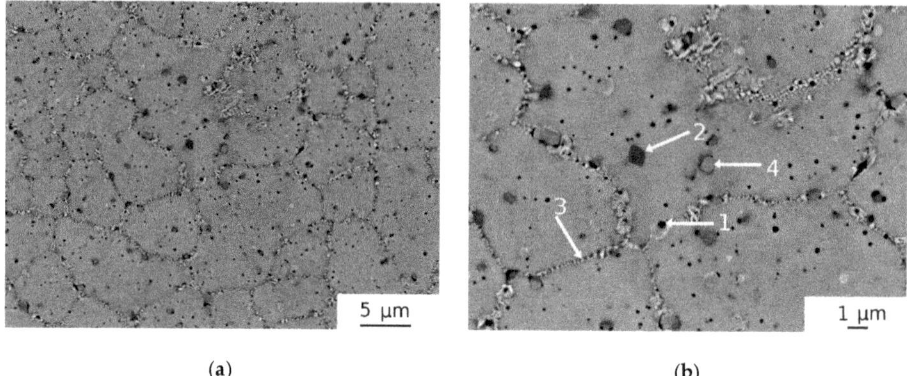

(a) (b)

Figure 6. SEM micrograph of MA-SPS-HPT specimen: (**a**) overview of microstructure; (**b**) detail of microstructure with the pores (marked "1") and three kinds of precipitates (marked "2, 3, 4").

The AT-SPS-HPT sample had an equiaxed microstructure with a pure bcc phase and the lattice parameter a = 3.406(4) Å. No residual porosity and no precipitates of secondary phases were observed, as seen in Figure 7. The mean grain size of sample AT-SPS (i.e., before HPT straining) was ≈50 μm, as seen in Figure 7. After HPT deformation (sample AT-SPS-HPT), the mean grain size decreased below 0.5 μm. The microstructure was highly deformed. The AT-SPS-HPT samples contained the lowest concentration of oxygen among the all samples studied (~0.12 wt %). Hence, the SPS treatment resulted only in a slight increase of oxygen content, if any.

One can see in Table 3 that both MA-SPS-HPT and AT-SPS-HPT samples exhibited a single component LT spectrum with lifetime $\tau_2 \approx 180$ ps. This testifies that virtually all positrons were annihilated in a trapped state in dislocations (so called saturated positron trapping). Hence, one can conclude that HPT processing resulted in formation of a high density of dislocations exceeding 5×10^{14} m^{-2}.

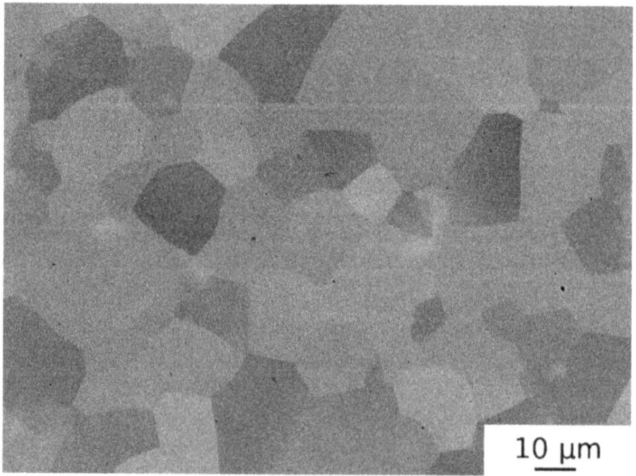

Figure 7. SEM micrograph of AT-SPS specimen.

4. Discussion

4.1. Residual Porosity

Results obtained for HEAP-S and HEAP-S-HIP samples (see Table 2) indicate that HIP processing had some potential to decrease residual porosity. However, this decrease was quite small and considerable porosity remained in the specimen. The fact that even HIP processing at 1400 °C/200 MPa/2 h did not remove the porosity completely can be ascribed to a high volume fraction of pores and to filling of the pores by Ar gas used as a pressing medium in HIP processing. The MA powder compacted by SPS contained residual porosity as well, and small pores remained in the sample even after HPT processing, as seen in Figure 6. It should be pointed out that the MA-SPS-HPT specimen was broken into several pieces during the HPT process and therefore the results of such specimen are disputable. In such cases, the HPT after cracking could not introduce full deformation (strain) to the specimen. Contrary to this, SPS processing at 1300 °C of AT powder resulted in porosity-free samples (see Figure 7).

It has to be mentioned that no microscopic porosity (vacancy clusters or voids) was found by LT investigations in the HEA samples studied in this work, which means that residual porosity in the present samples was mainly macroscopic porosity (large pores among grains).

4.2. Oxygen Content

The oxygen concentration in specimens fabricated by powder metallurgy techniques was determined not only by oxygen content in the initial powders, but also by the processing route. The highest oxygen concentration was found in the MA-SPS specimen. This was probably caused by a significant oxygen contamination of powders with large active surface area. This was possibly due to a long milling time and residual oxygen present in the milling atmosphere (Ar), and high reactivity of some elements with the oxygen (e.g., Ti, Hf) [40]. In the HEAP samples the oxygen content increased more significantly during HIP (Ar atmosphere) than during sintering (vacuum $<1.10^{-3}$ Pa). It also seems that the oxygen content was decisive for the plasticity of the alloy as the only specimen with enough ductility to undergo the cold plastic deformation was the AT-SPS specimen, which had the lowest oxygen content. Other specimens were damaged by crack formation during deformation at room temperature or even at higher temperature (swaging at 950 °C). It should be pointed out that the possibility of hot deformation of the alloy with increased oxygen content cannot be excluded under suitable conditions (i.e., low strain rate, or higher processing temperature) [41], but this possibility needs more detailed study.

The high oxygen content in the MA-SPS alloy well corresponds to the presence of numerous oxides (mainly HfO_2) in the microstructure detected by SEM and by XRD. They emerged probably during mechanical alloying and are also one of the reasons of the brittleness of the alloy [42]. The thermal stability of these oxides is very high and therefore they were not dissolved during the SPS process [43]. The second reason is probably the oxygen dissolved in the matrix [44]. The AT-SPS is the only sample with single phase bcc solid solution with no other phases. Single phase random solid solution is caused by relatively high cooling rate after SPS processing at high temperature of 1300 °C, where all secondary phases are supposed to be dissolved [2,22,26,45–48]. It was reported by Senkov et al. [22] that cooling rate higher than 15 °C/min is sufficient for retaining single phase random solid solution at this alloy. The cooling rate after SPS is very high as the sample was cooled from sintering temperature (1300 °C) to room temperature in 3–4 min. Therefore the achieved cooling rate around 300 °C/min is much higher than 15 °C/min. On the other hand the specimens after sintering or HIP were slowly cooled with the furnace. The estimated cooling rate (HIP and sintering had similar cooling rate) was lower than 12 °C/min under 800 °C (lowering with decreasing temperature). This cooling rate was not high enough to suppress formation of the bcc2 and hcp phases.

The grain size of the MA SPS specimen was lower than for the others (except that after HPT). This was caused by the intensive plastic deformation during high energy ball milling, which resulted in a high number of nucleation sites for new grains formed by recrystallization during SPS processing.

4.3. Hardness

Very high hardness of the MA-SPS-HPT specimen (584 HV10) can be ascribed to the high oxygen content in solid solution and also the presence of oxide particles that caused additional strengthening.

As seen in Figure 8 hardness was correlated with oxygen content in the sample. Hardness remained approximately unchanged with increasing oxygen content up to the oxygen concentration of ≈0.8 wt %. Above this value hardness strongly increased with oxygen concentration. This indicates that there was a certain threshold of oxygen concentration in the sample above which it had a significant hardening effect. It is likely that the remarkable hardening effect of oxygen above 0.8 wt % was connected with the formation of HfO_2 particles with a monoclinic structure, which were detected in MA-SPS and MA-SPS-HPT samples by SEM and X-ray diffraction. Additional strengthening (not connected with oxygen) was caused by dislocations introduced by severe (cold) plastic deformation. In Figure 8 one can see that HPT straining increased hardness of both MA-SPS-HPT and AT-SPS-HPT samples.

It has to be mentioned that HEAP-HIP and HEAP-HIP-S samples were characterized by surprisingly low hardness. Hardness measurement of these samples could be influenced by porosity. Moreover, the low hardness of the HEAP-HIP sample could be possibly attributed to different phase composition. One can see in Figure 4c that Ta and Nb rich particles were not fully dissolved during HIP and therefore this specimen was somewhat between a blend of powders and a high entropy alloy. This was proved also by the XRD pattern in Figure 2, which showed the highest difference between lattice parameters of the bcc1 and bcc2 phases. It was reported that the high entropy effect may result into higher hardness since various radii of atoms occupying randomly lattice sites cause random fluctuations of potential energy landscape, which hinders motion of dislocations [2]. This hardening effect was probably absent in the HEAP-HIP sample, as random solid solution was not completely formed. This picture is supported by the fact that both bcc1 (Nb, Ta rich) and bcc2 (Hf and Zr rich) phases have lower hardness than HfNbTaTiZr random solid solution [38]. Subsequent sintering improved the chemical homogeneity of the HEAP-HIP sample and increased hardness as well. However, the hardness of HEAP-HIP-S still did not reach the values of HEAP-S and HEAP-S-HIP. The cause of this effect remains unclear to the authors.

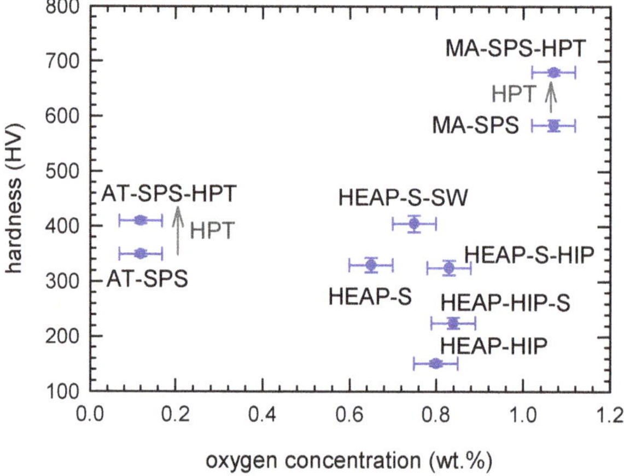

Figure 8. Correlation between hardness of samples studied and oxygen content. Grey arrows denote work hardening effect of HPT straining.

5. Conclusions

HfNbTaTiZr high entropy alloy was produced by various methods of powder metallurgy including mechanical alloying, sintering, hot isostatic pressing, spark plasma sintering, and high pressure torsion. The microstructure and mechanical properties of HfNbTaTiZr prepared by various methods were examined. It was found that the main problem in fabrication of HfNbTaTiZr alloy by powder metallurgy is oxygen contamination. Samples prepared by cold isostatic pressing followed by sintering and/or hot isostatic pressing contained oxygen in the concentration range of 0.55 wt %–0.85 wt %.

Oxygen contamination of samples prepared by mechanical alloying was even higher, and the oxygen concentration in these samples exceeded 1 wt %. Oxygen incorporated into the HfNbTaTiZr samples caused deterioration of mechanical properties (embrittlement) and prevented formation of a single phase random solid solution.

Investigations performed in the present work showed that the most promising powder metallurgy method of production of HfNbTaTiZr alloy was spark plasma sintering of gas atomized powder. HfNbTaTiZr alloy prepared using this route was found to be a single phase random solid solution, and oxygen content in the sample was ≈0.12 wt % only. Moreover, it was shown that high pressure torsion is a suitable method for grain refinement of high entropy alloys prepared by powder metallurgy.

Supplementary Materials: The following are available online at http://www.mdpi.com/1996-1944/12/23/4022/s1, Figure S1. (**a**) Nb; (**b**) Ti; (**c**) Zr; (**d**) Ta; (**e**): Hf.

Author Contributions: For research articles with several authors, a short paragraph specifying their individual contributions must be provided. The following statements should be used "conceptualization, F.L., M.V. and J.Č.; methodology, J.M. and O.M.; validation, J.Z., M.V., T.V., O.M., A.M. and H.-S.K.; formal analysis, J.M.; investigation, J.M., J.Z., F.L., T.V., J.Č. and H.-S.K.; resources, J.M., A.M. and H.-S.K.; writing—original draft preparation, J.M., F.L. and J.Č.; writing—review and editing, J.M., J.Z. and J.Č.; visualization, M.V. and O.M.; supervision, J.Č.; project administration, J.Č.; funding acquisition, J.Č.", please turn to the CRediT taxonomy for the term explanation. Authorship must be limited to those who have contributed substantially to the work reported.

Funding: Support by the Czech Science Foundation (project 17-17016S) is highly acknowledged.

Conflicts of Interest: The authors declare no conflict of interest.

References

1. Gorsse, S.; Miracle, D.B.; Senkov, O.N. Mapping the world of complex concentrated alloys. *Acta Mater.* **2017**, *135*, 177–187. [CrossRef]
2. Miracle, D.B.; Senkov, O.N. A critical review of high entropy alloys and related concepts. *Acta Mater.* **2017**, *122*, 448–511. [CrossRef]
3. Lu, Z.P.; Wang, H.; Chen, M.W.; Baker, I.; Yeh, J.W.; Liu, C.T.; Nieh, T.G. An assessment on the future development of high-entropy alloys: Summary from a recent workshop. *Intermetallics* **2015**, *66*, 67–76. [CrossRef]
4. Ye, Y.F.; Wang, Q.; Lu, J.; Liu, C.T.; Yang, Y. High-entropy alloy: Challenges and prospects. *Mater. Today* **2016**, *19*, 349–362. [CrossRef]
5. Zhang, Y.; Zuo, T.T.; Tang, Z.; Gao, M.C.; Dahmen, K.A.; Liaw, P.K.; Lu, Z.P. Microstructures and properties of high-entropy alloys. *Prog. Mater. Sci.* **2014**, *61*, 1–93. [CrossRef]
6. Yeh, B.J.; Chen, S.K.; Lin, S.J.; Gan, J.Y.; Chin, T.S.; Shun, T.T.; Tsau, C.H.; Chang, S.Y. Nanostructured High-Entropy Alloys with Multiple Principal Elements: Novel Alloy Design Concepts and Outcomes. *Adv. Eng. Mater.* **2004**, *6*, 299–303. [CrossRef]
7. Cantor, B.; Chang, I.T.H.; Knight, P.; Vincent, A.J.N. Microstructural development in equiatomic multicomponent alloys. *Mater. Sci. Eng. A* **2004**, *375–377*, 213–218. [CrossRef]
8. Fuerbacher, M.; Heidelmann, M.; Carsten, T. Hexagonal High-Entropy Alloys. *Mater. Res. Lett.* **2014**, *3*, 1–9. [CrossRef]
9. Yusenko, K.V.; Riva, S.; Carvalho, P.A.; Yusenko, M.V.; Arnaboldi, S.; Suknikh, A.S.; Hanfland, M.; Gromilov, S.A. First hexagonal close packed high-entropy alloy with outstanding stability under extreme conditions and electrocatalytic activity for methanol oxidation. *Scr. Mater.* **2017**, *138*, 22–27. [CrossRef]
10. Sheikh, S.; Shafeie, S.; Hu, Q.; Ahlstrom, J.; Persson, C.; Veselý, J.; Zýka, J.; Klement, U.; Guo, S. Alloy design for intrinsically ductile refractory high-entropy alloys. *J. Appl. Phys.* **2016**, *120*, 164902. [CrossRef]
11. Couzinié, J.; Dirras, G. Body-centered cubic high-entropy alloys: From processing to underlying deformation mechanisms. *Mater. Charact.* **2019**, *147*, 533–544. [CrossRef]
12. Li, Z.; Pradeep, K.G.; Deng, Y.; Raabe, D.; Tasan, C.C. Metastable high—Entropy dual—Phase alloys overcome the strength—Ductility trade-off. *Nature* **2016**, *534*, 227–230. [CrossRef] [PubMed]
13. Wang, L.; Fu, C.; Wu, Y.; Wang, Q.; Hui, X.; Wang, Y. Formation and toughening of metastable phases in TiZrHfAlNb medium entropy alloys. *Mater. Sci. Eng. A* **2019**, *748*, 441–452. [CrossRef]

14. Wang, D.; Lu, X.; Wan, D.; Li, Z.; Barnoush, A. In-situ observation of martensitic transformation in an interstitial metastable high-entropy alloy during cathodic hydrogen charging. *Scr. Mater.* **2019**, *173*, 56–60. [CrossRef]
15. Čapek, J.; Kyncl, J.; Kolařík, K.; Beránek, L.; Pitrmuc, Z.; Medřický, J.; Pala, Z. Grinding of Inconel 713 superalloy for gas turbines Grinding of Inconel 713 Superalloy for Gas Turbines. *Manuf. Technol.* **2016**, *16*, 38–45.
16. Praveen, S.; Kim, H.S. High-Entropy Alloys: Potential Candidates for High-Temperature Applications—An Overview. *Adv. Eng. Mater.* **2018**, *20*, 1700645. [CrossRef]
17. Couzinié, J.; Senkov, O.N.; Miracle, D.B.; Dirras, G. Comprehensive data compilation on the mechanical properties of refractory high-entropy alloys. *Data Br.* **2018**, *21*, 1622–1641. [CrossRef]
18. Senkov, O.N.; Scott, J.M.; Senkova, S.V.; Maisenkothen, F.; Micracle, D.B.; Woodward, C.F. Microstructure and elevated temperature properties of a refractory TaNbHfZrTi alloy. *J. Mater. Sci.* **2012**, *47*, 4062–4074. [CrossRef]
19. Biesiekierski, A.; Wang, J.; Abdel-Hady Gepreel, M.; Wen, C. A new look at biomedical Ti-based shape memory alloys. *Acta Biomater.* **2012**, *8*, 1661–1669. [CrossRef]
20. Donato, T.A.G.; Almeida, L.H.; Nogueira, R.A.; Niemeyer, T.C.; Grandini, C.R.; Caram, R.; Schneider, S.G.; Santos, A.R., Jr. Cytotoxicity study of some Ti alloys used as biomaterial. *Mater. Sci. Eng. C* **2009**, *29*, 1365–1369. [CrossRef]
21. Eisenbarth, E.; Velten, D.; Müller, M.; Thull, R.; Breme, J. Biocompatibility of β-stabilizing elements of titanium alloys. *Biomaterials* **2004**, *25*, 5705–5713. [CrossRef] [PubMed]
22. Senkov, O.N.; Semiatin, S.L. Microstructure and properties of a refractory high-entropy alloy after cold working. *J. Alloys Compd.* **2015**, *649*, 1110–1123. [CrossRef]
23. Maiti, S.; Steurer, W. Structural-disorder and its effect on mechanical properties in single-phase TaNbHfZr high-entropy alloy. *Acta Mater.* **2016**, *106*, 87–97. [CrossRef]
24. Juan, C.C.; Tsai, M.H.; Tsai, C.W.; Lin, C.M.; Wang, W.R.; Yang, C.C.; Chen, S.K.; Lin, S.J.; Yeh, J.W. Enhanced mechanical properties of HfMoTaTiZr and HfMoNbTaTiZr refractory high-entropy alloys. *Intermetallics* **2015**, *62*, 76–83. [CrossRef]
25. Chen, S.; Tseng, K.K.; Tong, Y.; Li, W.; Tsai, C.W.; Yeh, J.W.; Liaw, P.K. Grain growth and Hall-Petch relationship in a refractory HfNbTaZrTi high-entropy alloy. *J. Alloys Compd.* **2019**, *795*, 19–26. [CrossRef]
26. Senkov, O.N.; Pilchak, A.L.; Semiatin, S.L. Effect of Cold Deformation and Annealing on the Microstructure and Tensile Properties of a HfNbTaTiZr Refractory High Entropy Alloy. *Metall. Mater. Trans. A* **2018**, *49*, 2876–2892. [CrossRef]
27. Lukac, F.; Dudr, M.; Mušálek, R.; Klečka, J.; Cinert, J.; Čížek, J.; Chráska, T.; Čížek, J.; Melikhova, O.; Kuriplach, J.; et al. Spark plasma sintering of gas atomized high-entropy alloy HfNbTaTiZr. *J. Mater. Res.* **2018**, *33*, 3247–3257. [CrossRef]
28. Kang, B.; Lee, J.; Jin, H.; Hyung, S. Ultra-high strength WNbMoTaV high-entropy alloys with fi ne grain structure fabricated by powder metallurgical process. *Mater. Sci. Eng. A* **2018**, *712*, 616–624. [CrossRef]
29. Cao, Y.; Liu, Y.; Liu, B.; Zhang, W. Precipitation behavior during hot deformation of powder metallurgy Ti-Nb-Ta-Zr-Al high entropy alloys. *Intermetallics* **2018**, *100*, 95–103. [CrossRef]
30. Portnoi, V.K.; Leonov, A.V.; Gusakov, M.S.; Logachev, I.A.; Fedotov, S.A. Preparation of High-Temperature Multicomponent Alloys by Mechanochemical Synthesis from Refractory Elements. *Inorg. Mater.* **2019**, *55*, 219–223. [CrossRef]
31. Wang, S.; Xu, J. (TiZrNbTa)—Mo high-entropy alloys: Dependence of microstructure and mechanical properties on Mo concentration and modeling of solid solution strengthening. *Intermetallics* **2018**, *95*, 59–72. [CrossRef]
32. Lewis, G. Properties of open-cell porous metals and alloys for orthopaedic applications. *J. Mater. Sci. Mater. Med.* **2013**, *24*, 2293–2325. [CrossRef] [PubMed]
33. Guo, W.; Liu, B.; Liu, Y.; Li, T.; Fu, A.; Fang, Q.; Nie, Y. Microstructures and mechanical properties of ductile NbTaTiV refractory high entropy alloy prepared by powder metallurgy. *J. Alloys Compd.* **2019**, *776*, 428–436. [CrossRef]
34. Neslušan, M.; Minárik, P.; Čilliková, M.; Kolařík, K.; Rubešová, K. Barkhausen noise emission in tool steel X210Cr12 after semi-solid processing. *Mater. Charact.* **2019**, *157*, 109981. [CrossRef]

35. Valiev, R.Z.; Islamgaliev, R.K.; Alexandrov, I.V. Bulk nanostructured materials from severe plastic deformation. *Prog. Mater. Sci.* **2000**, *45*, 103–189. [CrossRef]
36. Bečvář, F.; Čížek, J.; Procházka, I.; Janotová, J. The asset of ultra-fast digitizers for positron-lifetime spectroscopy. *Nucl. Instrum. Methods Phys. Res. A* **2005**, *539*, 372–385. [CrossRef]
37. West, R.N. Positron studies of condensed matter. *Adv. Phys.* **1973**, *22*, 263–383. [CrossRef]
38. Čížek, J.; Haušild, P.; Cieslar, M.; Melikhova, O.; Vlasák, T.; Janeček, M.; Král, R.; Harcuba, P.; Lukáč, F.; Zýka, J.; et al. Strength enhancement of high entropy alloy HfNbTaTiZr by severe plastic deformation. *J. Alloys Compd.* **2018**, *768*, 924–937. [CrossRef]
39. Couzinie, J.P.; Dirras, G.; Perriere, L.; Chauveau, T.; Leroy, E.; Champion, Y.; Guillot, I. Microstructure of a near-equimolar refractory high-entropy alloy. *Mater. Lett.* **2014**, *126*, 285–287. [CrossRef]
40. Cichy, H.; Fromm, E. Oxidation kinetics of metal films at 300 K studied by the piezoelectric quartz crystal microbalance technique. *Thin Solid Films* **1991**, *195*, 147–158. [CrossRef]
41. Málek, J.; Zýka, J.; Lukáč, F.; Čížek, J.; Kunčická, L.; Kocich, R. Microstructure and mechanical properties of sintered and heat treated HfNbTaTiZr high entropy alloy. *Metals* **2019**, in press.
42. Piluso, P.; Ferrier, M.; Chaput, C.; Claus, J.; Bonnet, J. Hafnium dioxide for porous and dense high-temperature refractories (2600 °C). *J. Eur. Ceram. Soc.* **2009**, *29*, 961–968. [CrossRef]
43. Durov, A.V. Wetting oF Hafnium Dioxide by Pure Metals. *Powder Metall. Met. Ceram.* **2011**, *50*, 552–556. [CrossRef]
44. Lei, Z.; Liu, X.; Wu, Y.; Wang, H.; Jiang, S.; Wang, S.; Hui, X.; Wu, Y.; Gault, B.; Kontis, P.; et al. Enhanced strength and ductility in a high-entropy alloy via ordered oxygen complexes. *Nature* **2018**, *563*, 546–550. [CrossRef]
45. Juan, C.C.; Tsai, N.H.; Tsai, C.W.; Hsu, W.L.; Lin, C.M.; Chen, S.K.; Lin, S.J.; Yeh, J.W. Simultaneously increasing the strength and ductility of a refractory high-entropy alloy via grain refining. *Mater. Lett.* **2016**, *184*, 200–203. [CrossRef]
46. Schuh, B.; Volker, B.; Todt, J.; Schnell, N.; Perriere, L.; Li, J.; Couzinie, J.P.; Hohenwarter, A. Thermodynamic instability of a nanocrystalline, single-phase TiZrNbHfTa alloy and its impact on the mechanical properties. *Acta Mater.* **2018**, *142*, 201–212. [CrossRef]
47. Chen, S.Y.; Tong, Y.; Tseng, K.K.; Yeh, J.W.; Poplawsky, J.D.; Wen, J.G.; Gao, M.C.; Kim, G.; Chen, W.; Ren, Y.; et al. Phase transformations of HfNbTaTiZr high-entropy alloy at intermediate temperatures. *Scr. Mater.* **2019**, *158*, 50–56. [CrossRef]
48. Yao, J.Q.; Liu, X.W.; Gao, N.; Jiang, Q.H.; Li, N.; Liu, G.; Zhang, W.B.; Fan, Z.T. Phase stability of a ductile single-phase BCC Hf0.5Nb0.5Ta0.5Ti1.5Zr refractory high-entropy alloy. *Intermetallics* **2018**, *98*, 79–88. [CrossRef]

Article

Influence of Material Structure on Forces Measured during Abrasive Waterjet (AWJ) Machining

Libor M. Hlaváč [1,*], Adam Štefek [1], Martin Tyč [1] and Daniel Krajcarz [2]

[1] Department of Applied Physics, Faculty of Electrical Engineering and Computer Science, VSB–Technical University of Ostrava, 17. listopadu 2172/15, 70800 Ostrava–Poruba, Czech Republic; adam.stefek.st@vsb.cz (A.Š.); martin.tyc.st@vsb.cz (M.T.)

[2] Department of Materials Science and Materials Technology, Faculty of Mechatronics and Mechanical Engineering, Kielce University of Technology, al. Tysiąclecia Państwa Polskiego 7, 25-314 Kielce, Poland; d.krajcarz@wp.pl

* Correspondence: libor.hlavac@vsb.cz; Tel.: +420-59732-3102

Received: 27 July 2020; Accepted: 31 August 2020; Published: 2 September 2020

Abstract: Material structure is one of the important factors influencing abrasive waterjet (AWJ) machining efficiency and quality. The force measurements were performed on samples prepared from two very similar steels with different thicknesses and heat treatment. The samples were austenitized at 850 °C, quenched in polymer and tempered at various temperatures between 20 °C and 640 °C. The resulting states of material substantially differed in strength and hardness. Therefore, samples prepared from these material states are ideal for testing of material response to AWJ. The force measurements were chosen to test the possible influence of material structure on the material response to the AWJ impact. The results show that differences in material structure and respective material properties influence the limit traverse speed. The cutting to deformation force ratio seems to be a function of relative traverse speed independently on material structure.

Keywords: abrasive waterjet; machining; traverse speed; material structure; material properties; cutting force; deformation force

1. Introduction

Abrasive waterjet (AWJ) is a tool penetrating to machining technologies for 40 years demonstrated by Natarajan et al. in their brief review [1]. AWJ is applied in many manufacturing processes. Classical machining technics like cutting and milling are studied, e.g., by Axinte et al. [2] or Rabani et al. [3]. Turning is one of the unusual machining processes and it was studied by Zohourkari et al. [4]. Even less usual application studied by Schwartzentruber and Papini is piercing [5]. Grinding was studied by Liang et al. [6] or polishing, being studied by Loc and Shiou, are rarely used applications [7], but more widespread use of these new kinds of AWJ machining is anticipated in the near future. The AWJ is capable of machining almost every type of material, independently of its thickness, if the jet energy is sufficient for penetration through material. The accuracy of this technology is closely related to either regression or theoretical models, because no complex model for all types of materials has been prepared to date. Two basic groups of models are used in practice; the first are more suitable for ductile materials and the second ones are more appropriate for brittle materials. The efforts of many research teams all over the world are focused on better understanding the erosion process as one of the important bases for improvement of contemporary models or the creation of new ones. A better understanding of erosion processes can also lead to the design of some new AWJ tools. Therefore, information about quality of the AWJ machining process is very important for quality control and it is obtained through many online and offline measurements. Conventional online methods seem to be inappropriate due to the harsh environment during the machining process and strong sensitivity to irregularities present in jet flow or material properties. Therefore, usable methods for online monitoring of the AWJ processes are still being sought.

Investigation of the acoustic emissions studied (e.g., by Rabani et al. [8]) is one of methods tested for the monitoring of AWJ processes. Another type of AWJ cutting process monitoring was presented by Hreha et al., studying the relations between the vibrations of cut material measured by accelerometers and the surface roughness of cut wall [9]. However, the experiences of scientists described by Mikler show that more than 25% of the information from acoustic emissions or vibrations may be incorrect due to the misinterpretation of the complicated dynamic signal [10]. The attempts to determine the quality of the machining process by vibration signals measured on cutting heads were also presented in several research articles. Fabian and Salokyová [11–13] presented in these articles the measurements and analyses of signals from accelerometers placed on a cutting head (mixing chamber) and attempted to determine the relationship between some resonation frequencies and material cutting quality.

Research of new monitoring and control methods continues and research activities take various forms. Karas et al. [14], for example, tried to determine AWJ characteristics through mathematical modelling in Computational Fluid Dynamics CFD methods to prepare a tool to predict manufacturing processes results. Li et al. [15] studied deformation of stainless steel 304 and its mechanical properties after AWJ processing by using several methods. Pahuja et al. investigated the behaviour of the abrasive water jet during its passage through a system of plates made from various kinds of materials [16]. Their findings are very important for certain industrial applications, namely in aerospace. However, acoustic emissions, cutting head vibrations and other dynamic methods may be influenced by the dimensions of machines, their stiffness, or the dimensions of machined pieces of materials as mentioned by Mikler [10]. Searching for new ways of measurements applicable to new technologies has been mentioned by Królczyk et al. [17]. Therefore, some alternative to dynamic measurements have been investigated.

Forces can be used both for dynamic and static measurements. One of the first AWJ force measurements has been described by Li et al. in 1989 [18]. Authors tried to measure the force response of both the pure and abrasive water jet regarding several selected parameters—stand-off distance, water nozzle diameter, focussing tube diameter, abrasive mass flow rate and abrasive size. The results are inspirational for further research, but the basic relationships correspond with expectations based on theoretical analyses and contemporary measurements. In the past, the force sensors were also used for the determination of the pure waterjet velocity profile and diameter, as presented in Vala and Vala et al. in [19,20] or for the investigation of possible safety hazards of hand operating tools described by Hlaváčová and Vondra [21]. The problems with influencing the measured signal by dimensions of machined pieces of material could be partially be reduced using a measurement on samples with equal dimensions. Therefore, the force sensor measuring x-y-z forces during AWJ cutting has been designed and patented by Mádr et al. [22]. Here, the force sensor is tested in order to find out the limits of its use for the continuous monitoring of AWJ processes and their control.

The measurements presented in this article are aimed at the role of material structure impact on measured data. During the measurements of forces on various metal samples, it was mentioned that the most important parameter changing with this kind of material is the limit traverse speed. The mutual ratio of forces in the axes seemed to be influenced just by this parameter and respective "portion" of this value represented by the "actual traverse speed". The intent of the presented research was to eliminate the influence of other possible sources of material difference (elemental composition of the material, density). Therefore, the material enabling an extensive change of strength characteristics and thus the limit traverse speeds has been selected (in two very close modifications). Three different traverse speeds were used for testing. The experiments were performed on two similar steels with a broad scale of strength characteristics created by heat treatment of original raw materials. Therefore, the material structure and respective material characteristics differ significantly, while the density and elemental composition of material are identical. The results of force measurements confirm the influence of traverse speed on the material response against AWJ penetration, but the direct influence of material strength characteristics on the force response is not strong. On the contrary, the force measurement results indicate that the change of material properties influences the limit traverse speed of AWJ machining. This fact causes movement with identical relative traverse speeds that induce cutting forces in kerf (tangential and normal) in the same mutual ratio.

2. Theoretical Background

The theoretical model, used for the calculations necessary to determine the relative traverse speeds (the independent variables in this research), is based on the limit traverse speed $v_{P\lim}$ determined for each material by Hlaváč et al. [23] according to his equations. The respective jet velocity loss α_e is to be determined experimentally because of the interaction time uncertainty (explanation of this parameter and its determination was presented by Strnadel et al. in [24]). The respective equations are based on jet and material characteristic parameters:

$$v_{P\lim} = \left[\frac{C_A\, S_p\, \pi\, d_o\, \sqrt{2\rho_j p_j^3}\, e^{-5\xi_j L}\, (1 - \alpha_e^2)}{8\, H\left(p_j \rho_m \alpha_e^2\, e^{-2\xi_j L} + \sigma_m \rho_j\right)} \right]^{\frac{2}{3}} - v_{Pmin} \tag{1}$$

$$\alpha_e = 1 - \frac{\sqrt{2p_j^3}\, HV t_i}{8\, \sqrt{\rho_j}\, \sigma_m a_m} \tag{2}$$

The variables used in Equations (1) and (2) have the subsequent meaning: $v_{P\lim}$ the limit traverse speed; C_A the coefficient modifying abrasive water jet performance according to the changing content of abrasive below "saturation level" (above this level, the jet performance increases no more and $C_A = 1$); S_P-ratio between the quantity of non-damaged grains (i.e., not containing defects) and the total quantity of grains in the supplied abrasive material; d_o-diameter of the water nozzle (orifice); ρ_j-density of abrasive jet (conversion to homogeneous liquid); p_j-pressure obtained from Bernoulli's equation for liquid with density and velocity of abrasive jet; ξ_j-attenuation coefficient of abrasive jet in the environment between the focusing tube outlet and the material surface (usually air); L-stand-off distance (distance between the focusing tube outlet and the material surface); α_e-coefficient of abrasive water jet velocity loss in the interaction with material (experimentally determined); H-material thickness; ρ_m-density of material being machined; σ_m-strength of material being machined; v_{Pmin}-minimum limit traverse speed of cutting–correction for the zero traverse speed (usually the value $v_{Pmin} = a_n/60$ is used, where a_n is the average abrasive particle size after the mixing process inside the mixing head and focussing tube); HV-material hardness; t_i-interaction time; a_m-mean size of particles (elements) of material–grains or their chips.

The declination angle of striations measured at the bottom edge of a sample wall for a certain traverse speed introduced by Hlaváč [25] can be used for the calculation of the limit value according the equation

$$v_{P\lim} = v_P \left(\frac{\vartheta_{\lim}}{\vartheta} \right)^{\frac{2}{3}} \tag{3}$$

where the traverse speed v_P is the experimental one for which the declination angle ϑ is measured on the kerf wall [25]. The declination angle or respective relative traverse speed (the ratio of v_P and $v_{P\lim}$) can be utilized for determination of some quantities and process parameters regarding required quality of cutting, i.e., declination angle of striations on the kerf walls [23] and inclination of the walls (the taper) [26]. Because the traverse speed of the cutting head is the most easily controlled parameter of AWJ machining, it is very important to know the relationship between this speed and resulting quality, as it was presented by Hlaváč et al. [27]. However, the limit declination angle used for the calculation of the limit traverse speed depends on the jet energy and material parameters. If the energy of the AWJ is sufficient for full development of both types of material wear inside kerf, the limit declination angle is close to 45°. The limit declination angle value decreases, when the AWJ energy is decreasing regarding the necessary one for both types of wear, as indicated in [28]. Provided that only cutting wear can be utilized, the declination angle can decrease to 22.5° and for very hard or very thick materials the value 15° seems to be the limit one. These limits could be influenced not only by jet energy and macroscopic material properties, but also by abrasive grains rigidity and material structure.

The scheme of the impacting force decomposition inside the kerf into cutting forces (tangential), and the deformation (normal) and transverse (lateral) components is presented in Figure 1. Elements of the components are signed by the axes of the force sensor.

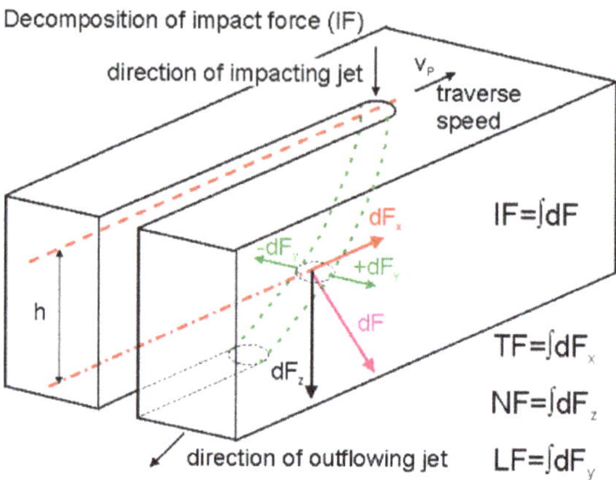

Figure 1. Decomposition of the total impact force acting on the kerf head into the cutting (TF), deformation (NF) and transverse (LF) forces; the forces elements are distinguished by subscripts corresponding with force sensor axes directions, measured values are integral values in the selected direction at a given moment.

Considering the force decomposition, the following is presumed of this behaviour of new quantities: "The cutting to deformation force ratio (CDFR) depends on the ratio of the cutting-to-deformation wear in the produced kerf". It is anticipated that an increase of the traverse speed from a zero value to the limit causes an increase of the CDFR up to a certain maximum value, achieved approximately for a half of the limit traverse speed. This maximum corresponds with the maximum portion of cutting force. Further increase of the traverse speed causes further increase of the deformation force. This is caused by the increasing portion of the deformation wear due to kerf head curvature, while the cutting wear cannot increase more. Therefore, the CDFR decreases for traverse speeds higher than a half of the limit traverse speed. This consideration is valid only in cases when the energy of the jet is sufficient for the full development of both types of wear. Lower energies than necessary (usually during the cutting of very thick materials) makes it so that only cutting wear can be used to penetrate through material. The residual jet energy inside the deep kerfs is insufficient for the proper development of the deformation wear and it reflects back from the kerf as a reverse flow when the conditions for cutting force are exceeded. This situation occurs when the outlet declination angle overcomes a value of approximately 22.5° for medium-thick materials (1.5–2.5 thicker than appropriate to jet energy) or 15° for very thick materials (for thickness 2.5 higher than the appropriate one).

The theoretical model presented in [23–26] makes it possible to limit the transfer values determined for one machining configuration to another, reducing the necessary experimental work. Equation (4) is based on functional dependences of traverse speed on respective changing variables-Equations (1) and (2). Therefore, the values ones determined for certain machine configuration like, e.g., in [27], need not be determined again making new samples and measuring respective angles on them. Provided that the nozzle diameter, pressure and abrasive type are changed, the limit traverse speed for configuration 2 is calculated from the value for configuration 1 through the equation

$$v_{Plim2} = \eta_{A21} \frac{d_2}{d_1} \sqrt{\frac{p_2^3}{p_1^3}} v_{Plim1} \tag{4}$$

where v_{Plim2} is the limit traverse speed calculated for machine configuration 2; v_{Plim1} is the limit traverse speed already known for machine configuration 1; η_{A21} is a ratio of abrasive qualities (the second to the first one), if they differ; d_1, d_2 are the respective nozzle diameters of configurations; p_1, p_2 are the respective pumping pressures of configurations. The relative traverse speed (the main quality indicator) is then determined as

$$v_R = \frac{v_P}{v_{Plim}} \tag{5}$$

To make the research as objective as possible, the experiments were performed on a different device than the one originally used to determine the limit traverse speeds and the first cutting studies. The limit traverse speeds were calculated from original values using Equation (4). The ratio of cutting and deformation forces were determined and it its dependence was studied on relative traverse speed to approve its relationship with material structure, because limit traverse speeds differ for individual samples due to the differing structures of the materials (see [24] for details).

3. Materials and Methods

Several metal samples with different structures and respective strength characteristics were used for experiments. The two similar steels 34CrMo4 (DIN norm) were used, differing namely by nickel ratio. The raw material for each one was in another thickness. The composition of both steels is presented in Table 1.

Table 1. Chemical composition of two modifications of 34CrMo4 steel in weight %, hydrogen content is in ppm.

Steel Kind	C	Mn	Si	Cr	Ni	Mo	Cu	V	Al	P	S	N	H
K	0.37	0.84	0.26	1.15	0.03	0.208	0.02	0.074	0.0113	0.012	0.004	0.0113	1.3
CK	0.36	0.84	0.29	1.13	0.24	0.200	0.04	0.093	0.0270	0.011	0.003	0.0120	1.1

The samples were prepared from material austenitized at 850 °C, quenched in polymer and tempered at various temperatures; their tempering and respective uniaxial tensile strengths (σ_m) and Vickers hardness (*HV10*) for individual samples are summarized in Table 2.

Table 2. Steel 34CrMo4 samples austenitized at 850 °C, quenched in polymer and tempered at various presented temperatures.

Sample Mark (10 mm Thick)	Tempering Temperature (°C)	Uniaxial Strength (MPa)	Vickers Hardness HV 10	Limit Traverse Speed v_{Plim1} (mm/min)	Limit Traverse Speed v_{Plim2} (mm/min)
K37	20	2230	589	180	121
K38	250	1865	521	192	129
K39	400	1560	459	203	136
K40	510	1340	391	211	141
K41	580	1190	363	222	148
K42	620	1060	330	235	157
Sample Mark (6 mm Thick)	**Tempering Temperature (°C)**	**Uniaxial Strength (MPa)**	**Vickers Hardness HV 10**	**Limit Traverse Speed v_{Plim1} (mm/min)**	**Limit Traverse Speed v_{Plim2} (mm/min)**
CK43	20	2160	581	278	186
CK44	250	1860	514	306	204
CK45	400	1550	464	330	221
CK46	510	1320	405	339	227
CK47	580	1230	372	341	228
CK48	640	970	314	378	253

The samples were previously used for research presented in [24], where the influence of tempering on the material structure, namely the amount of carbides, was largely discussed. Therefore, the limit traverse speeds v_{Plim1} are determined from the original cutting parameters used for sample preparation in laboratory at the VŠB–Technical University of Ostrava (see [24]). Some of the samples analysed in [24] are presented in Figure 2.

Figure 2. Samples used for analysis of material structure impact on declination angle (see [24]).

The limit traverse speeds v_{Plim2}, necessary for experiments in Kielce, are calculated from Equation (4) for respective values of parameters summarized in Table 3 (for workplace actually used for experimental work–in Kielce–and the one used for previous experiments–in Ostrava). Both limit traverse speeds (v_{Plim1}, v_{Plim2}) are presented in Table 2 for respective samples.

Table 3. Summary of factors and parameters used for experiments in Kielce and the values used for determination of the original limit traverse speeds v_{Plim1} in Ostrava (in brackets).

Variable (Unit)	Value
Pump pressure (MPa)	250 (380)
Water orifice diameter (mm)	0.33 (0.25)
Focusing tube diameter (mm)	1.02 (1.02)
Focusing tube length (mm)	76 (76)
Abrasive mass flow rate (g/min)	240 (240)
Abrasive material average grain size (mm)	0.177 (0.180)
Abrasive material type	Indian garnet (Australian garnet)
Abrasive quality ratio η_{A21}	0.95
Stand-off distance (mm)	2 (2)
Traverse speed (mm/min)	50–150

All cutting experiments were performed at the Faculty of Mechatronics and Mechanical Engineering, Department of Materials Science and Materials Technology, the Kielce University of Technology, Poland. Experimental factors and parameters of the abrasive waterjet are summarized in Table 3.

The cuts in steels were performed at fixed three traverse speeds for each thickness of tested steels. The respective relative traverse speeds were calculated from these three pre-set traverse speeds and the respective limit traverse speeds v_{Plim2} calculated from values v_{Plim1} listed in Table 2. The forces in x and z directions were measured by special measuring device being able to measure forces in three orthogonal axes [22]. Design of this device is shown in Figure 3.

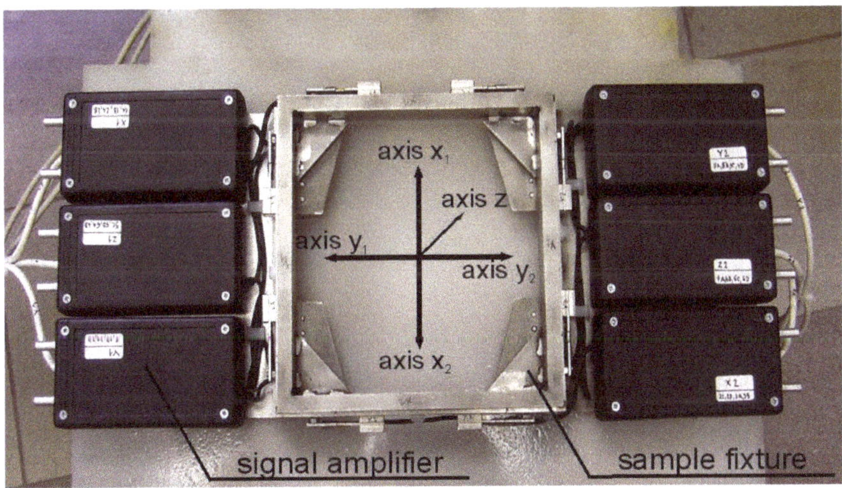

Figure 3. Overall view of the force sensor–black boxes contain amplifiers for strain gauges.

Signals from all measuring directions were recorded to a computer via the Signal Express program. After that, a raw voltage signal was processed with a program prepared in LabVIEW. The entire measuring system contains a DC source, the measuring part with deformation elements covered by extensometers, electric signal amplifiers (Figure 3), AD transducer, and computer with recording and processing software (Signal Express, LabVIEW, Austin, TX, USA).

The overall experimental procedure starts by placing the force sensor on the cutting table and fixing it to the grid, usually through putting heavy elements on the boxes with electronics, as presented in Figure 4. The sample is mounted to the removable plate with a central hole and placed into the frame signed "sample fixture" in Figure 3. The cutting machine and measuring system are switched on and prepared for starting the cut.

Figure 4. Force sensor on the cutting table in Kielce.

The recording on the measuring system is started first and then the cutting machine is put into operation. The movement of the cutting jet starts from outside the sample and ends approximately 20 mm inside the sample. The resulting shape of the samples after cutting is presented in Figure 5.

Figure 5. Samples after cutting for force measurement.

A typical measured signal is shown in Figure 6. There are electric signals from all Wheatstone bridges formed by extensometers placed on deformation elements of the force sensor. One bridge is for the x^+ axis (two deformation elements), the next one for x^- axis (two deformation elements). Two more bridges are for y^+ and y^- axes. The force in the selected direction is a mean of the absolute values in both directions of the respective axis. It was a different situation with z axis, because only the direction towards the ground is relevant. There are also two couples of deformation elements with respective bridges, but each of them registers only a part of the total force in the z direction. Therefore, the resulting force is a sum of these parts. The force signal after processing in LabVIEW program is presented in Figure 7.

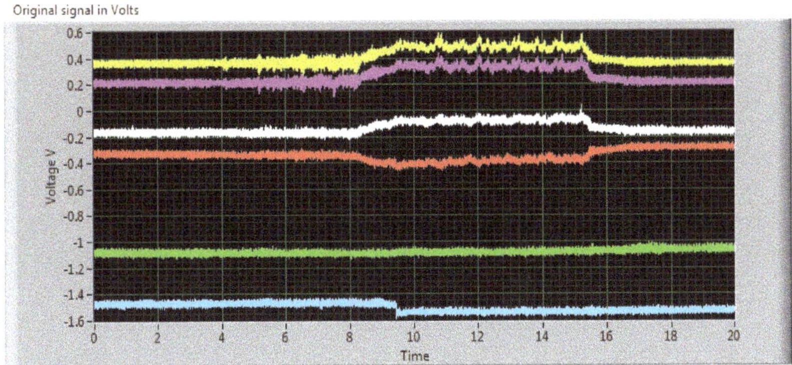

Figure 6. Electric signals measured by force sensor during cutting process of CK 44. Signal lines: yellow (1st from the top)–one part of the z axis, magenta (2nd from the top)–second part of the z axis, white (3rd from the top)–x^+ axis, red (4th from the top)–x^- axis, green (5th from the top)–y^+ axis, blue (the bottom line)–y^- axis.

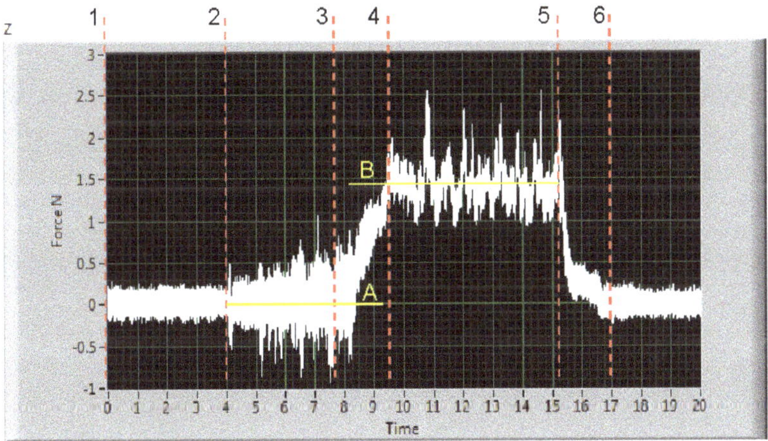

Figure 7. Force signal in z-axis after processing in LabVIEW program. Description of force signal parts: 1–2–signal before starting of the cutting process; 2–3–signal for the jet moving towards material (it is just impacting water in the waste energy attenuating vessel); 3–4–signal for jet starting to penetrate to material from the open side; 4–5–signal from the cutting through material; 5–6–signal from ending the cutting and stopping the machine; 6 to the end–signal from non-cutting system; A and B–lower and upper levels of force: the difference is force value in z-axis.

4. Results

The forces in the x-direction and z-direction (cutting and deformation ones respectively) were calculated from measured signals in the self-prepared program for signals processing in the LabVIEW™. The respective force values are the differences of the mean values of the signals in parts 4–5 and 2–3 (see Figure 7). The levels of average forces are represented by lines A (lower level) and B (upper level). The difference of these two values is the force value of the respective axis in a particular measurement. The relative traverse speeds were calculated from Equation (5) from the actual used traverse speed (the heading line of Table 4) and the respective limit traverse speed v_{Plim2} (see Table 2) calculated for each material from v_{Plim1} through Equation (4). CDFR values for various relative traverse speeds are summarized in Table 4 and graphically presented in Figure 8.

Table 4. CDFR values and the respective relative traverse speeds v_R (in brackets).

Samples 34CrMo4/10 mm	50 mm/min	75 mm/min	100 mm/min
K37	0.50 [0.41]	0.60 [0.62]	0.51 [0.83]
K38	0.60 [0.39]	0.70 [0.58]	0.61 [0.78]
K39	0.60 [0.37]	0.69 [0.55]	0.57 [0.73]
K40	0.50 [0.35]	0.61 [0.53]	0.52 [0.71]
K41	0.52 [0.34]	0.66 [0.51]	0.58 [0.67]
K42	0.50 [0.32]	0.65 [0.48]	0.59 [0.64]
Samples 34CrMo4/6 mm	**100 mm/min**	**125 mm/min**	**150 mm/min**
CK 43	0.60 [0.53]	0.68 [0.67]	0.56 [0.81]
CK 44	0.54 [0.48]	0.67 [0.61]	0.61 [0.73]
CK 45	0.59 [0.45]	0.66 [0.57]	0.57 [0.68]
CK 46	0.53 [0.44]	0.67 [0.55]	0.57 [0.66]
CK 47	0.51 [0.43]	0.65 [0.54]	0.56 [0.65]
CK 48	0.50 [0.40]	0.60 [0.49]	0.57 [0.59]

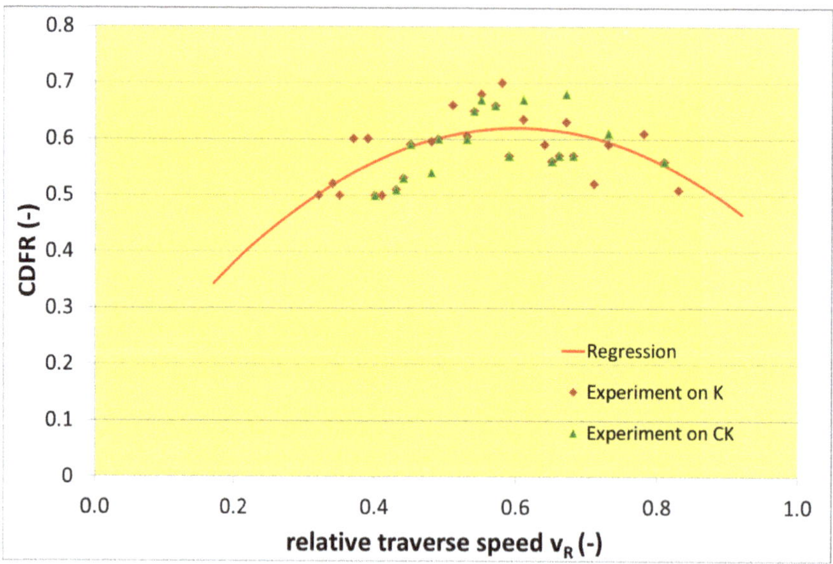

Figure 8. Graph of relationship between relative traverse speed and CDFR.

It can be seen that parabolic function is a good approximation of relationship between relative traverse speed and CDFR (representing cutting to deformation wear ratio inside the produced kerf). Nevertheless, no evident relationship on material structure is detected in the CDFR primarily. Quite the opposite—it seems like the CDFR is identical for a certain relative traverse speed independently of the material structure.

The respective regression equation for the calculation of the CDFR value from respective relative traverse speed value was determined in Excel for all measured values as.

$$CDFR = -1.5\,v_R^2 + 1.8\,v_R + 0.08 \tag{6}$$

Points (experimental results) presented in Figure 8 are determined as ratios from forces measured in x and z directions for respective relative traverse speeds (see Table 4). The regression equation and the respective curve in Figure 8 are presented just as a partial proof of the hypothesis presented in the theoretical portion and stating that the CDFR increases with increasing traverse speed up to the maximum at around the half of the limit traverse speed, and then it decreases.

5. Discussion

All experimental results indicate that material structure influences the material response to AWJ namely through changes of the limit traverse speeds. Material strength, hardness and structure, namely the amount and size of carbides [24], cause differences in absolute value of the limit traverse speed. This factor, the limit traverse speed, seems to be the main AWJ characteristic influenced by material structure directly, and probably the only one. The presented research was aimed at testing the influence of the material structure on the forces measured during AWJ machining, namely cutting. Therefore, two steels with a large change of structure depending on heat treatment were selected for tests. Each of the steels was of a different thickness. Heat treated samples with identical thickness were cut using the same scale of three traverse speeds. The different steel structures cause different limit traverse speeds. Therefore, it is evident that the ratio of the actual and limit traverse speed (relative traverse speed) should be the most appropriate independent variable for evaluation of the structure influence on the CDFR.

According to Hlaváč's theoretical model represented by Equations (1)–(3), it was anticipated that the identical relative traverse speeds produce similar cutting to deformation force ratios indicating so analogical ratios of cutting to deformation wear inside the produced kerfs. Differences in the CDFR are caused by a change of the relative traverse speed due to either the uncertainty of the material properties or the AWJ characteristics. Experimental results show independence of the CDFR on material thickness for AWJ energies sufficient for the full development of both types of material wear inside the produced kerf.

The contemporary research can be supported by conclusions based on results presented in [24]. The most important experimental and calculated results regarding size and volume percentage of carbides are summarized together with the respective limit traverse speeds in Table 5. Equation (7) for the calculation of carbide sizes d_c regarding the tempering temperature T_t has been determined from experimental results partially published in [24]

$$d_c = 0.0000004T_t^3 - 0.0003787T_t^2 + 0.1202T_t + 15 \tag{7}$$

Table 5. Tempering temperatures of the 34CrMo4 steel, respective experimental carbide sizes, carbide sizes calculated from experimentally based Equation (7), volume percentage of carbides and the limit traverse speeds set for the cutting machine in Kielce.

Sample Mark (10 mm Thick)	Tempering Temperature (°C)	Measured Size of Carbides d_c (m)	Calculated Carbide Size d_c (m)	Percentage of Carbides (%)	Limit Traverse Speed (mm/min)
K37	20				121
K38	250	27.54	27.63	2.20	129
K39	400	28.36	28.09	3.50	136
K40	510	31.01	30.86	5.37	141
K41	580	35.72	35.37	6.42	148
K42	620	39.26	39.28	7.53	157

The description of experimental materials presented in [24] states: "The increase in density of fine carbides at higher tempering temperatures is caused by the coarsening of carbides accompanied by a change in their coherence to non-coherent phase boundaries. The plastic deformation is easier in the presence of coarse carbides, primarily because dislocations may overcome fields of coarse carbides during plastic deformation by means of the Orowan mechanism, which requires relatively low energy [29]. The significant effect of fine coherent precipitates on hardening is replaced by hardening by means of non-coherent carbides, in which increasing carbide size is accompanied by a decline in hardening [30]." Therefore, it was necessary to test if the changing structure of material has the direct influence on the CDFR trends.

Figure 9 is showing a comparison of trends of the carbide size, volume percentage of carbides and the limit traverse speed dependence on tempering temperature. All three trends are identical, proving that the limit traverse speed clearly depends on the structure of the experimental material. Simultaneously, the difference between these trends and the trend of the CDFR dependence on the relative traverse speed confirms that the limit traverse speed is the proper variable influenced by the material structure. The CDFR is almost independent on the material structure, because it has a similar trend and values for similar or identical relative traverse speeds (portions of the respective limit traverse speed) independently on the absolute value of the limit traverse speed.

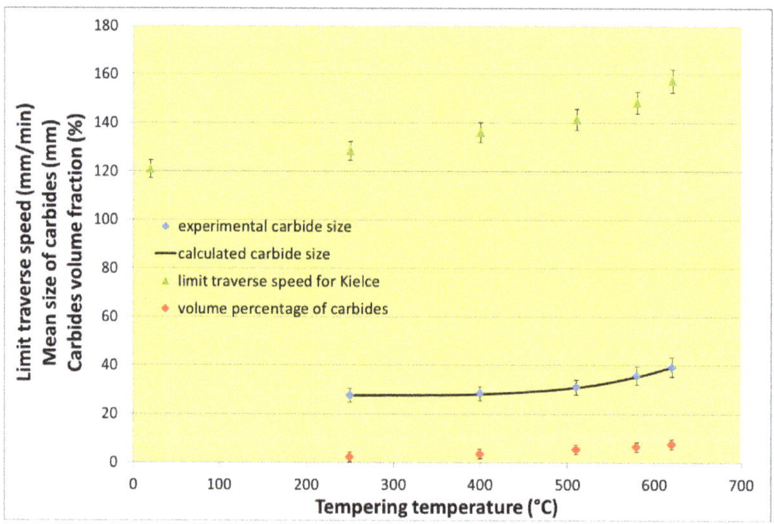

Figure 9. Trends of dependences of the carbide sizes, volume percentage of carbides and the limit traverse speeds on the tempering temperatures of samples used for research of material structure influence on AWJ cutting forces (the uncertainty magnitudes of experimental and calculated values are added).

Another way of explanation is based on the detail description of the cutting process with increasing traverse speed. When the traverse speed is close to zero, a very small part of the impacting jet is cutting material, i.e., the force acting towards the traverse speed direction (the tangential force) is small. Simultaneously, the part of the jet cross-section circumference projected to the plane perpendicular to the traverse speed is just a jet diameter. The force in the direction of jet flow (the normal one) is also small, but higher than cutting one due to the greater part of jet acting on the jet border (one half of the jet cross-section circumference, i.e., $\pi/2$ times longer than for the cutting force). By increasing the traverse speed the jet delay in the kerf makes the forehead of the kerf curve, increasing with the deformation wear of material. Simultaneously, the pushing of the jet towards traverse speed direction also increases. Up to a certain traverse speed, the tangential force pushing on the kerf forehead increases more rapidly than normal force increasing due to kerf forehead curving. However, for a certain traverse speed, the curvature of the kerf forehead is so high that tangential force cannot increase more, because part of the force pushing towards traverse speed direction projects to the normal direction. Therefore, the ratio of cutting to the deformation force decreases. This whole scale of ratios can be observed when the energy of the jet is sufficient for a full development of both wear types in the kerf. In the case of very thick material cutting, when only the cutting wear can be utilized, the CDFR relationship on the relative traverse speed either ends in the maximum value or at some point of the increasing part of the curve. This knowledge can be used for AWJ machining monitoring and control.

6. Conclusions

The subsequent conclusions resulting from presented research work were found out during the evaluation of data measured by the force sensor on the steel with a large scale of strength parameters prepared by various thermal processing of identical raw materials:

(1) Changes in material structure influence primarily the limit AWJ traverse speed. This finding is consistent with previously observed relationships between the material structure and the limit traverse speed published in prior articles. The benefit of the presented research is that relationship is confirmed independently on material composition and density.
(2) Differences in limit traverse speed caused by structure changes inflict different relative traverse speeds for the fixed testing traverse speed set. Therefore, it is possible to evaluate a large set of experimental results with a limited number of samples to yield broad range of relative traverse speeds. This is an important consequence of the application of the presented theoretical base for experimental planning and design.
(3) Dependence of CDFR on relative traverse speed has a parabolic relationship and the maximum is located near the value $0.5\,v_{\mathrm{Plim}}$. This finding is one of the important new results of the presented research. It is confirmed on steel samples 6 mm and 10 mm thick and AWJ power density higher than 8 $\mathrm{GW{\cdot}m^{-2}}$.
(4) The parabolic shape of the functional relationship between relative traverse speed and the CDFR is confirmed for AWJ characteristics being sufficient for a full development of both wear types (cutting and deformation ones) inside the produced kerf. An increase of material thickness or decrease of AWJ power may cause that only the cutting wear can be present and the deformation wear is either substantially suppressed or fully eliminated. Searching for respective limits is the aim of subsequent research.
(5) It can be anticipated that decreasing of the limit traverse speed for increasing material thickness causes shift of the CDFR maximum closer to the limit traverse speed, i.e., the maximum is either shifted to the relative traverse speed equal to one or no maximum occurs at all (the curve of CDFR relation on relative traverse speed has only increasing character).
(6) The dependence of the CDFR primarily on relative traverse speed and independence of the CDFR on the material structure are the most important new findings of presented research, because they make force measurements usable for monitoring and control of all kinds of materials, provided that AWJ power is sufficient for both kinds of material wear inside kerf.

Author Contributions: L.M.H. provided conceptualization, methodology, theoretical and formal analysis, writing—original draft preparation and supervision. A.Š. provided data measurement and processing. M.T. prepared software for measurement and data processing. D.K. performed sample preparation on AWJ cutting equipment and data curation. All authors have read and agreed to the published version of the manuscript.

Funding: This research was funded by Ministry of Education, Youths and Sports of the Czech Republic through projects SP2018/43, SP 2019/26 and SP 2020/45.

Conflicts of Interest: The authors declare no conflict of interest. The funders had no role in the design of the study; in the collection, analyses, or interpretation of data; in the writing of the manuscript, or in the decision to publish the results.

References

1. Natarajan, Y.; Murugesan, P.K.; Mohan, M.; Khan, S.A.L.A. Abrasive Water Jet Machining process: A state of art of review. *J. Manuf. Process.* **2020**, *49*, 271–322. [CrossRef]
2. Axinte, D.; Karpuschewski, B.; Kong, M.; Beaucamp, A.; Anwar, S.; Miller, D.; Petzel, M. High Energy Fluid Jet Machining (HEFJet-Mach): From scientific and technological advances to niche industrial applications. *CIRP Ann.* **2014**, *63*, 751–771. [CrossRef]
3. Rabani, A.; Madariaga, J.; Bouvier, C.; Axinte, D. An approach for using iterative learning for controlling the jet penetration depth in abrasive waterjet milling. *J. Manuf. Process.* **2016**, *22*, 99–107. [CrossRef]
4. Zohourkari, I.; Zohoor, M.; Annoni, M. Investigation of the Effects of Machining Parameters on Material Removal Rate in Abrasive Waterjet Turning. *Adv. Mech. Eng.* **2014**, *6*. [CrossRef]
5. Schwartzentruber, J.; Papini, M. Abrasive waterjet micro-piercing of borosilicate glass. *J. Mater. Process. Technol.* **2015**, *219*, 143–154. [CrossRef]
6. Liang, Z.; Xie, B.; Liao, S.; Zhou, J. Concentration degree prediction of AWJ grinding effectiveness based on turbulence characteristics and the improved ANFIS. *Int. J. Adv. Manuf. Technol.* **2015**, *80*, 887–905. [CrossRef]
7. Loc, P.H.; Shiou, F.-J. Abrasive Water Jet Polishing on Zr-Based Bulk Metallic Glass. *Adv. Mater. Res.* **2012**, *579*, 211–218. [CrossRef]
8. Rabani, A.; Marinescu, I.; Axinte, D. Acoustic emission energy transfer rate: A method for monitoring abrasive waterjet milling. *Int. J. Mach. Tools Manuf.* **2012**, *61*, 80–89. [CrossRef]
9. Hreha, P.; Radvanská, A.; Knapčíková, L.; Krolczyk, G.M.; Legutko, S.; Krolczyk, J.B.; Hloch, S.; Monka, P. Roughness Parameters Calculation By Means Of On-Line Vibration Monitoring Emerging From AWJ Interaction With Material. *Metrol. Meas. Syst.* **2015**, *22*, 315–326. [CrossRef]
10. Mikler, J. On use of acoustic emission in monitoring of under and over abrasion during a water jet milling process. *J. Mach. Eng.* **2014**, *142*, 104–115.
11. Fabian, S.; Salokyová, Š. AWJ Cutting: The Technological Head Vibrations with Different Abrasive Mass Flow Rates. *Appl. Mech. Mater.* **2013**, *308*, 1–6. [CrossRef]
12. Salokyová, Š. Measurement and analysis of technological head vibrations in hydro-abrasive cutting technology. *Acad. J. Manuf. Process.* **2014**, *12*, 90–95.
13. Salokyová, Š. Measurement and Analysis of Mass Flow and Feed Speed Impact on Technological Head Vibrations during Cutting Abrasion Resistant Steels with Abrasive Water Jet Technology. *Key Eng. Mater.* **2015**, *669*, 243–250. [CrossRef]
14. Karas, A.; Ameur, H.; Mazouzi, R.; Kamla, Y. Determination of the dynamic characteristics of the abrasive water jet for process manufacturing. *Int. J. Adv. Manuf. Technol.* **2020**, *110*, 101–112. [CrossRef]
15. Li, Y.; Lu, Z.; Li, T.; Li, D.; Lu, J.; Liaw, P.K.; Zou, Y. Effects of Surface Severe Plastic Deformation on the Mechanical Behavior of 304 Stainless Steel. *Meterials* **2020**, *10*, 831. [CrossRef] [PubMed]
16. Pahuja, R.; Ramulu, M.; Hashish, M. Surface quality and kerf width prediction in abrasive water jet machining of metal-composite stacks. *Compos. Part B Eng.* **2019**, *175*, 107134. [CrossRef]
17. Krolczyk, G.M.; Krolczyk, J.B.; Maruda, R.W.; Legutko, S.; Tomaszewski, M. Metrological changes in surface morphology of high-strength steels in manufacturing processes. *Measurement* **2016**, *88*, 176–185. [CrossRef]
18. Li, H.Y.; Geskin, E.S.; Chen, W.L. Investigation of forces exerted by an abrasive water jet on workpiece. In *Proceedings of the 5th American Water Jet Conference, Toronto, ON, Canada, 29–31 August 1989*; Vijay, M.M., Savanick, G.A., Eds.; National Research Council of Canada: Ottawa, QC, Canada; U.S. Water Jet Technology Association: St. Louis, MO, USA, 1989; pp. 69–77.
19. Vala, M. The measurement of the non-setting parameters of the high pressure water jets. In *Geomechanics 93*; Rakowski, Z., Ed.; Balkema: Rotterdam, The Netherlands, 1994; pp. 333–336.
20. Sitek, L.; Vala, M.; Vašek, J. Investigation of high pressure water jet behaviour using jet/target interaction. In *Proceedings of the 12th International Conference on Jet Cutting Technology: Applications and Opportunities, Rouen, France, 25–27 October 1994*; Allen, N.G., Ed.; Mech. Eng. Pub. Ltd.: London, UK, 1994; pp. 59–66.
21. Hlaváčová, I.M.; Vondra, A. Future in marine fire-fighting: High pressure water mist extinguisher with abrasive water jet cutting. *Nase More* **2016**, *63*, 102–107. [CrossRef]
22. Mádr, V.; Lupták, M.; Hlaváč, L. Force Sensor and Method of Force Sensing in the Process of Abrasive Water Jet Cutting. Patent 2011-82, 16 May 2012.

23. Hlaváč, L.; Strnadel, B.; Kaličinský, J.; Gembalová, L. The model of product distortion in AWJ cutting. *Int. J. Adv. Manuf. Technol.* **2011**, *62*, 157–166. [CrossRef]
24. Strnadel, B.; Hlaváč, L.; Gembalová, L. Effect of steel structure on the declination angle in AWJ cutting. *Int. J. Mach. Tools Manuf.* **2013**, *64*, 12–19. [CrossRef]
25. Hlaváč, L. Investigation of the abrasive water jet trajectory curvature inside the kerf. *J. Mater. Process. Technol.* **2009**, *209*, 4154–4161. [CrossRef]
26. Hlaváč, L.; Hlaváčová, I.M.; Geryk, V.; Plančár, Š. Investigation of the taper of kerfs cut in steels by AWJ. *Int. J. Adv. Manuf. Technol.* **2014**, *77*, 1811–1818. [CrossRef]
27. Hlaváč, L.; Hlaváčová, I.M.; Arleo, F.; Viganò, F.; Annoni, M.; Geryk, V. Shape distortion reduction method for abrasive water jet (AWJ) cutting. *Precis. Eng.* **2018**, *53*, 194–202. [CrossRef]
28. Hlaváčová, I.M.; Geryk, V. Abrasives for water-jet cutting of high-strength and thick hard materials. *Int. J. Adv. Manuf. Technol.* **2016**, *90*, 1217–1224. [CrossRef]
29. Cahn, R.W.; Haasen, P. *Physical Metallurgy*; North-Holland Physics Publishing: Amsterdam, The Netherlands, 1983; pp. 1376–1379.
30. Wilson, D.; Konnan, Y. Work hardening in a steel containing a coarse dispersion of cementite particles. *Acta Met.* **1964**, *12*, 617–628. [CrossRef]

Communication

Influence of Steel Structure on Machinability by Abrasive Water Jet

Irena M. Hlaváčová [1,*], Marek Sadílek [2], Petra Váňová [3], Štefan Szumilo [3] and Martin Tyč [1]

1 Department of Physics, Faculty of Electrical Engineering and Computer Science, VSB-Technical University of Ostrava, 708 00 Ostrava, Czech Republic; martin.tyc.st@vsb.cz
2 Department of Machining, Assembly and Engineering Metrology, Faculty of Mechanical Engineering, VSB-Technical University of Ostrava, 708 00 Ostrava, Czech Republic; marek.sadilek@vsb.cz
3 Department of Material Engineering, Faculty of Materials Science and Technology, VSB-Technical University of Ostrava, 708 00 Ostrava, Czech Republic; petra.vanova@vsb.cz (P.V.); stefanszum@gmail.com (Š.S.)
* Correspondence: irena.hlavacova@vsb.cz; Tel.: +42-072-893-1874

Received: 28 August 2020; Accepted: 30 September 2020; Published: 5 October 2020

Abstract: Although the abrasive waterjet (AWJ) has been widely used for steel cutting for decades and there are hundreds of research papers or even books dealing with this technology, relatively little is known about the relation between the steel microstructure and the AWJ cutting efficiency. The steel microstructure can be significantly affected by heat treatment. Three different steel grades, carbon steel C45, micro-alloyed steel 37MnSi5 and low-alloy steel 30CrV9, were subjected to four different types of heat treatment: normalization annealing, soft annealing, quenching and quenching followed by tempering. Then, they were cut by an abrasive water jet, while identical cutting parameters were applied. The relations between the mechanical characteristics of heat-treated steels and the surface roughness parameters *Ra*, *Rz* and *RSm* were studied. A comparison of changes in the surface roughness parameters and Young modulus variation led to the conclusion that the modulus was not significantly responsible for the surface roughness. The changes of *RSm* did not prove any correlation to either the mechanical characteristics or the visible microstructure dimensions. The homogeneity of the steel microstructure appeared to be the most important factor for the cutting quality; the higher the difference in the hardness of the structural components in the inhomogeneous microstructure was, the higher were the roughness values. A more complex measurement and critical evaluation of the declination angle measurement compared to the surface roughness measurement are planned in future research.

Keywords: abrasive water jet cutting; surface roughness; heat treatment; hardness; tensile strength

---

1. Introduction

The technology of Abrasive waterjet (AWJ) has been widely used in many areas of human activity for several decades. The physical principle of the process is the transfer of mechanical energy from the pump to material and its use for required operations, namely various types of material disintegration such as milling [1,2], machining of composites [3], rock breaking [4], wood cutting [5], and many others. One of the most frequent and important applications of this technology is material cutting. Its efficiency should be evaluated from two points of view: either by the amount of consumed energy along with the costs, or by the evaluation of the quality of the cut, namely the surface roughness and cutting accuracy [2,5–8]. In the case of plain dividing cuts, the quality is usually not the most important thing. However, it may become the key problem in complex engineering processes.

It is rational to suppose that the quality of the cut would be influenced by material characteristics. The father of the AWJ technology, Hashish, originally identified material strength as an important characteristic influencing the AWJ cutting process [9]. However, in his erosion model for predicting the depth of the cut of abrasive waterjets in different metals [10], he described the correlation with the elastic modulus and dynamic flow stress, one of the two main properties he needed for a metal cutting process characterization. Later on, Zeng and Kim studied the correlation of various material parameters: the material structure element size, flaw distribution, Young's modulus of elasticity, Poisson's ratio, fracture energy, density, flow stress, AWJ cutting efficiency and the quality of the surface [11]. In a model presented by Che et al. [12], the influence of the material hardness on the surface roughness was investigated. Vikram and Babu developed a model for the numerical simulation of the cutting process, introducing material density [13]; Deam et al. [14] based a theoretical 2D model on the cutting wear factor introduced by Bitter [15,16]. There has been a lot of other interesting research work since then [17]. Whereas in models published in [11–14], the data used for the proposed models' verification came from the cutting of nonmetal, rather brittle, materials, e.g., granite, perspex or ceramics, other authors focused their attention on metals, like Hashish did. Arola and Ramulu found that the depth of subsurface plastic deformation is inversely proportional to a metal's strength coefficient [18], Chen et al. improved the cutting quality of mild steel by controlled nozzle oscillations [19], Hascalik et al. studied the depth of cut and smooth cutting region on titanium [20], Hlaváč et al. proposed a method for the compensation of the jet retardation and the taper [21], and Monno et al. studied the influence of heat treatment on the kerf roughness of carbon steel cut by AWJ [22]. Some of these researchers prepared models for the process description, while others focused on specific problems, among which the geometric accuracy, surface integrity and kerf roughness are achieving much attention [22–24]. Roughness might be important not only for manufacturers but also for material engineers, due to its potential to reduce necessary final polishing in cases when AWJ is used for cold cutting in sample preparation [25]. Although a lot of research was done in the search for a direct and simple connection between the material parameters and the cutting quality, and some unsubstantiated theories were published, no reliable proofs of such a linkage has been brought forward up to now.

One of the most interesting and inspiring works is the research of Strnadel et al. [26], who studied the relationships between the declination angle introduced by Hlaváč [27] and selected material properties, namely the strength, hardness and material element size. Although their research was aimed at another problem, they announced that their results indicated the dependence of the AWJ declination angle not only on strength characteristics, but also on the microstructure of the steel—i.e., on the way in which plastic deformation occurred during the AWJ–material interaction. Mono et al. [22] tried to trace this dependence in more detail and studied carbon steel C40 subjected to two different heat treatments and cut with different traverse speeds. They found that workpiece hardness affected the surface finish in different ways depending on the AWJ cutting parameters.

Our research group investigated the changes in the interaction of AWJ and steels with different microstructures under standard cutting parameters usual for the chosen thickness of the steel material. Three different steel grades suitable for heat treatment were chosen, and experiments aiming to reach significant changes in the steel microstructure were designed.

Carbon steel C45 represents a structural carbon steel for quenching and tempering, surface hardening, and for smaller and large forgings. It is suitable for shafts of mining machines, turbochargers, compressor pistons, etc.; its weldability is very difficult [28].

Steel 37MnSi5 is a manganese-silicon steel for tempering, and for large and larger forgings. It is relatively easy to form and also easy to machine. It is very susceptible to tempering brittleness. The steel is used for medium-stress machine and motor vehicle components and is particularly wear-resistant. The optimal diameter for finishing is 50 mm; its weldability is also very difficult [29].

Steel 30CrV9 represents a low-alloy steel with good weldability and machinability, which is suitable for finishing, surface quenching, and chemical-thermal and nitriding treatment. It is used on heavily stressed hardened machine parts and on nitrided parts, including nitrided gears. Due to its high hardenability, it can be used for large forgings. It is also used for the crankshafts of aircraft engines, propeller heads, connecting rods, connecting rod bolts and similarly stressed machine parts like the drive axles of motor vehicles, steering levers, etc. [30].

Our research was aimed at testing the validity of the findings presented in [26] for other surface quality indicators than the declination angle, namely *Ra* and *Rz*, which are more generally applied in engineering. We also included the mean width of the assessed profile *RSm* into our surface roughness evaluation, as we supposed it to be potentially related to the internal structure of the material.

2. Materials and Methods

This paper deals with the influence of the microstructure of steels having passed through various types of heat treatment on the cutting properties of an abrasive water jet. For this experiment, three different types of steel, carbon steel C45 (W.Nr. 1.0503 or 1.1191 equivalent), micro-alloyed steel 37MnSi5 (W.Nr. 1.5122 equivalent) and low-alloy steel 30CrV9 (W.Nr. 1.7361 equivalent), were used. The materials were chosen so as to maximize the changes in their characteristics induced by heat treatment. As the amount of carbon was different in each of them, different changes in the steel microstructure could be expected. The chemical composition of steels determined by glow discharge optical emission spectrometry (GDOES) [31] on a GDA 750 instrument, produced and supplied by Spectruma analytic Gmbh, is given in Table 1.

Table 1. Chemical composition of examined steels in wt. %.

Steel Grade	C	Mn	Si	P	S	Cr	Ni	Mo	V
C45	0.42	0.66	0.29	0.02	0.01	0.08	0.05	0.01	0.00
37MnSi5	0.39	1.18	1.27	0.03	0.01	0.08	0.04	0.01	0.02
30CrV9	0.27	0.53	0.29	0.02	0.02	2.85	0.05	0.01	0.17

Originally, there were hot rolled round bars with a diameter of 200 mm. Discs 30 mm thick were cut from these bars and divided into quarters. Each quarter was subjected to a different heat treatment, namely: normalization annealing, soft annealing, quenching and quenching followed by high temperature tempering. The parameters recommended by the producer (the first three columns) and the real parameters used for the heat treatment are summarized in Table 2. All three quarters of different steels were placed in the furnace and processed simultaneously in order to save time and energy. Because the recommended ranges for heat treatment are different for the three grades of steel, the efficiency of the treatment was not the same.

Table 2. The heat treatment: recommended and real (last column) values.

Heat Treatment	C45	37MnSi5	30CrV9	Real Value
normalization annealing	840–870/air	850–890/air	860–900/air	865 [1]/air
soft annealing	680–720/furnace	680–720/furnace	700–740/furnace	710 [1]/furnace
quenching	800–830/water	820–850/water	830–880/water	830 [1]/water
quenching and tempering	530–670/water, air	530–680/water, air	550–660/water, air	600 [1]/ air

[1] The temperature within all intervals was chosen.

Both normalization and soft annealing lasted 4 h, and it was followed by cooling down either in the air (normalization) or in the furnace (soft). The quenching lasted 1 h in both cases and was followed afterwards by either cooling in water (pure quenching) or tempering (2 h in the furnace).

The change of the mechanical characteristics of the tested steels due to heat treatment was identified by a hardness measurement and tensile tests. The hardness was measured by a Vickers hardness test HV30 (under a load of 294.2 N) on transverse metallographic sections taken from the cut prisms (see below). The tensile tests were performed at room temperature on standardized test specimens [32] on the Multipurpose Dynamic and Fatigue System LFV (100 kN) produced by Walter+Bai AG, Switzerland. The loading speed was chosen to be 2 mm/min. Specimens for the tensile tests with a diameter d_0 = 6 mm were prepared on a lathe from the prisms cut from the bulk material by AWJ.

After the heat treatment, the parts were cut with a water jet into prism-shaped test specimens with a square base measuring 10 × 10 mm and a length of 100 mm. The cutting itself was performed in the liquid jet laboratory at VŠB-TUO with AWJ generated by a high pressure pump PTV 19/60 based on HSQ FlowX5. The pump was modified by the producer in order to have adjustable pressure with a pump with adjustable pressure from 50 MPa to 415 MPa; the maximum flow was 1.9 L per min. The parameters used in the water jet cutting are shown in Table 3.

Table 3. AWJ cutting parameters.

Cutting Parameter [unit]	Value
working pressure [MPa]	380
water flow rate [L/min]	1.9
water nozzle diameter [mm]	0.25
abrasive tube diameter [mm]	1.02
type of abrasive	Australian garnet
mesh	80 [1]
abrasive mass flow rate [g/min]	225
stand-off distance [mm]	2
traverse speed [mm/min]	100

[1] Mean diameter of particles should be approximately 180 μm.

A microstructure examination of the heat-treated samples was carried out on an optical microscope Olympus GX51. The samples were grinded with 60, 80, 160, 320, 600, 800, 1200, 2400 SiC papers, then polished with a diamond suspension (crystals 1 μm) and finally etched with Nital (4% solution of HNO_3 in ethanol) long enough to make the expected microstructures be well-distinguishable.

The surfaces cut by the above defined technology were then measured in three lines (upper line 1 mm below the upper edge, the central line, and bottom line 1 mm above the bottom edge of the sample cut surface) by a contact tester SurfTest SJ-400 (Mitutoyo) using a diamond stylus tip with radius 2 μm and cone 60°, and three required surface roughness parameters were evaluated for each line: the arithmetical mean deviation *Ra*, maximum height *Rz* and mean width *RSm*.

The arithmetical mean deviation of the assessed profile *Ra* [33] is one of the most widely used parameters. It represents the arithmetic mean of the absolute ordinate *Z(x)* within the sampling length, and it therefore provides for stable results, as the parameter is not significantly influenced by scratches, contamination and measurement noise; furthermore, it is hardly affected by individual peaks or valleys because it is the mean value of the whole profile [34]:

$$Ra = \frac{1}{l}\int_0^l |Z(x)|dx \quad (1)$$

On the contrary, the maximum height *Rz* [33] represents the sum of the maximum peak height *Zp* and the maximum valley depth *Zv* of a profile within the reference length and so is significantly influenced by scratches, contamination, and measurement noise due to its reliance on peak values. Nevertheless, it is widely used and sometimes preferred in engineering applications [35]:

$$Rz = Rp + Rv = max(Zp) + |min(Zv)| \quad (2)$$

The third parameter that was selected to be potentially related to the internal structure of the material was the mean width *RSm* [33]. This parameter is used to evaluate the horizontal size of parallel grooves and grains instead of the height parameters; it represents the mean value of the lengths of profile elements within the sampling length (Equation (3)):

$$RSm = \frac{1}{N}\sum_{i=1}^{N} X_{si} \tag{3}$$

where X_{si} means the length of the *i*-th profile element.

3. Results

The results are presented in three sections, introducing each of the three materials individually. Each section contains a table summarizing the measured mechanical parameters of the respective steel grade (hardness HV30 results and results of the tensile tests, i.e., lower R_{eL} and upper R_{eH} yield strength—or, if these are not significant, proof strength $Rp_{0.2}$—ultimate tensile strength R_m and percentage elongation after fracture A). Another table provides the surface roughness parameters: the arithmetical mean deviation *Ra*, maximum height *Rz* and mean width *RSm*; all three lines are presented, without an average evaluation, because the interaction of the AWJ with the material traced on the bottom line is distorted by the pressure oscillations; therefore, the lines should be discussed individually.

Finally, photos of the steel microstructure after the respective heat treatments are presented and discussed. There are two photos of the same treatment; the second one with a five-time bigger magnification highlights the details of the microstructure.

3.1. Steel C45

3.1.1. Mechanical Parameters

Various types of heat treatment induced more or less significant changes of steel mechanical parameters (Table 4). As the tensile tests of the quenched C45 steel did not reveal a clear value of lower and upper yield strength, the proof strength $R_{p0.2}$ was evaluated.

Table 4. Vickers hardness HV30, lower R_{eL} and upper R_{eH} yield strength, ultimate tensile strength R_m, percentage elongation after fracture A.

Steel C45	HV30	R_{eL} [MPa]	R_{eH} [MPa]	R_m [MPa]	A [%]
normalization annealing	184 ± 1	372.5 ± 2.5	381.8 ± 2.3	636.5 ± 2.9	19.5 ± 4.8
soft annealing	153 ± 4	274.1 ± 1.4	278.9 ± 2.4	577.9 ± 0.4	25.2 ± 2.8
quenching	458 ± 14	739.5 ± 9.8 [1]	- [1]	1145.2 ± 7.8	2.87 ± 1.9
quenching and tempering	152 ± 2	492.6 ± 22.1	506.5 ± 24.7	732.7 ± 13.3	22.1 ± 1.6

[1] The proof strength $R_{p0.2}$ had to be evaluated for quenched material.

3.1.2. Microstructure

The microstructure of C45 steel after normalization annealing was fine-grained ferritic-pearlitic (Figure 1a) with unevenly precipitated pearlite (Figure 1b). According to [36], the steels with a lower hardness (below 270 HV) are, to a great extent, affected by other factors than hardness, namely by the pearlite nodular size. A larger pearlite size leads to a decrease of machinability in the form of an increased wear of the classic tool. Several earlier studies regarding steel machinability proved that the best machinability was provided by a spheroidized microstructure, as the amount of proeutectoid ferrite was increased in such a microstructure. The same should be expected with a smaller pearlite nodular size.

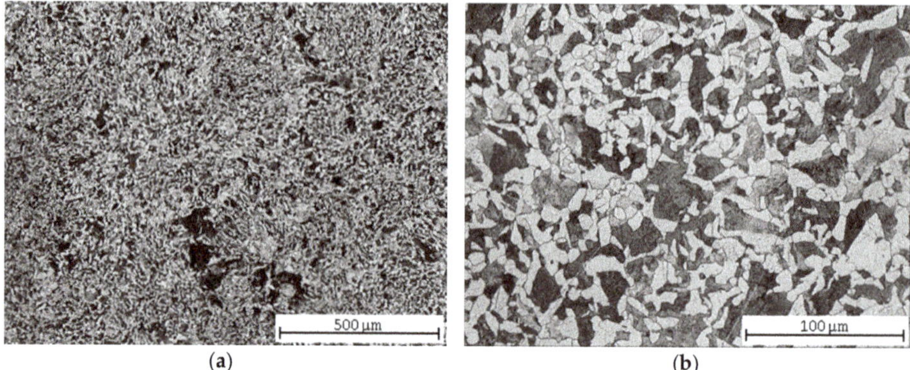

Figure 1. C45 steel microstructures after normalization annealing: (**a**) central part; (**b**) detail.

After soft annealing, partially spheroidized pearlite and Widmannstätten morphology ferrite (Figure 2a), i.e., needle-like growths of cementite within the crystal boundaries of the martensite, became visible in the microstructure (Figure 2b). Widmanstätten structures tend to form when the coarse-grained steel is rapidly cooled, which might have occurred with the hot rolled material during production. The structure increases the brittleness of the steel, and it can only be relieved by recrystallization above A_{c3} (the final critical temperature at which free ferrite is completely transformed into austenite during heating). The soft annealing temperature of 710 °C, lying below the eutectoid temperature A_{c1} = 727 °C (the critical temperature at which pearlite transforms into austenite), notwithstanding the holding time of 4 h, is ineffective in the removal of this morphology; a complete spheroidization of cementite could not be realized; therefore, a higher brittleness of the material should be expected [37].

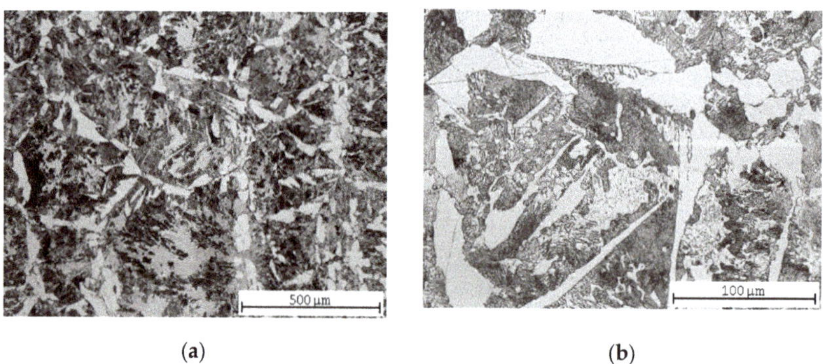

Figure 2. C45 steel microstructures after soft annealing: (**a**) central part; (**b**) detail.

As the hardenability of this steel was only 7 mm, the structure was not completely hardened. The evaluated sample might have been taken too far from the original hardened surface, and therefore the structure in Figure 3a consists mainly of pearlite and a ferritic network along the grain boundaries, while the proportion of martensite, resp. bainite is very low. The heterogeneity of the microstructure is also evident in this case (Figure 3b).

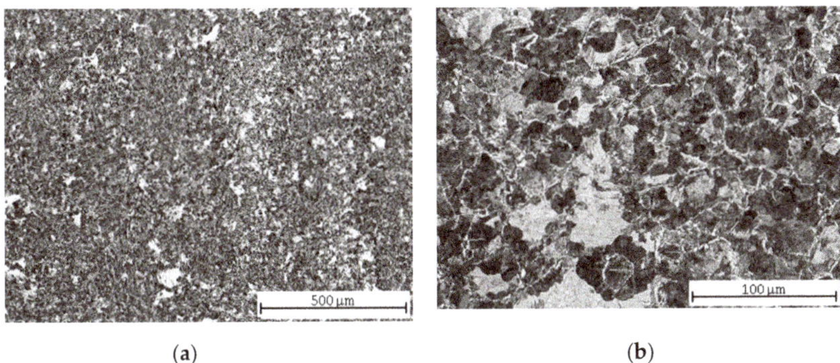

Figure 3. C45 steel microstructures after quenching: (**a**) central part; (**b**) detail.

After quenching and high-temperature tempering, the microstructure was formed mainly by bainite with a ferritic network along the grain boundaries (Figure 4a). The sample was taken closer to the surface in this case, and therefore the incidence of the tempered turbid microstructure was higher (Figure 4b).

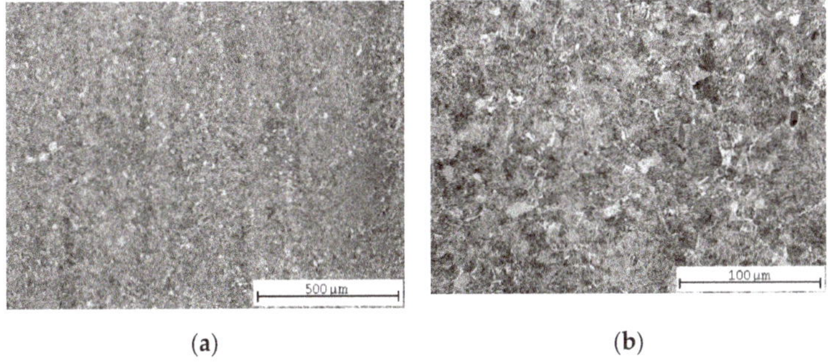

Figure 4. Quenched and tempered C45 steel microstructures: (**a**) central part; (**b**) detail.

3.1.3. Surface Roughness Measurement

The results of surface roughness measurement provided by contact tester are summarized in Table 5. The results of the all three lines are presented, named "upper", "central", and "bottom" in this table. The most significant of them is the central line; the values measured on bottom lines might be distorted by AWJ pressure fluctuations.

Table 5. Surface roughness characteristics after abrasive waterjet cutting.

Roughness Parameter		Normalization Annealing	Soft Annealing	Quenching	Quenching and Tempering
Ra	upper	7.69 ± 1.12	7.67 ± 0.81	9.56 ± 0.82	8.38 ± 1.21
	central	10.97 ± 1.05	10.35 ± 1.53	15.42 ± 1.88	13.51 ± 1.66
	bottom	21.04 ± 2.25	16.58 ± 2.89	22.20 ± 2.52	20.65 ± 4.88
Rz	upper	43.72 ± 4.23	44.17 ± 3.78	50.42 ± 5.28	46.70 ± 4.57
	central	56.43 ± 6.86	52.22 ± 7.63	75.87 ± 10.66	64.78 ± 6.22
	bottom	97.03 ± 12.07	80.57 ± 7.63	102.17 ± 11.10	96.43 ± 20.31
RSm	upper	713.33 ± 113.16	723.85 ± 142.80	1025.58 ± 186.18	748.88 ± 192.38
	central	860.18 ± 79.28	862.72 ± 134.19	1338.57 ± 196.84	937.60 ± 81.18
	bottom	1249.15 ± 142.80	1056.90 ± 220.03	1165.18 ± 115.64	1054.15 ± 218.15

3.2. Steel 37MnSi5

3.2.1. Mechanical Parameters

The heat treatment of manganese steel led to different changes of mechanical parameters than it was with the carbon steel (Table 6). After quenching, this steel grade did not reveal a clear value of lower and upper yield strength, therefore, proof strength $R_{p0.2}$ had to be evaluated both for quenched and tempered material. The changes are displayed graphically in the discussion section.

Table 6. Vickers hardness HV30, lower R_{eL} and upper R_{eH} yield strength, ultimate tensile strength R_m, percentage elongation after fracture A.

Steel 37MnSi5	HV30	R_{eL} [MPa]	R_{eH} [MPa]	R_m [MPa]	A [%]
normalization annealing	163 ± 2	546.1 ± 40	562.5 ± 38.7	864.3 ± 43.8	22.6 ± 1.3
soft annealing	139 ± 2	495.6 ± 3.8	499.2 ± 5.2	782.0 ± 6.3	27.5 ± 2.6
quenching	306 ± 8	912.0 ± 29.5 [1]	- [1]	1940.4 ± 21.7	3.86 ± 1.5
quenching and tempering	181 ± 1	724.4 ± 12.4	-	922.1 ± 8.0	19.8 ± 0.3

[1] The proof strength $R_{p0.2}$ was evaluated for hardened material.

3.2.2. Microstructure

The microstructure of 37MnSi5 steel in the state after normalization annealing was fine-grained ferrite-pearlitic (Figure 5a) with uneven pearlite precipitation associated with low-degree annealing (Figure 5b).

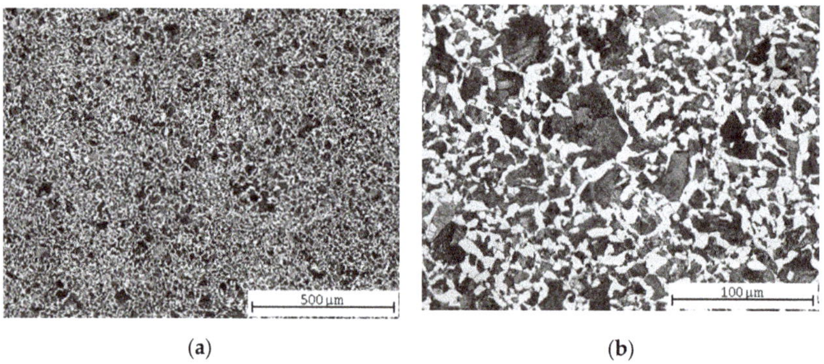

Figure 5. 37MnSi5 steel microstructures after normalization annealing: (**a**) central part; (**b**) detail.

After soft annealing, an almost 100% spheroidization of the pearlite occurred (Figure 6a). Compared to C45 steel, the spheroidization rate is higher due to the lower carbon content. Ferrite is precipitated along the grain boundaries (Figure 6b). The microstructure after quenching was martensitic with a low proportion of bainite and ferrite (Figure 7a). Although the individual martensite needles are not clearly observed (they are too small), thanks to their tendency to become aligned parallel to one another in a large region of the austenite grain, their characteristic microstructure can be recognized quite well (Figure 7b). After refining, the microstructure is tempered martensitic-bainitic (Figure 8a,b).

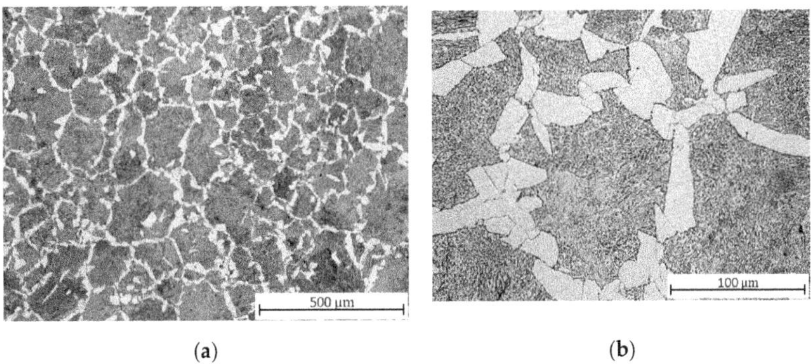

(a) (b)

Figure 6. 37MnSi5 steel microstructures after soft annealing: (**a**) central part; (**b**) detail.

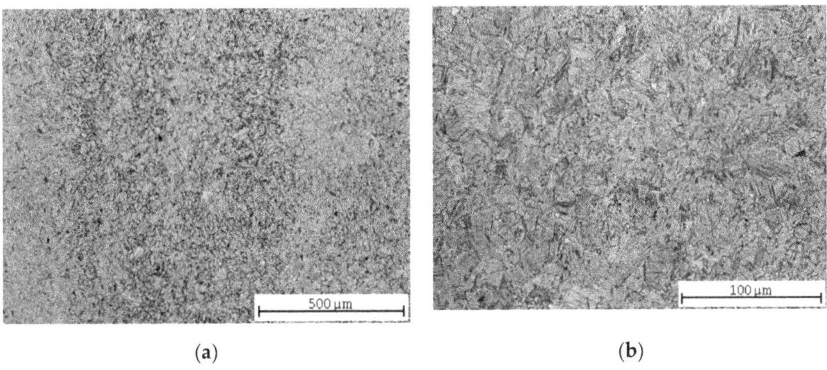

(a) (b)

Figure 7. 37MnSi5 steel microstructures after quenching: (**a**) central part; (**b**) detail.

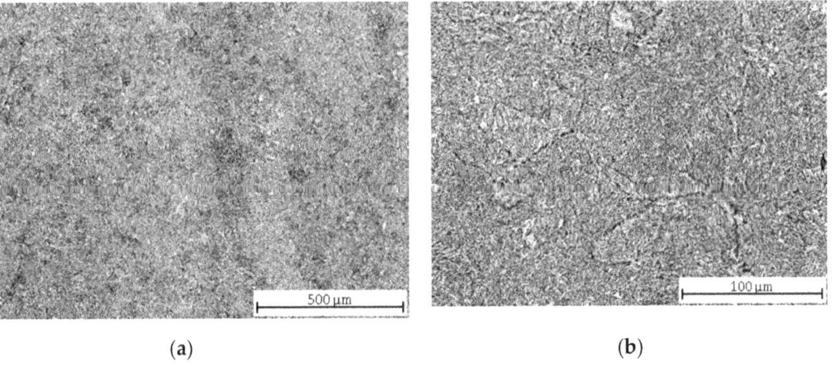

(a) (b)

Figure 8. Quenched and tempered 37MnSi5 steel microstructures: (**a**) central part; (**b**) detail.

3.2.3. Surface Roughness Measurement

The Table 7 presents the results of surface roughness measurement on the three lines in similar manner as it was done for the carbon steel. They are analyzed in detail in the discussion section.

Table 7. Surface roughness characteristics after abrasive waterjet cutting.

Roughness Parameter	Normalization Annealing	Soft Annealing	Quenching	Quenching and Tempering
	3.57 ± 0.32	3.63 ± 0.30	1.86 ± 0.14	2.00 ± 0.22
Ra	6.63 ± 0.63	9.20 ± 1.45	3.22 ± 0.45	2.56 ± 0.24
	17.39 ± 3.27	21.41 ± 2.65	5.56 ± 0.82	3.61 ± 0.45
	25.13 ± 2.16	24.50 ± 6.78	10.92 ± 1.13	13.68 ± 1.04
Rz	38.70 ± 4.22	49.57 ± 7.28	15.57 ± 1.45	13.68 ± 1.04
	84.13 ± 15.23	99.63 ± 11.50	23.18 ± 2.31	17.47 ± 1.83
	262.72 ± 30.01	239.73 ± 33.98	182.68 ± 30.49	148.03 ± 29.96
RSm	659.73 ± 98.06	791.65 ± 95.27	379.53 ± 69.09	242.63 ± 20.46
	951.48 ± 142.95	1131.05 ± 100.84	1054.53 ± 276.95	851.50 ± 367.08

3.3. Steel 30CrV9

3.3.1. Mechanical Parameters

The low-alloy steel 30CrV9 was the least hard steel included to this research. Its mechanical parameters after heat treatment are summarized in Table 8. Opposite to the other two steel grades it had no lower and upper yield strength, its proof strength was evaluated instead.

Table 8. Vickers hardness HV30, proof strength $R_{p0.2}$ [MPa], ultimate tensile strength R_m [MPa], percentage elongation after fracture A [%].

Steel 30CrV9	HV30	$R_{p0.2}$ [MPa]	R_m [MPa]	A [%]
normalization annealing	216 ± 0	745.9 ± 6.8	1203.1 ± 0.2	13.3 ± 1.2
soft annealing	120 ± 2	430.5 ± 5.3	682.0 ± 5.0	22.7 ± 1.2
quenching	292 ± 8	957.3 ± 80.1	1751.7 ± 48.3	3.7 ± 0.8
quenching and tempering	196 ± 1	920.3 ± 9.1	1032.9 ± 3.4	16.2 ± 0.4

3.3.2. Microstructure

After normalization annealing of 30CrV9 steel, the microstructure was formed by upper bainite and ferrite (Figure 9a), although the cooling was performed in air. Due to the high hardenability of this steel grade, the microstructure was partially quenched (Figure 9b).

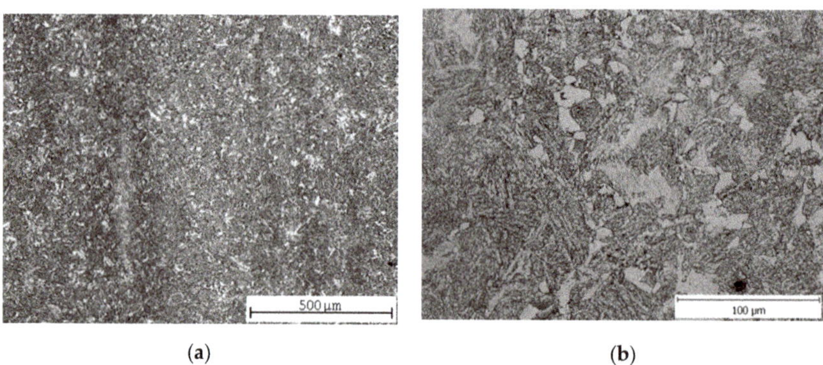

Figure 9. 30CrV9 steel microstructures after normalization annealing: (**a**) central part; (**b**) detail.

Soft annealing led to a partial spheroidization of the pearlite (Figure 10a). Due to presence of Cr and V carbides (the steel is alloyed with chromium and vanadium), spheroidization was slowed down [38]. Ferrite was precipitated along the grain boundaries (Figure 9b).

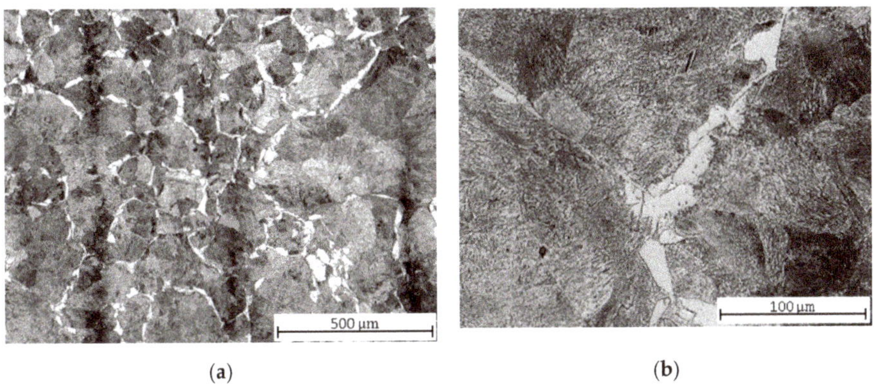

(a) (b)

Figure 10. 30CrV9 steel microstructures after soft annealing: (**a**) central part; (**b**) detail.

The microstructure after quenching consists of low-carbon martensite and a small proportion of bainite (Figure 11a,b), and after tempering it corresponds to tempered martensite (Figure 12a,b) [39].

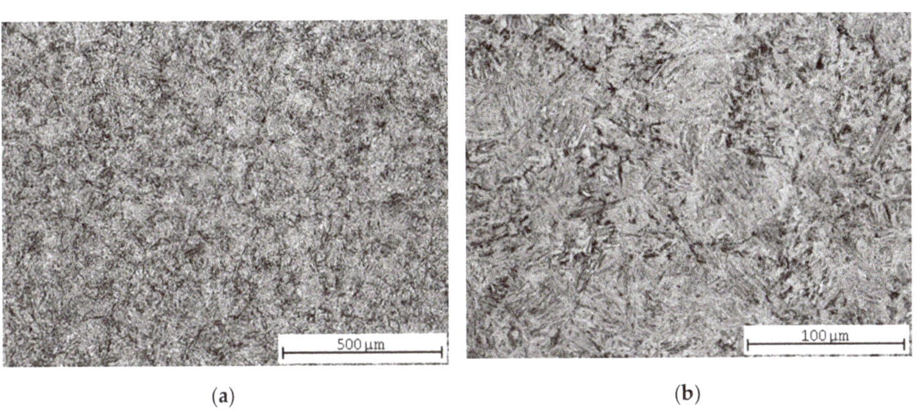

(a) (b)

Figure 11. 30CrV9 steel microstructures after quenching: (**a**) central part; (**b**) detail.

(a) (b)

Figure 12. Quenched and tempered 30CrV9 steel microstructures: (**a**) central part; (**b**) detail.

3.3.3. Surface Roughness Measurement

The Table 9 presents the results of surface roughness measurement on the three lines in the same manner as it was for the carbon steel. They are analyzed in detail in the discussion section. The horizontal parameter the mean width *RSm* shows such a large variance of values, that it cannot be taken to any trustworthy consideration.

Table 9. Surface roughness characteristics after abrasive waterjet cutting.

Roughness Parameter	Normalization Annealing	Soft Annealing	Quenching	Quenching and Tempering
	10.03 ± 1.24	8.10 ± 1.42	10.73 ± 1.24	7.17 ± 0.93
Ra	13.65 ± 1.63	11.08 ± 1.70	16.26 ± 1.38	9.15 ± 1.13
	31.51 ± 2.27	26.47 ± 3.63	42.04 ± 3.75	17.60 ± 2.69
	51.27 ± 6.49	46.60 ± 6.77	56.67 ± 5.38	40.95 ± 3.66
Rz	66.72 ± 5.42	56.60 ± 5.74	79.32 ± 6.63	51.08 ± 8.08
	174.60 ± 13.61	144.48 ± 34.30	230.38 ± 25.25	82.50 ± 11.98
	967.97 ± 182.43	697.45 ± 142.47	995.27 ± 220.03	765.17 ± 121.69
RSm	1315.48 ± 131.10	1012.73 ± 185.23	1116.10 ± 112.33	904.85 ± 138.45
	2025.05 ± 292.21	2125.05 ± 340.08	1985.23 ± 172.60	1121.83 ± 140.32

4. Discussion

A set of 36 triplets of surface roughness parameters, which were assigned to the respective values of four different material characteristics (Young's modulus of elasticity, hardness, ultimate tensile strength and ductility), was examined, in search of some dependency or correlation. The observed findings will be described and their possible connection with the AWJ operation considered.

The first finding was obvious and undisputable: the range of changes in *Ra* (in the same line position) varied from 41% to 54% for the same material; it is clear that such changes cannot correspond to Young's modulus values, which are never influenced by heat treatment to such an extent This finding corresponds to a discrepancy between the modulus values and respective mean declination angles referred to in [26]. It can, therefore, reasonably be stated that the modulus is not significantly responsible for the surface roughness.

As for other mechanic characteristics, all measured values were studied and compared, and various dependencies were analyzed. No obvious conclusions were discovered, although it is evident that there should be some correlation. The mechanism of the AWJ operation is inherently different from that of solid tools, and so macroscopic mechanical properties may exhibit contradictory effects from the point of view of AWJ machinability. The steel microstructure can then represent the decisive factor for a final classification. For example, a hard material may induce particle breakage followed by a decrease of jet efficiency, but at the same time it should prevent the abrasives from making scratches in the material and should result in a lower roughness of the cut if the material microstructure is homogenous.

The abrasive waterjet operation can be classified as a form of impact abrasive machining; it uses a mixture of fluid and solid particles, impacts the surface of a target workpiece and causes either permanent deformation or material removal [40]. In our AWJ machine, the abrasive mass flow rate is 32.5 g/s and the nominal size of particles is 0.18 mm; this amounts to ~4×10^6 particles per second, which is only four particles per microsecond. At an impact velocity of several hundreds of m/s, the full process of particle indentation and rebound lasts a fraction of μs [40]. Considering that the jet spreads over a ~1 mm spot size, i.e., ~40× the area for a single particle impact, we realize that it is reasonable to treat AWJ as a sequence of single particle impacts: each particle indents, removes material and rebounds away from the target as an independent event without interference from other particles or from workpiece motion [40]. The material response to the impact is described either as ductile or brittle. Generally, ductile erosion is relevant to metals and similar materials that are capable of significant plastic deformation. Brittle erosion applies to materials that crack and fragment under impact. However, from a microscopic point of view, the type of erosion is to a great extent influenced by the microstructure of the material, namely by the fact that it is either homogeneous or that it is a mixture of several components with different characteristics [40].

In the design of our experiment, it was expected that the roughness characteristic that might trace the individual particle impact on the material would be the horizontal one: the mean width *RSm*. This expectation, however, appeared to be false: the changes of *RSm* did not prove any correlation to either the mechanical characteristics or the visible microstructure dimensions. Moreover, it is evident that the values of *RSm* at the bottom lines are burdened with large uncertainty. When material is cut with a traverse speed that ensures just a medium quality, as was the case in our experiments, the bottom lines show considerable waviness rising from both pressure fluctuations and a decrease of jet energy along the sample profile. This leads to a distortion of *RSm* values that is so extensive that on the bottom lines they do not have any informative value.

Having compared the results of the vertical roughness parameters of all three grades of steel, some interesting findings appeared. *Ra* reached its maximum value after pure quenching for two steels: carbon steel C45 and low-alloy steel 30CrV9 (Figures 3 and 11, Tables 5 and 9). The manganese-silicon steel exhibited a different behavior: the surface roughness dropped down significantly after pure quenching, although both the hardness and ultimate tensile strengths increased (Figure 7, Table 7). The quenched and tempered 37MnSi5 sample was cut with the lowest surface roughness (Figure 8, Table 7).

The changes in *Rz* are similar, with the range of changes being just a bit lower. This result could be anticipated; it corresponded to the fact that the samples were handled with care so that no additional scratches or damage were generated. Although *Rz* is preferred to *Ra* by the contact tester manufacturer, in the case of normal AWJ cutting *Ra* is usually more important.

Another interesting finding seems to be the difference of the roughness results for the carbon steel and the two remaining ones. The quenched and tempered carbon steel had worse surface parameters than both types of annealed carbon steel (Table 5); in the case of the steels with lower contents of carbon, it was the other way around (Tables 7 and 9). The Vickers hardness HV30 measured on carbon steel was lower, but the ultimate tensile strength was higher (Table 4, Figure 13). The reason for the worse AWJ machining may, therefore, rather correlate with the steel microstructure. The steel microstructure analysis revealed that the final microstructure of the carbon steel was formed mainly by bainite with a ferritic network along the grain boundaries (Figure 4). This microstructure is likely to contain more internal stresses [41], which tend to release during the cutting and cause a higher surface roughness after AWJ cutting.

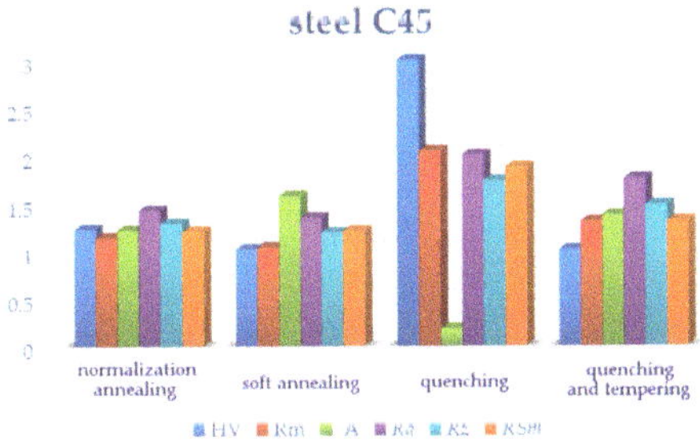

Figure 13. Relative changes of mechanical characteristics and measured surface roughness parameters for C45. All values were normalized, being divided by respective reference values (150 HV, 560 MPa, 16%—values for raw C45 steel [37]; *Ra, Rz, RSm* values for the upper line of the normalization annealing sample).

Quenched and high temperature tempered 37MnSi5 and 30CrV9 were both martensitic with a low proportion of bainite and ferrite (Figures 8 and 12); however, their elemental composition is different, and this leads to a different way of precipitating carbon. Both manganese and silicon tend to deposit in the solid phase and strengthen it. In contrast, both chromium and vanadium are elements that induce carbide formation in the material; therefore, the carbides are present in this steel grade, although the carbon content is lower than for manganese steel.

Both normalization and soft annealing led to a decrease of both the hardness and ultimate tensile strength; therefore, it is not surprising that *Ra* and *Rz* decreased as well (Figures 14 and 15). The microstructure of individual steel grades was identified in comparison with similar grades published in the literature, such as [42].

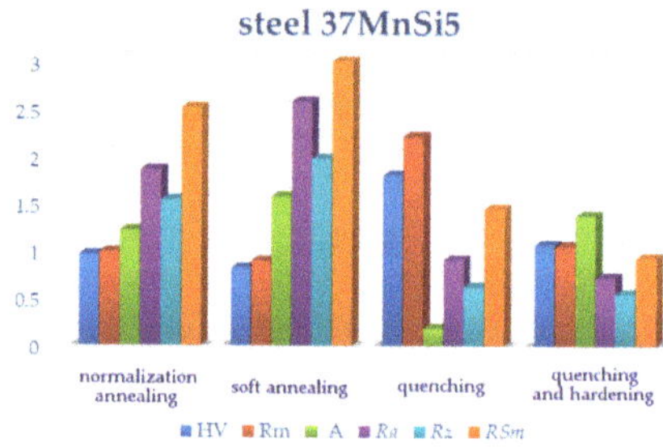

Figure 14. Relative changes of mechanical characteristics and measured surface roughness parameters for 37MnSi5. All values were normalized, being divided by respective reference values (170 HV, 880 MPa, 19%; *Ra*, *Rz*, *RSm* values for the upper line of the normalization annealing sample).

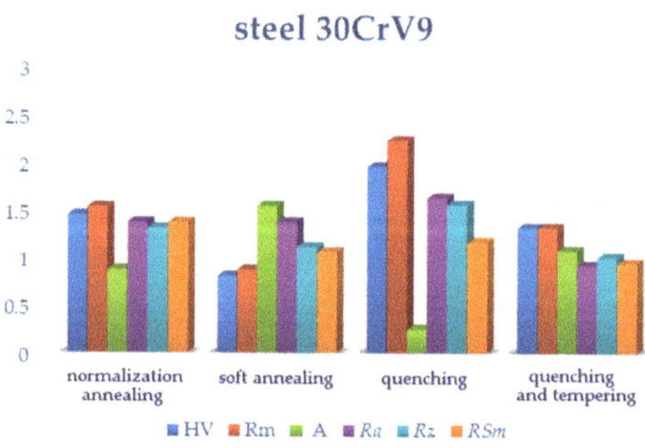

Figure 15. Relative changes of mechanical characteristics and measured surface roughness parameters for 30CrV9. All values were normalized, being divided by respective reference values (150 HV, 790 MPa, 15%; *Ra*, *Rz*, *RSm* values for the upper line of the normalization annealing sample).

The abovementioned findings can be summarized, specifying the individual behavior of the examined steel grades subjected to four types of heat treatment:

(1) C45 steel exhibits the effect of a higher carbon content, which increases the hardness and yield strength R_m. The roughness values also increase for this steel grade, due to the presence of different structural components; the inhomogeneous microstructure after normalization annealing consists of ferrite (F)—soft phase and perlite (P)—harder phase. After hardening, there are three structural components with different hardness in the microstructure: martensite (M)—the hardest, P—softer, F—the softest. Two main phases (tempered M and B, softer than M but harder than P) appear again after hardening and tempering.

(2) 37MnSi5 steel contains less carbon; moreover, the presence of Mn and Si strengthen the solid solution. The roughness values are lower than for C45 steel. After normalization and soft annealing, two structural components occur in the microstructure: ferrite and perlite; meanwhile, after quenching and quenching and tempering, the microstructure is quite homogeneous, formed by M and B with a small proportion of ferrite, and therefore the roughness values are low when compared to those of carbon steel, although the strength limits are relatively higher than for C45 steel.

(3) Although 30CrV9 steel has a low carbon content (see Table 1), the roughness values are relatively high, probably due to the content of Cr and V-based carbides, which precipitate the matrix in the state after hardening and hardening and tempering. In the normalized state and the state after soft annealing, the influence of the inhomogeneous microstructure is again manifested, which will have an effect on the impact of abrasive particles on the individual structural components, and thus on the increase of the roughness value.

5. Conclusions

This paper presents findings of a new research topic that represents an interdisciplinary theme requiring the cooperation of researchers from several fields. They can be summarized as follows:

1. Young's modulus is not significantly responsible for the surface roughness of steel materials cut with an abrasive water jet.
2. Pure quenching lowers the roughness of the AWJ cut if it is accompanied with the substantial growth of the ultimate tensile strength and homogeneous steel microstructure.
3. The carbon contents in steel appear to be the primary factor influencing the behavior of the steel grade in terms of AWJ machinability.
4. The homogeneity of the steel microstructure is another important factor. The higher the difference in hardness of structural components (martensite, bainite, lamellar pearlite, spheroidized pearlite, ferrite) in the inhomogeneous microstructure, the higher the roughness values.

In our future research, these findings will be verified on other steel grades, and the correlation between the surface roughness parameters and the declination angle will be studied. Measurements with various traverse speeds and sample thicknesses will enable us to study in detail the mechanism of AWJ interaction with material and to derive cutting parameter optimization proposals for heat-treated steels.

Author Contributions: Conceptualization, I.M.H.; methodology, M.S. and P.V.; validation, M.S. and M.T.; formal analysis, I.M.H.; investigation, P.V. and Š.S.; data curation, I.M.H., Š.S. and M.T.; writing—original draft preparation, I.M.H.; writing—review and editing, I.M.H., M.S. and P.V. All authors have read and agreed to the published version of the manuscript.

Funding: This research was funded by Ministry of Education, Youths and Sports of the Czech Republic through projects Student Grant Competition SP2020/39, SP 2020/45 and SP2020/58.

Conflicts of Interest: The authors declare no conflict of interest.

References

1. Hashish, M. AWJ Milling of Gamma Titanium Aluminide. *J. Manuf. Sci. Eng.-Trans. ASME* **2010**, *132*, 041005. [CrossRef]
2. Yuvaraj, N.; Pavithra, E.; Shamli, C.S. Investigation of Surface Morphology and Topography Features on Abrasive Water Jet Milled Surface Pattern of SS 304. *J. Test. Eval.* **2020**, *48*, 2981–2997. [CrossRef]
3. Thakur, R.K.; Singh, K.K. Abrasive waterjet machining of fiber-reinforced composites: A state-of-the-art review. *J. Braz. Soc. Mech. Sci. Eng.* **2020**, *42*, 381. [CrossRef]
4. Liu, S.Y.; Zhou, F.Y.; Li, H.S.; Chen, Y.Q.; Wang, F.C.; Guo, C.W. Experimental Investigation of Hard Rock Breaking Using a Conical Pick Assisted by Abrasive Water Jet. *Rock Mech. Rock Eng.* **2020**, *53*, 4221–4230. [CrossRef]

5. Pelit, H.; Yaman, O. Influence of Processing Parameters on the Surface Roughness of Solid Wood Cut by Abrasive Water Jet. *BioResources* **2020**, *15*, 6135–6148. [CrossRef]
6. Pahuja, R.; Ramulu, M.; Hashish, M. Surface quality and kerf width prediction in abrasive water jet machining of metal-composite stacks. *Compos. Part B-Eng.* **2019**, *175*, UNSP 107134. [CrossRef]
7. Ipar, C.E.D.E.L.; Neis, P.D.; Ferreira, N.F.; Lasch, G.; Zibetti, T.F. Analysis of the initial damage region in agate plates cut by abrasive waterjet (AWJ) process. *Int. J. Adv. Manuf. Technol.* **2020**, *109*, 2629–2638. [CrossRef]
8. Modica, F.; Basile, V.; Viganò, F.; Arleo, F.; Annoni, M.; Fassi, I. Micro-Abrasive Water Jet and Micro-WEDM Process Chain Assessment for Fabricating Microcomponents. *J. Micro Nano-Manuf.* **2019**, *7*, 010903. [CrossRef]
9. Hashish, M. A Modeling Study of Metal Cutting with Abrasive Waterjets. *J. Eng. Mater. Technol.-Trans. ASME* **1984**, *106*, 88–100. [CrossRef]
10. Hashish, M. A Model for Abrasive-Waterjet (AWJ) Machining. *J. Eng. Mater. Technol.-Trans. ASME* **1989**, *111*, 154–162. [CrossRef]
11. Zeng, J.; Kim, T.J. An Erosion Model of Polycrystalline Ceramics in Abrasive Waterjet Cutting. *Wear* **1996**, *193*, 207–217. [CrossRef]
12. Che, C.L.; Huang, C.Z.; Wang, J.; Zhu, H.T.; Li, Q.L. Theoretical Model of Surface Roughness for Polishing Super Hard Materials with Abrasive Waterjet. *Key Eng. Mater.* **2008**, *375–376*, 465–469. [CrossRef]
13. Vikram, G.; Babu, N.R. Modelling and Analysis of Abrasive Water Jet Cut Surface Topography. *Int. J. Mach. Tools Manuf.* **2002**, *42*, 1345–1354. [CrossRef]
14. Deam, R.T.; Lemma, E.M.; Ahmed, D.H. Modelling of the Abrasive Water Jet Cutting Process. *Wear* **2004**, *257*, 877–891. [CrossRef]
15. Bitter, J.G.A. A Study of Erosion Phenomena Part I. *Wear* **1963**, *7*, 5–21. [CrossRef]
16. Bitter, J.G.A. A Study of Erosion Phenomena Part II. *Wear* **1963**, *7*, 169–190. [CrossRef]
17. Liu, X.C.; Liang, Z.W.; Wen, G.L.; Yuan, X.F. Waterjet machining and research developments: A review. *Int. J. Adv. Manuf. Technol.* **2019**, *102*, 1257–1335. [CrossRef]
18. Arola, D.; Ramulu, M. Material Removal in Abrasive Waterjet Machining of Metals—Surface Integrity and Texture. *Wear* **1997**, *210*, 50–58. [CrossRef]
19. Chen, F.L.; Siores, E.; Patel, K. Improving the Cut Surface Qualities Using Different Controlled Nozzle Oscillation Techniques. *Int. J. Mach. Tools Manuf.* **2002**, *42*, 717–722. [CrossRef]
20. Hascalik, A.; Caydas, U.; Gurun, H. Effect of traverse speed on abrasive waterjet machining of Ti-6Al-4V alloy. *Mater. Des.* **2007**, *28*, 465–469. [CrossRef]
21. Hlaváč, L.M.; Strnadel, B.; Kaličinský, J.; Gembalová, L. The model of product distortion in AWJ cutting. *Int. J. Adv. Manuf. Technol.* **2012**, *62*, 1257–1335. [CrossRef]
22. Monno, M.; Pellegrini, G.; Ravasio, C. Effect of workpiece heat treatment on surface quality of AWJ kerf. *Int. J. Mater. Prod. Technol.* **2015**, *51*, 345–358. [CrossRef]
23. Kong, M.C.; Srinivasu, D.; Axinte, D.; Voice, W.; McGourlay, J.; Hon, B. On geometrical accuracy and integrity of surfaces in multi-mode abrasive waterjet machining of NiTi shape memory alloys. *CIRP Ann.-Manuf. Technol.* **2013**, *62*, 555–558. [CrossRef]
24. Xu, D.D.; Liao, Z.R.; Axinte, D.; Hardy, M. A novel method to continuously map the surface integrity and cutting mechanism transition in various cutting conditions. *Int. J. Mach. Tools Manuf.* **2020**, *151*, 103529. [CrossRef]
25. Kunčická, L.; Macháčková, A.; Lavery, N.P.; Kocich, R.; Cullen, J.C.T.; Hlaváč, L.M. Effect of thermomechanical processing via rotary swaging on properties and residual stress within tungsten heavy alloy. *Int. J. Refract. Met. Hard Mater.* **2020**, *87*, 105120. [CrossRef]
26. Strnadel, B.; Hlaváč, L.M.; Gembalová, L. Effect of steel structure on the declination angle in AWJ cutting. *Int. J. Mach. Tools Manuf.* **2013**, *64*, 12–19. [CrossRef]
27. Hlaváč, L.M. Investigation of the abrasive water jet trajectory curvature inside the kerf. *J. Mater. Process. Technol.* **2009**, *209*, 4154–4161. [CrossRef]
28. JKZ Bučovice. Available online: https://www.jkz.cz/en/products/structural-steel/csn-12-050-11191-c45/ (accessed on 4 September 2020).
29. Ferona Online Materiálové Normy (in Czech). Available online: https://online.ferona.cz/materialove-normy/ (accessed on 4 September 2020).
30. ARET STEEL Hutní Materiál. Available online: http://www.aretsteel.com/en/steel-grades (accessed on 7 September 2020).

31. GDOES Theory, SPECTRUMA Analytik GmbH. Available online: https://www.spectruma.de/en/gdoes-theory.html (accessed on 5 September 2020).
32. *Metallic Materials—Tensile Testing—Part 1: Test Method at Room Temperature;* ČSN EN ISO 6892-1 (420310); International Organization for Standardization: Geneva, Switzerland, 2019.
33. *Geometrical Product Specifications (GPS)—Surface Texture: Profile Method—Terms Definitions and Surface Texture Parameters;* ČSN EN ISO 4287; International Organization for Standardization: Geneva, Switzerland, 1997.
34. Mitutoyo Quick Guide to Surface Roughness Measurement. Reference Guide for Laboratory and Workshop. Available online: https://www.mitutoyo.com/wp-content/uploads/2012/11/1984_Surf_Roughness_PG.pdf (accessed on 9 September 2020).
35. Surface Roughness Measurement. Practical Tips for Laboratory and Workshop. p. 5. Available online: https://www.mitutoyo.com/wp-content/uploads/2015/04/Surface_Roughness_Measurement.pdf (accessed on 9 September 2020).
36. Björkeborn, K.; Klement, U.; Oskarson, H.B. Study of microstructural influences on machinability of case hardening steel. *Int. J. Adv. Manuf. Technol.* **2010**, *49*, 441–446. [CrossRef]
37. Blaoui, M.M.; Zemri, M.; Brahami, M. Effect of Heat Treatment Parameters on Mechanical Properties of Medium Carbon Steel. *Mech. Mech. Eng.* **2018**, *22*, 909–918. [CrossRef]
38. Korda, A.; Miyashita, Y.; Mutoh, Y.; Sadasue, T. Fatigue crack growth behavior in ferritic–pearlitic steels with networked and distributed pearlite structures. *Int. J. Fatigue* **2007**, *29*, 1140–1148. [CrossRef]
39. Bowen, L.; Qin, T.; Wei, X.; Chengchang, J.; Qiuchi, W.; Mingying, C.; Zhe, L. Effect of Tempering Conditions on Secondary Hardening of Carbides and Retained Austenite in Spray-Formed M42 High-Speed Steel. *Materials* **2019**, *12*, 3714. [CrossRef]
40. Ali, Y.M.; Wang, J. Impact Abrasive Machining. In *Machining with Abrasives*, 2nd ed.; Jackson, M.J., Davim, J.P., Eds.; Springer Science+Business Media: Cham, Switzerland, 2011; pp. 385–419. [CrossRef]
41. Holmberg, J.; Steuwer, A.; Stormvinter, A.; Kristoffersen, H.; Haakanen, M.; Berglund, J. Residual stress state in an induction hardened steel bar determined by synchrotron—And neutron diffraction compared to results from lab-XRD. *Mater. Sci. Eng. A* **2016**, *667*, 199–207. [CrossRef]
42. Zhou, T.P.; Wang, C.Y.; Wang, C.; Cao, W.Q.; Chen, Z.J. Strong Interactions between Austenite and the Matrix of Medium-Mn Steel during Intercritical Annealing. *Materials* **2020**, *13*, 3366. [CrossRef] [PubMed]

Article

Effects of Sintering Conditions on Structures and Properties of Sintered Tungsten Heavy Alloy

Lenka Kunčická [1,2,*], Radim Kocich [2] and Zuzana Klečková [2]

1 Institute of Physics of Materials, CAS, 61662 Brno, Czech Republic
2 Faculty of Materials Science and Technology, VŠB-Technical University of Ostrava, 70833 Ostrava 8, Czech Republic; radim.kocich@vsb.cz (R.K.); zuzana.kleckova@vsb.cz (Z.K.)
* Correspondence: kuncicka@ipm.cz; Tel.: +420-532-290-471

Received: 27 April 2020; Accepted: 18 May 2020; Published: 19 May 2020

Abstract: Probably the most advantageous fabrication technology of tungsten heavy alloys enabling the achievement of required performance combines methods of powder metallurgy and processing by intensive plastic deformation. Since the selected processing conditions applied for each individual processing step affect the final structures and properties of the alloys, their optimization is of the utmost importance. This study deals with thorough investigations of the effects of sintering temperature, sintering time, and subsequent quenching in water on the structures and mechanical properties of a 93W6Ni1Co tungsten heavy alloy. The results showed that sintering at temperatures of or above 1525 °C leads to formation of structures featuring W agglomerates surrounded by the NiCo matrix. The sintering time has non-negligible effects on the microhardness of the sintered samples as it affects the diffusion and structure softening phenomena. Implementation of quenching to the processing technology results in excellent plasticity of the green sintered and quenched pieces of almost 20%, while maintaining the strength of more than 1000 MPa.

Keywords: tungsten heavy alloy; powder metallurgy; sintering; quenching; microstructure

1. Introduction

Given by the requirements on high density, high strength, and favorable toughness and ductility, tungsten heavy alloys (THAs) are advantageously used for demanding applications, such as for radiation shielding, space industry components, therapeutic devices in oncology, aircraft counterbalances, or kinetic penetrators [1–3]. THAs can be fabricated via modern technologies, such as spark plasma sintering (SPS), or selective laser melting (SLM) [4–8]. However, the processing parameters, which can be varied during such procedures, are limited. Probably the most advantageous THA fabrication technology enabling the achievement of required performance of the final product combines methods of powder metallurgy and processing by intensive plastic deformation, which can preferably be performed by severe plastic deformation (SPD) methods [9–14] imparting significant grain refinement, down to the ultra-fine scale (0.1–1.0 μm, e.g., equal channel angular pressing—ECAP [15]), or even to nano-scale (≤100 nm, e.g., high pressure torsion—HPT [16]), resulting in the enhancement of mechanical and utility properties via imposing high shear strain.

THAs typically consist of 90 to 97 wt. % of tungsten plus a mixture of other relatively low melting elements, such as Fe, Ni, Co, and Cu, the combination and volume fractions of which significantly influence the strength and plastic properties of the final product [17–20]. The tungsten content can be increased up to 98 wt. % to improve the strength, however, the strength usually increases at the expense of ductility; the content of matrix-forming elements lower than 3 wt. % usually supports brittleness of the final product [21]. On the other hand, increasing the addition of alloying, i.e., matrix-forming, elements generally decrease strength, but increase ductility [22]; high contents of matrix-forming elements contribute to uneven shape of the cross-section of the sintered piece due to gravity sedimentation during sintering, and consequently to non-uniformity of mechanical properties.

THAs are mostly consolidated from initial powder mixtures through powder metallurgy techniques; preferably via liquid phase sintering (LPS) [23], the boundaries of the grains during which melt and ensure binding of the tungsten particles and elimination of porosity via diffusion [24]. LPS offers the advantages of a relatively low processing temperature, favorable densification and structure homogenization, high productivity, and minimum production waste (<5%) [25]. Under optimized sintering conditions, the W and NiCo phases are homogenously distributed and the sintered piece contains no visible pores [7]. Sintering is typically performed at temperatures between 1000 and 1500 °C to ensure melting of the matrix-forming binding elements [22,26,27]. The green sintered THA work pieces typically feature high density (16–18 $g{\cdot}cm^{-3}$), and relatively high strength and plasticity. Nevertheless, besides the chemical composition, the absolute values of the mechanical properties after sintering depend on the character of the sintered structure, and, last but not least, also on the subsequent processing steps (the mechanical properties of THAs can also be enhanced by post-sintering deformation processing).

The presented work deals with the effects of sintering conditions, sintering temperature, and time in particular, and possible quenching on the structure and basic mechanical properties of the studied WNiCo tungsten heavy alloy. The primary goal is to optimize the sintering procedure in order to provide the best possible starting conditions for subsequent deformation processing.

2. Materials and Methods

The investigated 93W6Ni1Co tungsten heavy alloy samples were prepared at ÚJP Praha a.s. company (Praha, Czech Republic) from a homogeneous mixture of individual powders prepared by mechanical alloying (impurities ~<13 ppm of Fe, Cr, Mo, Al, and Ca). The granulometric distribution of the powder particles within the mixture was between 2 and 4 μm, and the mean particle size was 2.78 μm. The scanning electron image of the powder is depicted in Figure 1. After powders mixing, the mixture was cold isostatically pressed at 400 MPa, and subsequently prepared in the following steps:

(1) sintering under a protective atmosphere (hydrogen) at temperatures varying between 1450 and 1550 °C, with the step of 25 °C, for 30 min each (the temperature range was suggested by the ÚJP Praha a.s. company based on their previous experience);

(2) optional quenching in water-a set of samples was quenched after the mentioned sintering procedures in order to observe possible changes in structures and properties;

(3) based on the acquired data, additional experiments with variable sintering time were performed-sintering for 60 to 180 min, with the step of 30 min, at the sintering temperature of 1500 °C, and sintering for 180 min at the sintering temperature of 1525 °C.

Figure 1. SEM image of original powder mixture.

The following structure analyses performed on cross-sectional cuts of the sintered (and possibly quenched) pieces primarily focused on structure character and possible changes in distribution of the individual elements within the structure. The transversally cut cross-sectional samples were ground mechanically and polished vibrationally using a colloidal silica suspension (VibroMet 2 Vibratory Polisher, Buehler, Esslingen, Germany) and subjected either to optical microscopy using an Olympus DSX1000 digital microscope (Olympus Czech Group, s.r.o., Prague, Czech Republic), or to scanning electron microscopy (SEM) using a Tescan Lyra 3 device (TESCAN Brno s.r.o, Brno, Czech Republic). TEM images were acquired on ion polished thin foils using a JEOL 2100F (JEOL, Akishima, Tokio prefecture, Japan) device. Porosity of the prepared samples was detected by an image analysis using the ImageJ software; for each sample, four individual random scans the average porosity value from which was subsequently calculated were observed.

The microhardness was measured via the Vickers microhardness method with the load of 1 kg and loading time of 10 s using a Zwick-Roell machine (Zwick/Roell—Messphysik KAPPA LA, Zwick/Roell CZ s.r.o., Brno, Czech Republic). The average value for each sample was calculated from a line scan measured through the sintered sample cross-section with 0.5 mm spacing of the individual indentations. As the diameter of the sintered samples was approximately 12 mm (up to ±3 mm), the final average microhardness value was calculated from 13 indents for each sample. The tensile tests were performed on 150 mm long testing samples with circular cross-sections of the diameters of 10 mm at the strain rate of 10^{-3} using a Testometric M500-50CT testing machine (Testometric Co. Ltd., Rochdale, UK). In order to eliminate errors caused by possible inhomogeneities occurring during sintering, four tensile tests were performed for each sintering (and quenching) regime. The reported stress–strain curves in the relevant section then show the ones corresponding the most to the average values for each processing regime.

3. Results

3.1. Effect of Sintering Temperature on Structures

The results of the analyses investigating the effects of sintering temperature on the structures of the pieces sintered for 30 min revealed significant change in the structure character at the temperature of 1525 °C. Figure 2a shows the optical microscopy structure scan of the sintered-piece prepared at 1450 °C. The scan clearly shows structure inhomogeneity, as well as the presence of voids/gaps; the average porosity for this sample was 4.86%. The optical microscopy scan of structure of the piece sintered at 1475 °C is shown in Figure 2b depicting structure inhomogeneity comparable to the 1450 °C sintered piece. However, the presence of voids is significantly lower when compared to the 1450 °C sample (average porosity of 1.45%). SEM scan of the sample sintered at 1500 °C is shown in Figure 2c. As can be seen, this sintering regime leads to certain structure homogenization and almost complete elimination of porosity, which was as low as 0.23%. As depicted in the SEM scan in Figure 2d, the sintering temperature of 1525 °C results in the formation of the microstructure consisting of tungsten agglomerates and the NiCo matrix. Finally, Figure 2e shows the SEM scan of the structure of the 1550 °C sintered piece, the tungsten agglomerates surrounded by the NiCo matrix in which can clearly be seen, too. Both the 1525 and 1550 °C samples exhibited no residual porosity.

In order to examine the effects of the sintering temperature on chemical composition, i.e., distribution of the individual elements within the structures, SEM-EDX analyses were performed for selected sintered pieces. Figure 3a shows the chemical composition for the sample sintered at 1500 °C for 30 min depicted via distributions of the individual elements within the scanned area. Such distribution was typical for the samples sintered at the temperature of 1500 °C and lower. Figure 3c then shows quantitative evaluation of the individual elements within the scanned map. The chemical composition depicted via distributions of the individual elements within the sample sintered at 1525 °C for 30 min is shown in Figure 3b, corresponding quantitative evaluation of the individual elements within the scanned area of the sample is then depicted in Figure 3d. Such distribution of the individual elements was typical for the samples sintered at and above the temperature of 1525 °C. The maps in Figure 3b show that tungsten is primarily concentrated in the agglomerates surrounded by the NiCo matrix.

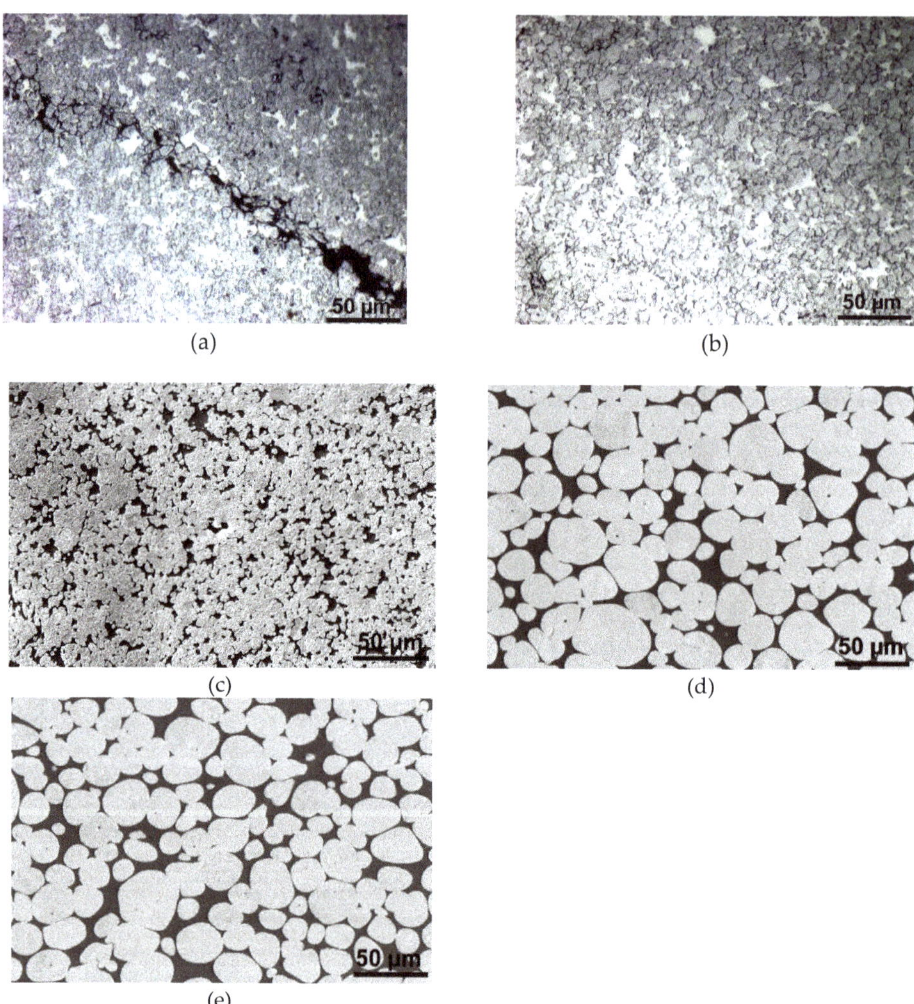

Figure 2. Structure scans of individual sintered pieces: 1450 °C/30 min (**a**); 1475 °C/30 min (**b**); 1500 °C/30 min (**c**); 1525 °C/30 min (**d**); 1550 °C/30 min (**e**). In OM images, white/light areas depict the NiCo matrix, grey areas depict tungsten, and black areas depict voids/cracks.

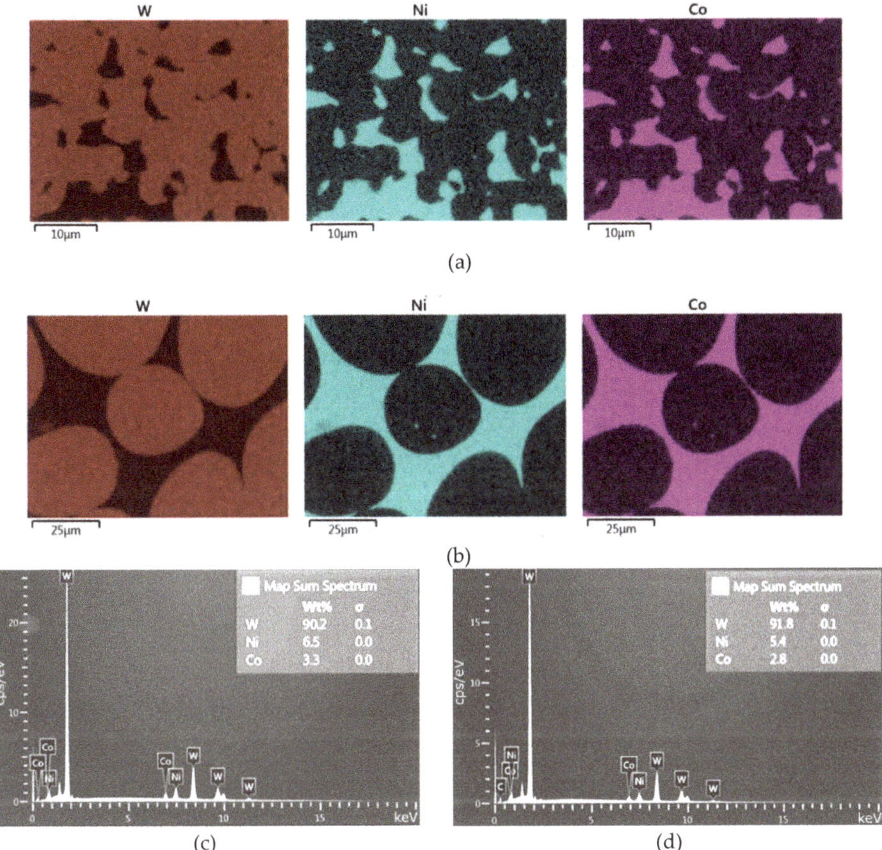

Figure 3. SEM-EDX maps of distribution of individual elements (W-red, Ni-blue, Co-violet) for sintered pieces: 1500 °C/30 min (**a**); 1525 °C/30 min (**b**); quantitative evaluation of individual elements within maps depicted in (**a**) and (**b**) in investigated structures, in wt. %: 1500 °C/30 min (**c**); 1525 °C/30 min (**d**).

3.2. Effect of Sintering Time on Structures

The subsequent research step was the investigation of the effects of variable sintering time at the critical sintering temperature of 1500 °C. The results of the analyses are depicted in Figure 4a–e, the SEM scans of samples sintered for 60, 90, 120, 150, and 180 min in which are depicted, respectively. Evidently, increasing the sintering time at the temperature of 1500 °C to at least 60 min leads to the formation of the microstructures consisting of tungsten agglomerates surrounded by the NiCo matrix.

Despite the fact that increasing the sintering time has favorable effects on the structure character when examined via SEM, more detailed TEM investigations showed that increasing the time dwell on the high temperature leads to inter-diffusion of tungsten into the NiCo matrix, as documented by Figure 5a,b. Figure 5a shows a TEM scan of a W-NiCo-W interface within the sample sintered at 1500 °C for 180 min, Figure 5b then shows the chemical composition (in wt. %) along a line scan measured across the mentioned interface. As can be seen, the content of tungsten increased to about 50 wt. % in the inter-agglomerate region by the effect of increased sintering time (negligible tungsten content within the NiCo matrix was observed for the sample sintered at 1500 °C for 60 min (not shown here)).

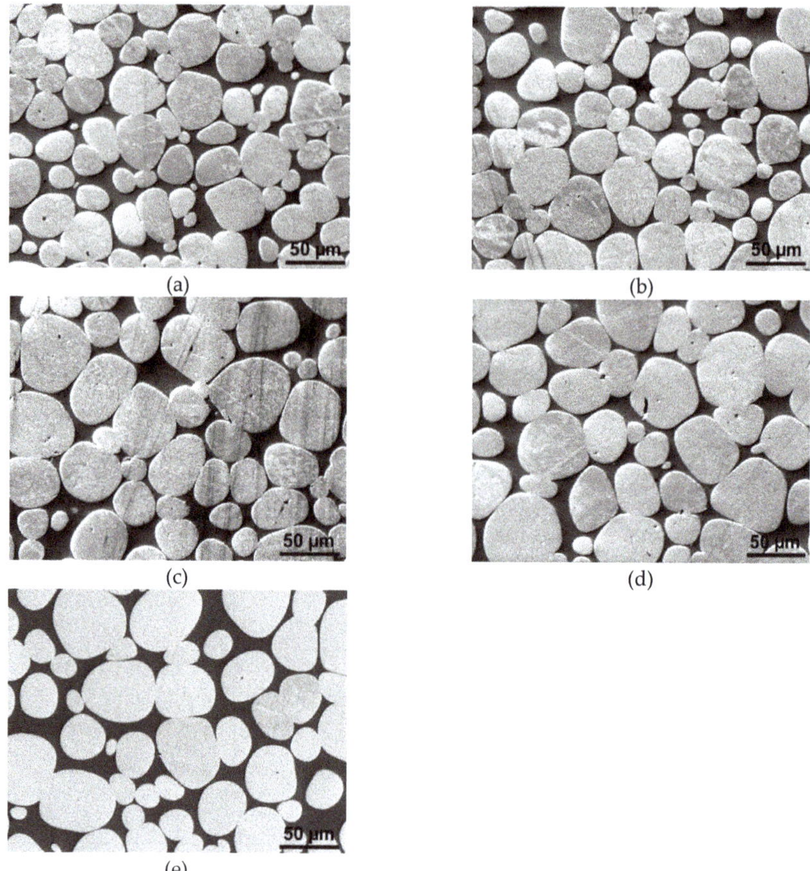

Figure 4. Structure scans of individual pieces sintered at 1500 °C for: 60 min (**a**); 90 min (**b**); 120 min (**c**); 150 min (**d**); 180 min (**e**).

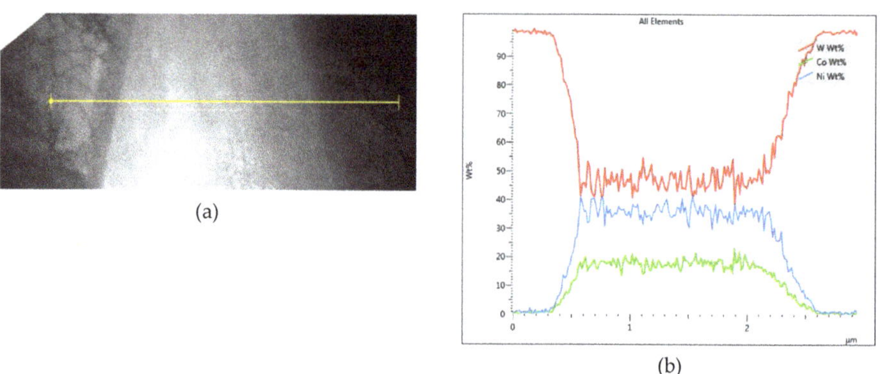

Figure 5. TEM scan showing a W-NiCo-W interface and measured line scan area for sample sintered at 1500 °C for 180 min (**a**); chemical composition of measured line scan (**b**).

3.3. *Effect of Quenching on Structures*

Investigation of the effects of quenching into room temperature water right after sintering was performed for all the samples sintered in the temperature range from 1450 to 1550 °C for 30 min, as well as the sample sintered at 1500 °C for 180 min. As shown by the structure analyses the results of which are depicted in Figure 6a–f, quenching does not have any significant effect on the structure character, as the quenched structures were similar to the non-quenched ones; quenched samples sintered at/above the temperature of 1525 °C for 30 min, and the sample sintered at 1500 °C for 180 min, exhibited the structure consisting of W agglomerates and the NiCo matrix, while the other analyzed samples featured more or less randomly distributed W/NiCo locations.

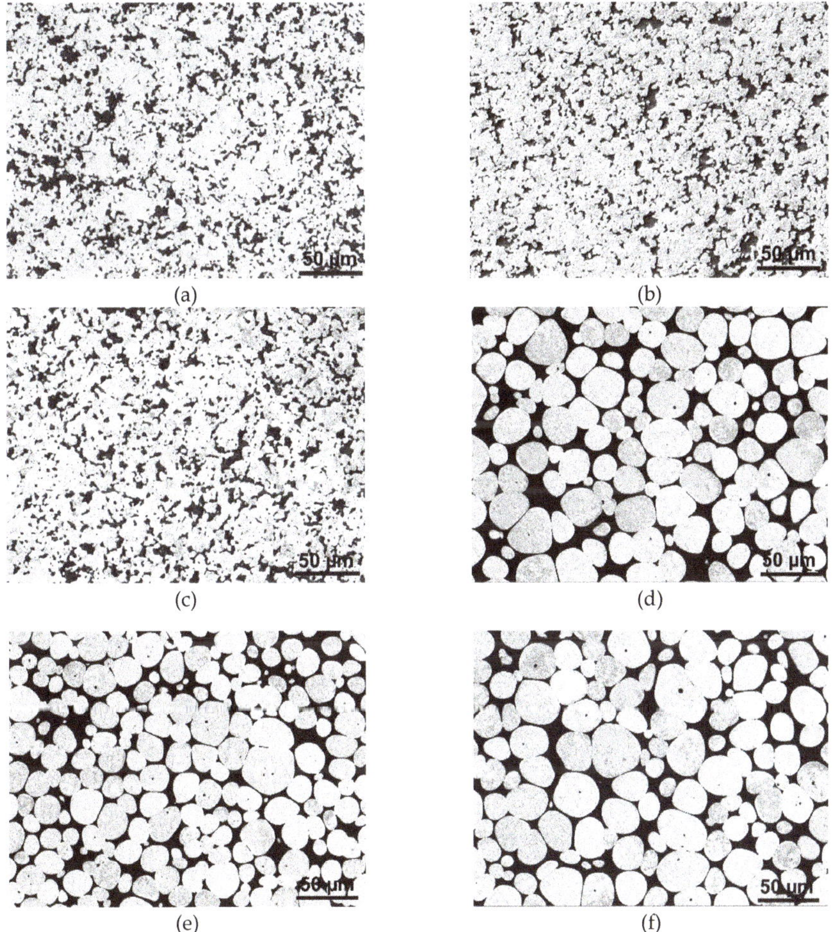

Figure 6. Structure scans of individual quenched pieces sintered at: 1450 °C/30 min (**a**); 1475 °C/30 min (**b**); 1500 °C/30 min (**c**); 1500 °C/180 min (**d**); 1525 °C/30 min (**e**); 1550 °C/30 min (**f**).

3.4. *Mechanical Properties*

Microhardness was measured for all the sintered pieces in order to analyze the effects of the sintering temperatures on their properties. The results of the measurements are summarized in Table 1. As regards to the samples sintered for 30 min, the highest average microhardness value was recorded for the sample sintered at 1550 °C. However, the highest microhardness value of all was measured for the sample sintered at 1500 °C for 180 min. This finding led to the performance of a supplementary experiment, in which a sample was sintered at the temperature of 1525 °C for 180 min. However, for the elevated temperature, the effect of increasing sintering time is not positive, as the microhardness decreased dramatically when compared to the sample sintered at 1500 °C for 180 min.

Table 1. Average microhardness for sintered pieces.

Sample Sintering Regime	Average Microhardness [HV]	Standard Deviation
1450 °C/30 min	378.54	14.86
1475 °C/30 min	389.75	13.40
1500 °C/30 min	376.61	10.54
1500 °C/180 min	483.99	5.16
1525 °C/30 min	377.97	9.67
1525 °C/180 min	337.56	7.64
1550 °C/30 min	392.83	11.42

Last but not least, the tensile tests were performed for the sintered and subsequently quenched pieces; the resulting stress–strain curves are depicted in Figure 7. Ass can be seen from the results, the samples sintered at/below 1500 °C for 30 min, featuring structures with randomly distributed W/NiCo locations and voids, exhibit very low plasticity (lower than 2.5%). The 1450 and 1757 °C samples exhibited the lowest plasticity of less than 1%. On the other hand, structures featuring the homogeneous distribution of tungsten agglomerates within a NiCo matrix exhibit a very high strength of more than 1000 MPa, together with excellent plasticity (elongation to failure exceeding 20%).

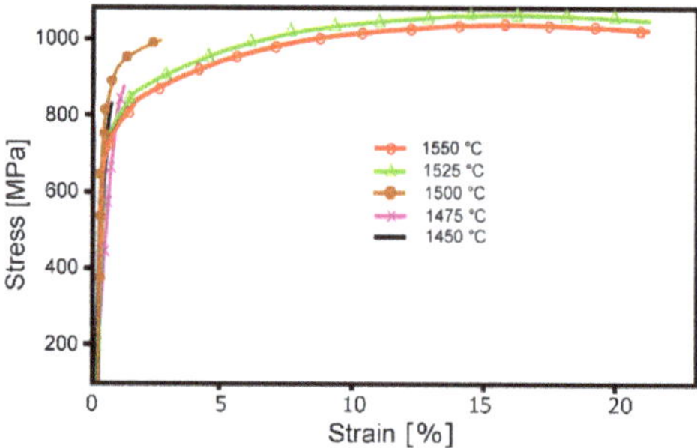

Figure 7. Stress–strain curves for sintered and subsequently quenched pieces.

4. Discussion

Microstructure observations showed that, during sintering at temperatures above the critical value of 1525 °C, the original powder particles consolidated and the structures exhibited the desired distribution of tungsten agglomerates with the sizes of several dozens of micrometres surrounded by a matrix primarily consisting of Ni and Co, whereas the samples sintered below this temperature featured more or less randomly distributed locations with high tungsten or nickel-cobalt contents. Moreover, the structures of pieces sintered below 1525 °C also featured voids and gaps. Sintering at low temperatures thus does not impart the desired structure homogenization and spheroidization of tungsten powder particles into agglomerates. In other words, sintering at temperatures at/above 1525 °C leads to local melting of the alloying elements, the melting temperatures of which are lower than the sintering one (1455 °C for nickel and 1495 °C for cobalt [28]), which supports the development of the desired structure character via diffusion. By this reason, especially the samples sintered below 1500 °C exhibited the presence of voids/gaps, as documented in Figure 2a,b.

Having performed the first part of the study, we found that the optimum sintering regime among the investigated ones, from the viewpoint of achievement of the required structure character, is the temperature of 1525 °C and time of 30 min. However, as the sintering temperature is definitely higher than the melting temperatures of both the alloying elements, the work pieces exhibit the tendency to gravity sedimentation during sintering, which increases the cross-sectional ovality. In other words, increasing the sintering temperature causes the shape of the cross-section of the sintered pieces to be oval, more than circular. By this reason, an additional experiment with increasing the sintering time at the sintering temperature of 1500 °C was performed.

Generally, increasing the sintering time was found to affect the mean size of the tungsten agglomerates—it increased with increasing sintering time [29], as well as to support inter-diffusion of the individual elements. Previous neutron powder diffraction measurements identified two individual phases in the sintered pieces featuring the required structure character, the main one of which was α-W featuring the B2 structure (W-B2 phase) [29]. The second phase, $NiCo_2W$, with the weight fraction of 6–7% had a pure Ni-like structure (FCC) with the lattice parameter of about 3.60 Å, which is, however, slightly larger than for pure nickel (3.55 Å). This fact indicated alloying of the matrix with larger atoms (atomic radius of W, Ni, and Co is 139, 124, and 125 pm, respectively), and thus revealed the presence of a certain amount of diffused W in the NiCo-based matrix. This phenomenon was explained by Chuvil'deev et al. [30], who documented that introducing distortions to the tungsten lattice, which can be performed by introducing energy via high sintering temperature and/or time, or via subsequent plastic deformation, leads to formation of strong fields of internal stress, which decreases the activation energy necessary for decoupling of tungsten atoms and supports their diffusion to the NiCo matrix. These phenomena are behind the increase in microhardness for the 1500 °C/180 min sintered piece. On the other hand, the 1525 °C/180 min sintered piece exhibited rapid decrease in microhardness when compared to the 1500 °C/180 min one. This can be explained via softening processes occurring within the matrix [17].

The subsequent quenching does not affect the structure character of the sintered pieces. However, it has positive effects on the mechanical properties, testing of which revealed excellent plasticity exceeding 20%, together with the ultimate tensile strength (UTS) of more than 1000 MPa for the pieces sintered at 1525 and 1550 °C for 30 min and quenched. The fact that the UTS of the 1550 °C sintered piece is slightly lower than of the 1525 °C one can be attributed to the occurring restoration phenomena supported by the increased sintering temperature (further temperature increase from 1525 °C introduces similar effects as increasing the sintering time at 1525 °C) [17,28].

5. Conclusions

The study focused on the investigations of the effects of sintering temperature, sintering time, and optional subsequent quenching in water on the structures and mechanical properties of a 93W6Ni1Co tungsten heavy alloy. The results revealed the following:

- 30-min-sintering at temperatures below 1525 °C does not provide sufficient homogenization as the structures feature more or less random locations with high W or NiCo concentrations, as well as voids and gaps. The microhardness is also low for such samples;
- sintering at or above 1525 °C leads to formation of homogeneous structures featuring W agglomerates surrounded by the NiCo matrix, which also increases the mechanical properties;
- sintering at 1500 °C for 30 min leads to inhomogeneous structure, but increasing the time to at least 60 min results in comparable effects as increasing the sintering temperature, however, increasing the sintering time provokes diffusion of W into the matrix which, on the other hand, increases the microhardness;
- implementation of quenching results in excellent mechanical properties of the quenched pieces sintered at/above 1525 °C—plasticity of almost 20% and strength of more than 1000 MPa.

The optimized preparation regime providing the 93W6Ni1Co alloy with structure and properties being the most favorable for subsequent processing is sintering at 1525 °C for 30 min and subsequent quenching in water.

Author Contributions: The contributions of the individual authors were the following: methodology, R.K., L.K. and Z.K.; experimental validation, R.K., L.K.; microscopy investigation and evaluation, L.K.; writing-original draft preparation, L.K., R.K.; writing—review and editing, L.K.; project administration, Z.K.; funding acquisition, R.K., Z.K. All authors have read and agreed to the published version of the manuscript.

Funding: The authors acknowledge the financial support of the FV40286 TRIO programme by ministry of Industry and Trade, and the 19-15479S Project of the Grant Agency of the Czech Republic.

Acknowledgments: The experiments were performed using equipment available at the Technical University of Ostrava. We also acknowledge the help of IPMinfra research infrastructure and working team of Institute of Physics of Materials, CAS.

Conflicts of Interest: The authors declare no conflict of interest. The funders had no role in the design of the study; in the collection, analyses, or interpretation of data; in the writing of the manuscript, or in the decision to publish the results.

References

1. Liu, J.; Li, S.; Zhou, X.; Wang, Y.; Yang, J. Dynamic recrystallization in the shear bands of tungsten heavy alloy processed by hot-hydrostatic extrusion and hot torsion. *Rare Met. Mater. Eng.* **2011**, *40*, 957–960. [CrossRef]
2. Guo, W.; Liu, J.; Yang, J.; Li, S. Effect of initial temperature on dynamic recrystallization of tungsten and matrix within adiabatic shear band of tungsten heavy alloy. *Mater. Sci. Eng. A* **2011**, *528*, 6248–6252. [CrossRef]
3. Kunčická, L.; Kocich, R.; Hervoches, C.; Macháčková, A. Study of structure and residual stresses in cold rotary swaged tungsten heavy alloy. *Mater. Sci. Eng. A* **2017**, *704*, 25–31. [CrossRef]
4. Senthilnathan, N.; Annamalai, A.R.; Venkatachalam, G. Microstructure and mechanical properties of spark plasma sintered tungsten heavy alloys. *Mater. Sci. Eng. A* **2018**, *710*, 66–73. [CrossRef]
5. Iveković, A.; Montero Sistiaga, M.L.; Vanmeensel, K.; Kruth, J.-P.; Vleugels, J. Effect of processing parameters on microstructure and properties of tungsten heavy alloys fabricated by SLM. *Int. J. Refract. Met. Hard Mater.* **2019**, *82*, 23–30. [CrossRef]
6. Tkachenko, S.; Cizek, J.; Mušálek, R.; Dvořák, K.; Spotz, Z.; Montufar, E.B.; Chráska, T.; Křupka, I.; Čelko, L. Metal matrix to ceramic matrix transition via feedstock processing of SPS titanium composites alloyed with high silicone content. *J. Alloys Compd.* **2018**, *764*, 776–788. [CrossRef]
7. Montufar, E.B.; Horynová, M.; Casas-Luna, M.; Diaz-de-la-Torre, S.; Celko, L.; Klakurková, L.; Spotz, Z.; Diéguez-Trejo, G.; Fohlerová, Z.; Dvořák, K.; et al. Spark plasma sintering of load-bearing iron–carbon nanotube-tricalcium phosphate CerMets for orthopaedic applications. *JOM* **2016**, *68*, 1134–1142. [CrossRef]
8. Ročňáková, I.; Slámečka, K.; Montufar, E.B.; Remešová, M.; Dyčková, L.; Břínek, A.; Jech, D.; Dvořák, K.; Čelko, L.; Kaiser, J. Deposition of hydroxyapatite and tricalcium phosphate coatings by suspension plasma spraying: Effects of torch speed. *J. Eur. Ceram. Soc.* **2018**, *38*, 5489–5496. [CrossRef]
9. Valiev, R.; Islamgaliev, R.; Alexandrov, I. Bulk nanostructured materials from severe plastic deformation. *Prog. Mater. Sci.* **2000**, *45*, 103–189. [CrossRef]
10. Kunčická, L.; Kocich, R.; Drápala, J.; Andreyachshenko, V.A. FEM simulations and comparison of the ecap and ECAP-PBP influence on Ti6Al4V alloy's deformation behaviour. In Proceedings of the 22nd International Conference on Metallurgy and Materials, Brno, Czech, 15–17 May 2013; pp. 391–396.
11. Thomasová, M.; Seiner, H.; Sedlák, P.; Frost, M.; Ševčík, M.; Szurman, I.; Kocich, R.; Drahokoupil, J.; Šittner, P.; Landa, M. Evolution of macroscopic elastic moduli of martensitic polycrystalline NiTi and NiTiCu shape memory alloys with pseudoplastic straining. *ACTA Mater.* **2017**, *123*, 146–156. [CrossRef]
12. Kocich, R.; Kunčická, L.; Macháčková, A. Twist Channel Multi-Angular Pressing (TCMAP) as a method for increasing the efficiency of SPD. *IOP Conf. Ser. Mater. Sci. Eng.* **2014**, *63*, 012006. [CrossRef]
13. Kocich, R.; Kunčická, L.; Král, P.; Strunz, P. Characterization of innovative rotary swaged Cu-Al clad composite wire conductors. *Mater. Des.* **2018**, *160*, 828–835. [CrossRef]
14. Kunčická, L.; Kocich, R.; Král, P.; Pohludka, M.; Marek, M. Effect of strain path on severely deformed aluminium. *Mater. Lett.* **2016**, *180*, 280–283. [CrossRef]

15. Hlaváč, L.M.; Kocich, R.; Gembalová, L.; Jonšta, P.; Hlaváčová, I.M. AWJ cutting of copper processed by ECAP. *Int. J. Adv. Manuf. Technol.* **2016**, *86*, 885–894. [CrossRef]
16. Kunčická, L.; Lowe, T.C.; Davis, C.F.; Kocich, R.; Pohludka, M. Synthesis of an Al/Al2O3 composite by severe plastic deformation. *Mater. Sci. Eng. A* **2015**, *646*, 234–241. [CrossRef]
17. Kunčická, L.; Macháčková, A.; Lavery, N.P.; Kocich, R.; Cullen, J.C.T.; Hlaváč, L.M. Effect of thermomechanical processing via rotary swaging on properties and residual stress within tungsten heavy alloy. *Int. J. Refract. Met. Hard Mater.* **2020**, *87*, 105120. [CrossRef]
18. Yu, Y.; Zhang, W.; Chen, Y.; Wang, E. Effect of swaging on microstructure and mechanical properties of liquid-phase sintered 93W-4.9(Ni, Co)-2.1Fe alloy. *Int. J. Refract. Met. Hard Mater.* **2014**, *44*, 103–108. [CrossRef]
19. Wu, Y.; German, R.M.; Marx, B.; Bollina, R.; Bell, M. Characteristics of densification and distortion of Ni–Cu liquid-phase sintered tungsten heavy alloy. *Mater. Sci. Eng. A* **2003**, *344*, 158–167. [CrossRef]
20. Shen, J.; Campbell, L.; Suri, P.; German, R.M. Quantitative microstructure analysis of tungsten heavy alloys (W-Ni-Cu) during initial stage liquid phase sintering. *Int. J. Refract. Met. Hard Mater.* **2005**, *23*, 99–108. [CrossRef]
21. Ravi Kiran, U.; Panchal, A.; Sankaranarayana, M.; Nageswara Rao, G.V.S.; Nandy, T.K. Effect of alloying addition and microstructural parameters on mechanical properties of 93{%} tungsten heavy alloys. *Mater. Sci. Eng. A* **2015**, *640*, 82–90. [CrossRef]
22. Ryu, H.J.; Hong, S.H.; Lee, S.; Baek, W.H. Microstructural control of and mechanical properties of mechanically alloyed tungsten heavy alloys. *Met. Mater.* **1999**, *5*, 185–191. [CrossRef]
23. Jiao, Z.; Kang, R.; Dong, Z.; Guo, J. Microstructure characterization of W-Ni-Fe heavy alloys with optimized metallographic preparation method. *Int. J. Refract. Met. Hard Mater.* **2019**, *80*, 114–122. [CrossRef]
24. Bose, A.; Schuh, C.A.; Tobia, J.C.; Tuncer, N.; Mykulowycz, N.M.; Preston, A.; Barbati, A.C.; Kernan, B.; Gibson, M.A.; Krause, D.; et al. Traditional and additive manufacturing of a new Tungsten heavy alloy alternative. *Int. J. Refract. Met. Hard Mater.* **2018**, *73*, 22–28. [CrossRef]
25. Upadhyaya, A. Processing strategy for consolidating tungsten heavy alloys for ordnance applications. *Mater. Chem. Phys.* **2001**, *67*, 101–110. [CrossRef]
26. Kocich, R.; Kunčická, L.; Dohnalík, D.; Macháčková, A.; Šofer, M. Cold rotary swaging of a tungsten heavy alloy: Numerical and experimental investigations. *Int. J. Refract. Met. Hard Mater.* **2016**, *61*, 264–272. [CrossRef]
27. Ryu, H.J.; Hong, S.H.; Kim, D.-K.; Lee, S. Impact and dynamic deformation behavior of mechanically alloyed tungsten heavy alloy. *Met. Mater.* **1998**, *4*, 367–371.
28. Russell, A.; Lee, K.L. *Structure-Property Relations in Nonferrous Metals*, 1st ed.; John Wiley & Sons, Inc.: Hoboken, NJ, USA, 2005; ISBN 978-0-471-64952-6.
29. Strunz, P.; Kunčická, L.; Beran, P.; Kocich, R.; Hervoches, C. Correlating microstrain and activated slip systems with mechanical properties within rotary swaged WNiCo pseudoalloy. *Materials* **2020**, *13*, 208. [CrossRef]
30. Chuvil'deev, V.N.; Nokhrin, A.V.; Boldin, M.S.; Baranov, G.V.; Sakharov, N.V.; Belov, V.Y.; Lantsev, E.A.; Popov, A.A.; Melekhin, N.V.; Lopatin, Y.G.; et al. Impact of mechanical activation on sintering kinetics and mechanical properties of ultrafine-grained 95W-Ni-Fe tungsten heavy alloys. *J. Alloys Compd.* **2019**, *773*, 666–688. [CrossRef]

Article

Affecting Structure Characteristics of Rotary Swaged Tungsten Heavy Alloy Via Variable Deformation Temperature

Adéla Macháčková, Ludmila Krátká *, Rudolf Petrmichl, Lenka Kunčická and Radim Kocich *

Faculty of Materials Science and Technology, VŠB–Technical University of Ostrava, 17. listopadu 15, 708 00 Ostrava, Czech Republic; adela.machackova@vsb.cz (A.M.); rudolf.petrmichl@vsb.cz (R.P.); lenka.kuncicka@vsb.cz (L.K.)

* Correspondence: ludmila.kratka@vsb.cz (L.K.); radim.kocich@vsb.cz (R.K.); Tel.: +420-596-99-44-33 (L.K.)

Received: 2 November 2019; Accepted: 12 December 2019; Published: 13 December 2019

Abstract: This study focuses on numerical prediction and experimental investigation of deformation behaviour of a tungsten heavy alloy prepared via powder metallurgy and subsequent cold (20 °C) and warm (900 °C) rotary swaging. Special emphasis was placed on the prediction of the effects of the applied induction heating. As shown by the results, the predicted material behaviour was in good correlation with the real experiment. The differences in the plastic flow during cold and warm swaging imparted differences in structural development and the occurrence of residual stress. Both the swaged pieces exhibited the presence of residual stress in the peripheries of W agglomerates. However, the NiCO matrix of the warm-swaged piece also exhibited the presence of residual stress, and it also featured regions with increased W content. Testing of mechanical properties revealed the ultimate tensile strength of the swaged pieces to be approximately twice as high as of the sintered piece (860 MPa compared to 1650 MPa and 1828 MPa after warm and cold swaging, respectively).

Keywords: tungsten heavy alloy; rotary swaging; finite element analysis; deformation behaviour; residual stress

---

1. Introduction

For their exceptional mechanical (high strength) and physical (high density, melting point, etc.) properties, tungsten heavy alloys (THAs) are primarily used to shield radiation or to block kinetic energy [1,2]. Nevertheless, they can also be used advantageously for other demanding applications, such as for production of therapeutic devices in oncology, for kinetic penetrators in the military, or as aircraft counterbalances [3,4]. THAs are typically produced via powder metallurgy—they are isostatically pressed and subsequently sintered at temperatures between 1000 and 1500 °C—from powder mixtures containing a majority (over 90 wt.%) of tungsten, in addition to other alloying elements featuring lower melting temperatures (e.g., Ni, Co, Fe), the boundaries of the grains of which melt during sintering and ensure binding of the tungsten particles and elimination of porosity via diffusion [5].

To enhance the strength of THAs, the tungsten content can be increased to 98 wt.%; however, the increase in strength is typically at the expense of plasticity. Additions of alloying (i.e., binding) elements generally decrease the strength, but increase plasticity. Contents of the alloying elements (i.e., matrix) lower than 3 wt.% can cause brittleness of the final product [6]. On the other hand, high contents of alloying elements contribute to inhomogeneity of the mechanical properties and usually also to an uneven shape of the sintered piece cross-section due to gravity sedimentation during sintering. Under optimized conditions, the matrix is homogenously distributed in the gaps between the tungsten agglomerates [7].

In addition to modification/optimization of their chemical composition, the mechanical properties of THAs can also be enhanced by post-sintering deformation processing, which can advantageously be performed via methods of severe plastic deformation [8–10]. Technologies applying intensive plastic deformation have recently gained attention primarily due to their ability to effectively refine the structural units and consequently enhance the mechanical and utility properties of the processed materials. Severe plastic deformation (SPD) methods, such as equal channel angular pressing (ECAP) [11–13], twist channel (multi) angular pressing (TCAP, TCMAP) [14,15], or high pressure torsion (HPT) [16], have been proven to be very effective; however, they are discontinuous and are only applicable for limited volumes of material. On the other hand, methods such as ECAP-conform [17,18], accumulative roll bonding (ARB) [19,20], or rotary swaging (RS) [21,22] are continuous and can favourably be used to manufacture relatively large (semi)products. Given its versatility, the RS technology is applicable in various industrial branches.

Rotary swaging is an incremental process, i.e., it is characterized by gradual increments of the imposed strain provided by high-frequency strokes of the swaging dies. The method can be applied under cold, as well as warm and hot conditions and is used to process solid work pieces, but also to produce hollow tubes and shaped axisymmetric products [23–25]. The processed materials are typically conventionally cast alloys, such as magnesium [26] and aluminium alloys [27]. Nevertheless, the favourable stress state and incremental character enables the method to advantageously be applied also to process composites, the example of which can be mechanically bonded Cu/Al clad composites for electrotechnic applications [28,29], powder-based materials, demanding compounds based on zirconium [30], or tungsten heavy alloys [31].

The aim of this work was to investigate material behaviour of the studied WNiCo tungsten heavy alloy during cold and warm rotary swaging, and the effects of the individual processing steps on its structure and properties. To better characterize the influence of the used technology, especially the effect of the applied pre-swaging induction heating, experimental investigations were supplemented with finite element analyses. The predicted results were verified by real swaging experiments. Attention was also given to characterization of residual stress within the structure.

2. Materials and Methods

2.1. Numerical Prediction

The first step of the performed study was the numerical prediction of the applied induction heating process, which was performed via Finite Element Method (FEM) using FORGE (NxT version, Transvalor, Sophia Antipolis, France) software. The used computational model was a model based on the Maxwell's equations available in the software. The prediction was implemented primarily to optimize control of the heating process and determine the heating time necessary to achieve homogeneous temperature distribution throughout the work piece cross-section before the warm rotary swaging, i.e., to predict behaviour of the studied THA during induction heating. The investigated work piece, as well as the used inductor, were modelled according to the experimentally used pieces (Figure 1a).

The simulation was designed considering two main aims. Firstly, minimisation of the computational time. For this reason, meshing was performed thoughtfully to ensure both reliable results and reasonable computational time. The mesh of the work piece consisted of 2700 volume tetrahedral elements, whereas the mesh of the inductor consisted of 21,314 nodes in total (Figure 1b).

The second aim was minimisation of the real heating time. For this reason, the simulation (and the following experiment) involved two-step heating. The first step, defined with the current of 68 A and frequency of 16.5 kHz, was designed to provide the quickest possible achievement of the swaging temperature of 900 °C, whereas the second step ensured homogenisation of the temperature throughout the cross-section, i.e., minimised temperature gradient from the surface to the axis of the work piece. Once the surface temperature of 900 °C had been reached, the second heating step, defined by a current of 50 A and a frequency of 15.8 kHz, began. However, possible development of residual stress resulting from the variable heating rate and occurring temperature gradients was also considered. The entire heating process was optimized for the maximum temperature of 900 °C not to be exceeded by more than 20 °C. To ensure thorough evaluation and control of the process, three locations at the axial longitudinal cut through the work piece, depicted as sensors 1, 2, and 3 (Figure 1c), were selected and monitored during the entire heating process.

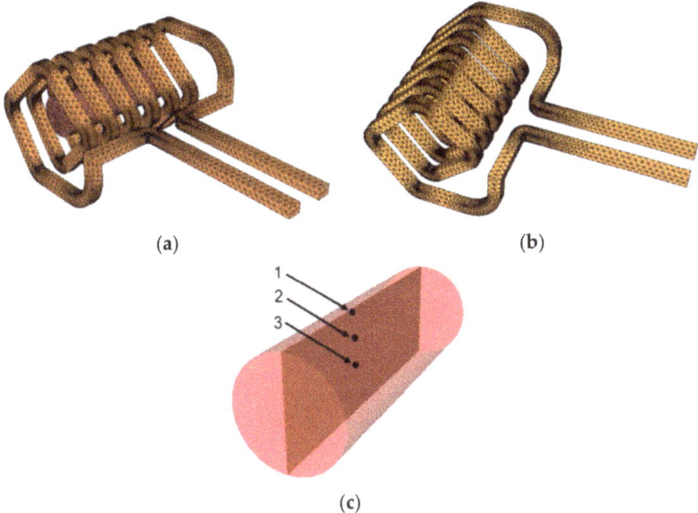

Figure 1. Schematic depiction of setup of induction heating, meshed (**a**); inductor with mesh (**b**); axial longitudinal cut through work piece with depicted monitored sensors 1, 2, and 3 (**c**).

The FORGE NxT software was also used to perform numerical prediction of deformation behaviour of the investigated 93W6Ni1Co alloy. The conditions for the simulation were selected to correspond to the real ones, and the entire simulation consisted of two simulation steps, corresponding to the real two swaging passes on which were modelled. The assembly for the simulation consisted of four swaging dies and the work piece, which was fed continuously towards the swaging head with the dies (Figure 2). The dies were created as shells, while the work piece was designed as a full deformable body with the mesh element size of 0.3 mm. Deformation behaviour of the material was characterized via the Haensel-Spittel equation, presented as Equation (1),

$$\sigma = A\exp(m_1 T)T^{m_9}\varepsilon^{m_2}\exp\left(\frac{m_4}{\varepsilon}\right)(1+\varepsilon)^{m_5 T}\exp(m_7\varepsilon)\dot{\varepsilon}^{m_3}\dot{\varepsilon}^{m_8 T}, \tag{1}$$

where $\dot{\varepsilon}$ is the equivalent strain rate (s^{-1}), ε is the equivalent strain (-), T is the temperature (°C), and A, m_1, m_2, m_3, m_4, m_5, m_7, m_8, and m_9 are regression coefficients, the values of which were set based on regression analyses of plastometric test results (A_1 = 1447.002738, m_1 = −0.009, m_2 = 0.0895, m_3 = 0.0044, m_4 = −0.0069, m_5, m_7, m_8, and m_9 were equal to 0). The dimensions of the test specimens for the plastometric tests were 10 mm in diameter and 150 mm in length. Friction was defined via the Coulomb law, μ = 0,1, and the specific parameters for the simulation were the following: Poisson coefficient 0.3, specific heat 130 $J \cdot kg^{-1}K^{-1}$, density 18.5 $g \cdot cm^{-3}$, and thermal conductivity (λ_{20} = 149.3 $W \cdot m^{-1}K^{-1}$, λ_{900} = 85.7 $W \cdot m^{-1}K^{-1}$).

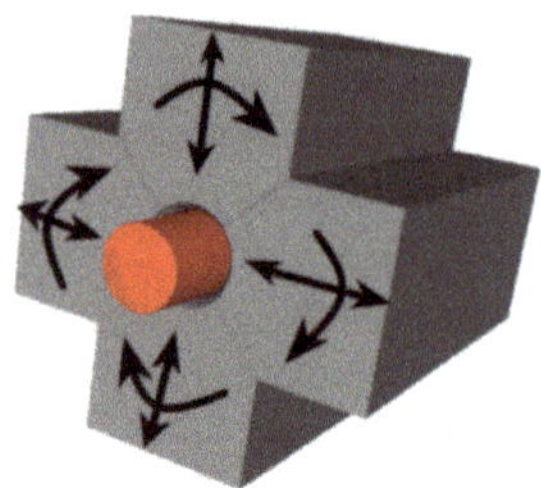

Figure 2. Schematic depiction of setup of rotary swaging.

2.2. Experimental Verification

The selected 93W6Ni1Co pseudo-alloy with the chemical composition of 92.6 wt.% W, 5 wt.% Ni and 2.40 wt.% Co (determined by SEM-EDX analysis) was prepared from powders with the mean grain size of 2.78 μm, containing ~<13 ppm of Fe, Mo, Cr, Al, Ca and other impurities. The process involved mixing of powders, cold isostatic pressing at 400 MPa, and subsequent sintering at 1525 °C for 20 min under a protective atmosphere followed by quenching in water. The protective atmosphere is hydrogen during sintering and argon during cooling. This preparation procedure was performed at ÚJP Praha a.s.

The sintered pieces were subsequently processed via rotary swaging, the initial temperatures for which were 20 °C (cold swaging) and 900 °C (warm swaging). The swaging temperatures were selected based on our previous experimental studies [3,31]. The room temperature was selected in order to maximize the possible achievable strength and to investigate, whether the material would maintain reasonable plasticity. The temperature of 900 °C was selected as the highest applicable temperature, since our previous studies showed that processing at temperatures higher than 900 °C resulted in massive oxidation causing rapid embrittlement during processing of the WNiCo. The temperature control during induction heating was provided by optical pyrometer (surface temperature), and a couple of thermocouples of type K (one on the surface, another in the work piece axial region). The initial diameter of the sintered pieces was 30 mm, the length was 100 mm. Both were gradually swaged down in two consequent reduction steps to the final diameter of 20 mm. Such swaged pieces are suitable, e.g., for the fabrication of kinetic penetrators [26]. The following structural analyses performed on cross-sectional cuts of the swaged pieces focused primarily on the influence of the applied deformation temperature on (sub)structure development and residual stress. The observations were performed via scanning electron microscopy (SEM EBSD analyses). The cut samples were mechanically ground on SiC papers and subsequently polished using Eposil F substance (Saphir 520 device, ATM, Germany). The EBSD analyses were performed in the sub-surface sample region, 1 mm from the outer rim of the swaged-rod, with the scan step of 0.25 μm using a Tescan Lyra 3 equipment (TESCAN Brno s.r.o, Brno, CZ) with a NordlysNano EBSD detector (Oxford Instruments, Abingdon-on-Thames, UK). The scans were evaluated using the ATEX (Win10 version) software [32]. Residual stress was characterized via analyses of internal grains misorientations in the scale from 0° to 15° (rainbow colour distribution). TEM images were acquired on ion polished thin foils with a JEOL 2100F (JEOL, Akishima, Tokio prefecture, Japan) device. The tensile tests were performed with the strain rate of 10^{-3} using a Testometric M500-50CT testing machine (Testometric Co. Ltd., UK) on 150 mm long testing samples with circular cross-sections.

3. Results

3.1. FE Analyses

3.1.1. Induction Heating

As shown by the FEA results, induction heating imparted the development of zones featuring various temperatures, similar to conventional heating in furnaces. As documented by Figure 3a, depicting the temperature field throughout the work piece in the moment of achieving the surface temperature of 900 °C (heating time 12 s), the difference between the temperatures in the surface and axial work piece regions was up to 350 °C in this moment. Nevertheless, the temperature gradient decreased rapidly as heating continued. The results showed that the heating time necessary to heat the THA work piece to the required temperature of 900 °C with the maximum allowable deviation of 20 °C and, at the same time, to provide homogeneous temperature distribution throughout the entire work piece, was 24 s (Figure 3b,c).

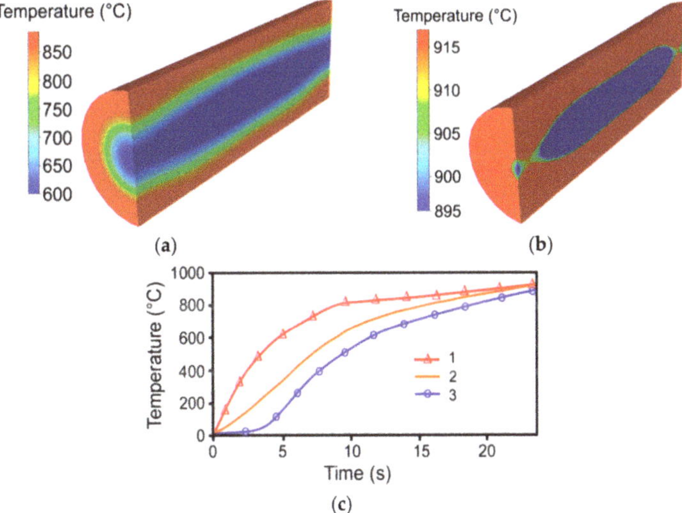

Figure 3. Temperature field during induction heating in time: 12 s (**a**); 24 s (**b**); distribution of temperature in monitored sensors (**c**).

The temperature distribution depicted in Figure 3b shows that the temperature field across the cross-section was not homogeneous along the work piece, especially during the second heating step. This is also confirmed by Figure 3c, which shows the differences between the temperatures in the individually monitored sensors. This phenomenon was a result of the geometry of the used inductor. The temperature increase was the highest in the surface work piece region up to the heating time of 10 s; the heating rate was lower in the other two monitored locations during that time. The lowest heating rate was monitored in the axial work piece area. The second heating step then imparted rather unexpected material behaviour, the heating rate in the peripheral work piece area decreased, while the other two monitored areas kept more or less the same heating rate. This behaviour finally introduced a decrease in the temperature gradient, i.e., homogenisation of the temperature field throughout the work piece cross-section, which is documented by Figure 3c.

3.1.2. Deformation Behaviour

The predicted developments of temperature in the surface and axial regions during processing of both the work pieces, swaged at 20 °C and 900 °C, are depicted in Figure 4a,b, respectively. As shown in the Figures, the initial temperatures for both the swaged pieces were higher than the original swaging temperatures due to the previous deformation history (first swaging pass).

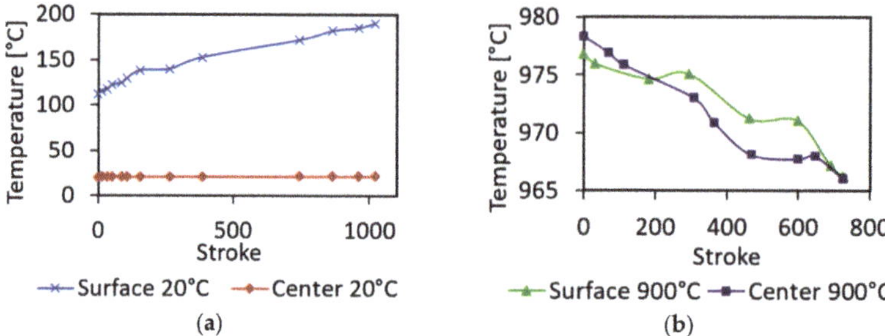

Figure 4. Temperature developments in surface and axial regions of the cold-swaged piece (**a**) and the warm-swaged piece (**b**).

As can be seen from Figure 4a, the temperature in the surface area increased gradually to almost 200 °C during cold swaging, whereas the temperature in the axial area remained more or less constant. On the other hand, the warm-swaged piece exhibited a decrease in temperature during swaging; the temperature decrease rate was comparable for both the surface and the axial areas (Figure 4b). However, multiple affecting phenomena need to be considered for the surface areas of the swaged pieces. The surface temperature increases due to the effect of friction with the rotating dies being in contact with the surface. On the other hand, the dies are cooler than the swaged piece, which causes a certain portion of the generated heat to dissipate via the swaging dies.

Figure 5a depicts the predicted developments of the imposed effective strain in the surface and axial regions of both the swaged pieces. The comparison of both the curves for the cold-swaged piece shows that the imposed strain was substantially higher in the surface region of the swaged piece, where it reached a maximum value of almost 4. The figure also shows that the maximum effective strain in the axial region of the cold-swaged piece reached to the value of 1.2. On the other hand, warm swaging imparted an increase in the imposed effective strain, which reached and even slightly exceeded the value of 2, approximately in the middle of the swaging pass. Mutual comparison of the effective strain developments for both the swaged pieces reveals that the strain gradient from the surface towards the swaged piece axis was substantially lower for the warm-swaged piece. Nevertheless, the maximum imposed effective strain was comparable for both.

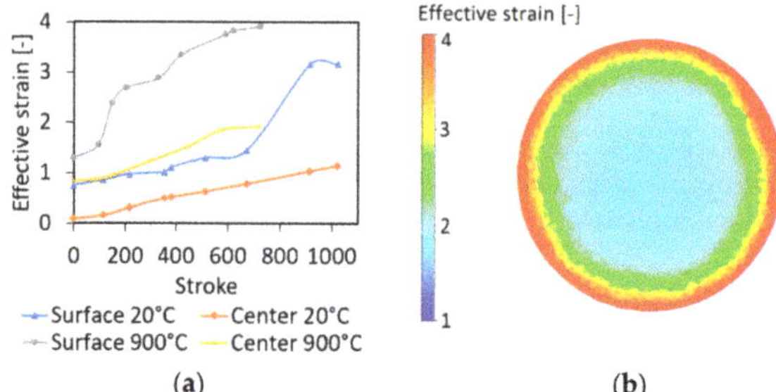

Figure 5. Developments of imposed effective strain in surface and axial regions of both swaged pieces (**a**); effective strain across cross-section of warm-swaged piece (**b**).

The distribution of the effective strain across the cross-sectional cut through the warm-swaged piece is depicted in Figure 5b. The figure shows that the imposed effective strain reached a value of 4 at the very periphery of the swaged piece. The imposed strain then gradually decreased towards the axial region where it reached the values of approximately 2.

Figure 6a,b depicts the material flow during swaging of the cold and warm swaged piece, respectively. Given by the incremental nature of the swaging process and rotary movement of the swaging dies, material flow during swaging is variable and quite complex. The material flow was different for both the swaged pieces, and also changed its character during the individual swaging pass. At the beginning of swaging, the axial material flow dominated for both the cold and warm swaged pieces; however, the plastic flow vectors were flexed for both the pieces. With continued swaging, the material flow changed its character, i.e., it rotated backwards in the neutral zone, due to the effect of the already swaged deformation strengthened material volume. For the cold-swaged piece, the neutral zone was located in the reduction zone (Figure 6a), while for the warm-swaged piece, the position of the neutral zone shifted towards the feeding zone of the swaging dies (Figure 6b).

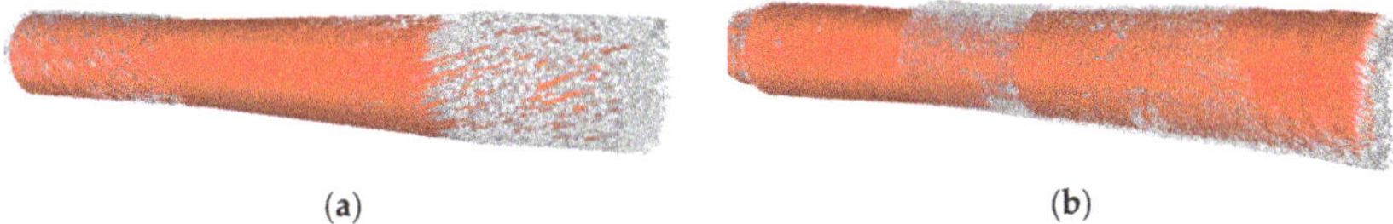

(**a**) (**b**)

Figure 6. Vectors of material flow for the cold-swaged piece (**a**) and the warm-swaged piece (**b**).

The finally evaluated predicted parameters were stress-related ones, stress intensity during swaging, and residual stress. Figure 7a,b depicts stress intensity via Von Mises stress throughout the axial longitudinal cut for the cold and warm swaged piece, respectively. As can be seen in Figure 7a, the stress intensity during cold-swaging increased to 750 MPa, while for the warm-swaged piece, the maximum stress intensity reached approximately 350 MPa.

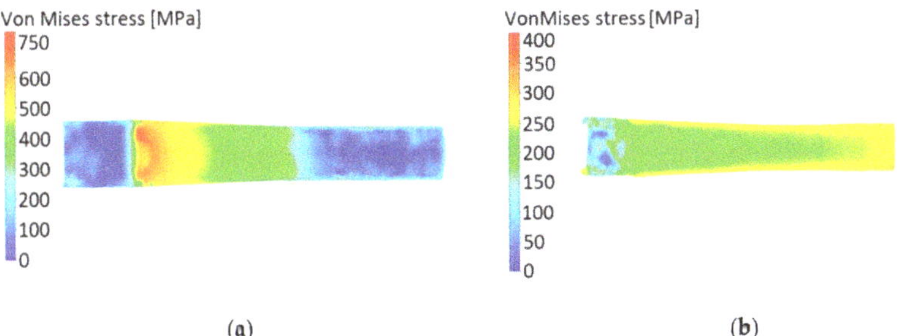

(**a**) (**b**)

Figure 7. Stress intensity in axial longitudinal cut depicted via Von Mises stress for the cold-swaged piece (**a**) and the warm-swaged piece (**b**).

Stress can be present in the material not only during the actual processing, but residual stress can also be present after the material left the swaging dies' reduction zone. Figure 8a,b depicts the predicted distribution of residual stress across the cross-section of the cold and warm swaged piece, respectively. Mutual comparison of the figures reveals similar distribution of residual stress within both of the swaged pieces, its intensity corresponding to the intensity of the imposed effective strain for both pieces. Deeper insight into the residual stress distribution was subsequently provided via experimental analysis.

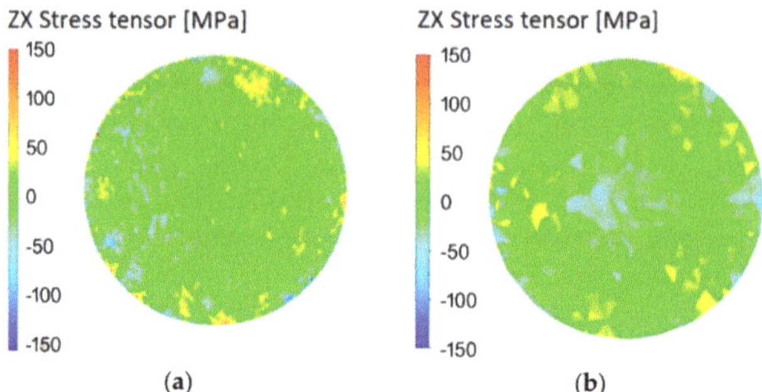

Figure 8. Predicted residual stress intensity across cross-section of the cold-swaged piece (**a**) and the warm-swaged piece (**b**).

3.2. Structure Analyses

Figure 9a–c shows the SEM-BSE scans of the tungsten pseudo-alloy after sintering, cold-swaging, and warm-swaging, respectively. As can be seen, the sintered state exhibited individual round-shaped W agglomerates surrounded by the NiCo matrix (Figure 9a). The structures after cold and warm swaging (Figure 9b,c) exhibited visible differences in the shapes of the agglomerates. While the grains of both the matrix and tungsten deformed significantly during cold swaging due to the substantial imposed shear strain, the elevated temperature during warm swaging decreased the matrix flow stress, which caused the softer matrix to preferentially consume the imposed strain. For this reason, the tungsten agglomerates remained more or less of round shapes.

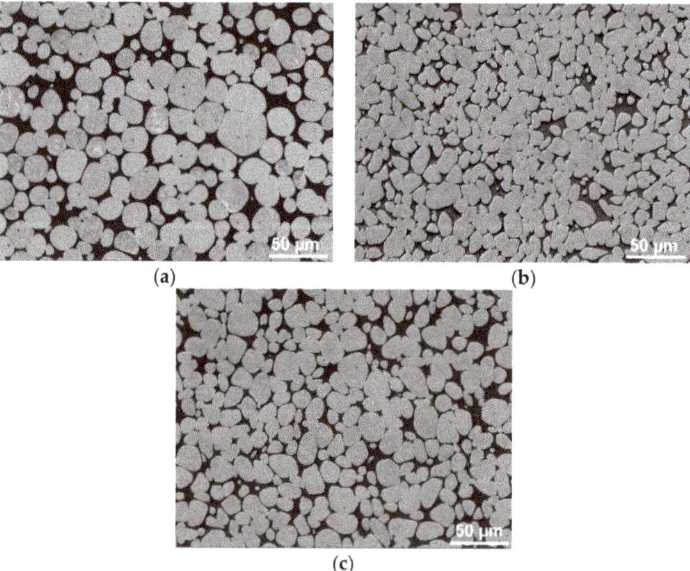

Figure 9. SEM-BSE structure scans for sintered state (**a**); cold-swaged piece (**b**); warm-swaged piece (**c**).

Figure 10a,b shows residual stress depicted via internal misorientations from 0° to 15° for the cold and warm swaged pieces, respectively. For both pieces, the presence of residual stress was detected in the peripheral regions of the tungsten agglomerates. This phenomenon can be attributed primarily to substructure development and accumulations of dislocations, as proven by the further discussed TEM structure characterisation.

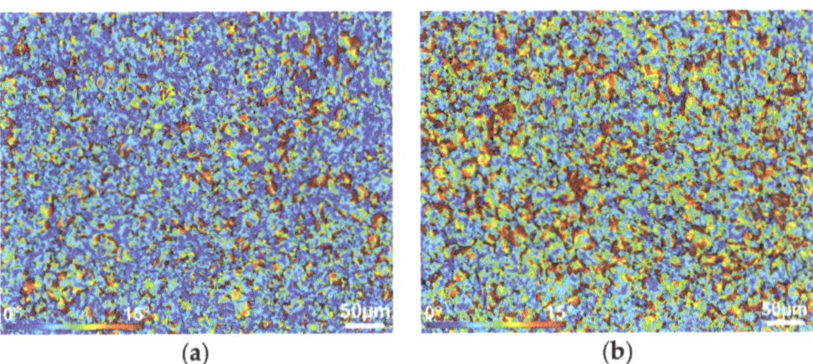

(a) (b)

Figure 10. Integral grain misorientations for ChemEngineering-610706 cold-swaged piece (**a**) and the warm-swaged piece (**b**).

As shown by the computational analyses of material flow, the increased temperature of 900 °C imparted a more intense plastic flow, which imparts a higher probability of fixation of elastic strain within the warm-swaged material. In other words, the warm-swaged piece (Figure 10b) exhibited a higher volume of locations featuring high internal misorientations, i.e., internal stress, than the cold-swaged piece (Figure 10a). As can be seen from Figure 9b, the misorientations occurred not only at the peripheries of the agglomerates, but also within the more ductile NiCo matrix.

Figure 11a,b depicts TEM scans of the cold- and warm-swaged pieces, respectively. As can be seen, the cold-swaged piece exhibited distinctive interfaces between the tungsten agglomerates (dark area) and the NiCo matrix (light area). On the other hand, the warm-swaged piece exhibited diffusion of tungsten to the NiCo matrix; the areas within the matrix with increased tungsten content can be seen in the figure as "dark shadows" within the light matrix area. To confirm the presence of tungsten within the matrix, a line scan measuring chemical composition through such a darker shadow within the matrix was carried out. Figure 11c depicts the area in which the line scan was carried out (depicted by the straight line). Figure 11d then shows the results of the line scan as regards the chemical composition of the investigated area. As is evident, the size of the interface between the agglomerate and the matrix after warm swaging and the content of tungsten within the matrix in the vicinity of the agglomerates reached up to 50 wt.%.

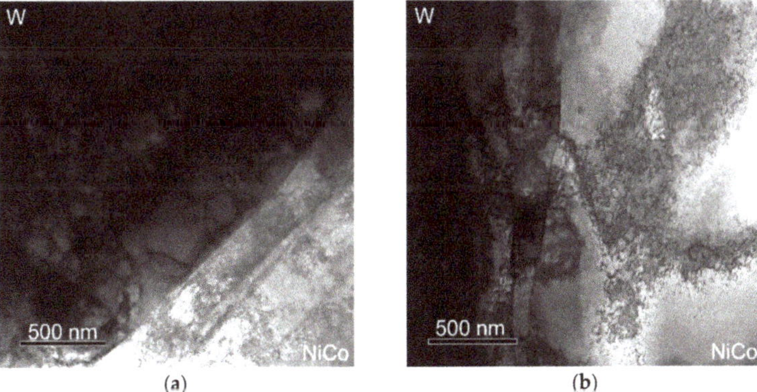

(a) (b)

Figure 11. *Cont.*

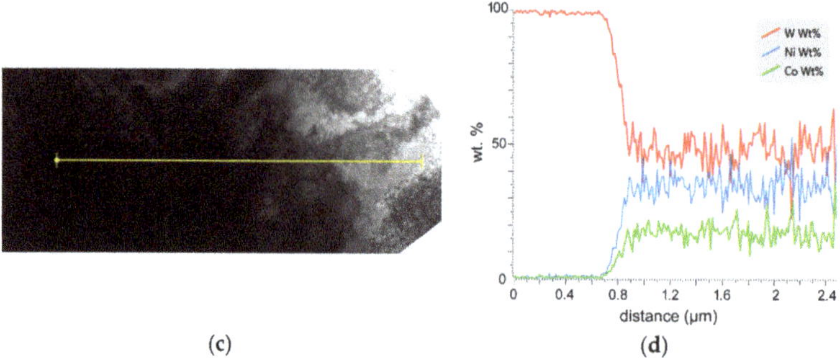

Figure 11. TEM scan showing agglomerate/matrix interface for the cold-swaged piece (**a**) and the warm-swaged piece (**b**); measured line scan area within the warm-swaged piece (**c**); and chemical composition of investigated line scan (**d**).

3.3. Mechanical Properties

The mechanical properties of both the swaged pieces were evaluated via tensile tests; the ultimate tensile strength (UTS) and elongation till failure for all the investigated pieces are summarized in Table 1.

Table 1. Mechanical properties resulting from tensile tests.

Swaging Pass	Swaging Temperature (°C)	UTS (MPa)	Elongation (%)
0	-	857 ± 12	18.3 ± 3.5
1	20	1412 ± 21	9.2 ± 2.4
2	20	1828 ± 17	5.6 ± 1.3
1	900	1010 ± 19	23.4 ± 3.2
2	900	1650 ± 15	7.8 ± 1.5

As can be seen, the powder-based sintered material exhibited the lowest recorded UTS of approximately 860 MPa; however, the elongation till failure reached up to ~20%. The highest strength was recorded for the cold-swaged piece, which it exhibited a UTS of 1828 MPa after the second swaging pass. The UTS after warm swaging increased too, but the UTS value was slightly lower than after cold swaging (1650 MPa vs. 1828 MPa). The fact that the increase in UTS occurred at the expense of plasticity for both the swaged pieces needs to be stressed. In other words, the decrease in plastic properties after swaging was evident in both cases (especially with higher imposed strain). However, the more favourable mutual combination of strength and plastic properties was achieved after warm swaging.

4. Discussion

The numerical simulations revealed certain differences in the material behaviours of both the work pieces during swaging. For successful swaging of the cold-swaged piece, no preheating was necessary, which makes the swaging process more effective from the viewpoint of processing time. On the other hand, induction heating of the warm-swaged piece is considerably quicker when compared to heating in a furnace and homogeneous heating of the work piece to the swaging temperature only took 24 s. Nevertheless, certain temperature gradient between the surface and axial regions of the work piece developed during heating. The gradient was primarily dependent on the heating time; it was most significant at the beginning of heating and diminished by the time heating was finished.

Among the advantages of induction heating is also that it can favourably be placed directly in front of the swaging machine, which, however, directly affects the material characteristics [33,34], especially the complex material flow. Continuous heating keeps the unswagged work piece in front of the swaging head at the swaging temperature of 900 °C, which introduces gradients in mechanical properties, especially flow stress, between the swaged and unswaged material volumes. As shown by the predicted plastic flow vectors, gradients in temperature-related material properties influenced significantly localization of the neutral zone, the plastic flow vectors in which changed their orientations.

Regardless the effect of induction heating, the predictions revealed temperature variations during processing for both the swaged pieces. The cold-swaged piece exhibited gradually increasing difference in temperatures in the surface and axial swaged piece regions (up to almost 200 °C). Such behaviour is typical for cold swaging technology; the surface temperature of the swaged piece increases primarily due to the effect of direct contact of the surface with the swaging dies, i.e., friction effect [35]. Nevertheless, this effect is reduced by the heat transfer through the cold swaging dies. Warm swaging also introduced a certain increase in temperature, primarily in the work piece surface region, where the maximum temperature increased to 990 °C. In addition to the effect of friction, the temperature increase during swaging can also be attributed to deformation heat generated by the influence of the imposed shear strain [36].

The material characteristics during swaging are primarily influenced by the effect of the swaging force, featuring two main components: tangential and axial [29]. Whereas the axial swaging force component contributes to the gradual elongation of the swaged piece and provides the swaged product with its final shape, the tangential swaging force component imparts intensive shear strain, resulting in structure changes, introducing an increase in the actual processing temperature, imparting flexure of the plastic flow vectors, but also imparting the gradient of the imposed effective strain from the swaged piece surface region towards its axial region; the effective strain is the highest in the surface swaged piece region, the effect of the rotating dies introducing shear strain, which is the most intense. This phenomenon was evident especially for the cold-swaged piece, the flow stress for which was aggravated. During warm swaging, the flow stress of the THA decreased, reducing the effective strain gradient across the swaged piece cross-section, but also resulted in the higher total imposed effective strain.

The flow stress, together with the differences in the plastic flows of both the swaged pieces, also non-negligibly influenced the stress intensity, and consequently the load on the swaging machine. The predicted stress intensity, pointing to the resistance of the swaged material against deformation, was significantly higher for the cold-swaged piece. The predicted distribution of residual stress was more or less comparable for the cold and warm swaged piece. Nevertheless, the experimental observations showed micro-differences in the distribution of residual stress within the NiCo matrix. The results showed that both the swaged pieces exhibited the presence of residual stress in the peripheral areas of tungsten agglomerates. As proven by the transmission electron microscopy analyses, this phenomenon could primarily be attributed to substructure development within the agglomerates. However, the more intense plastic flow supported by the elevated swaging temperature introduced the presence of residual stress, i.e., unrelaxed elastic deformation, within the matrix of the warm-swaged piece. The primary reason for such behaviour lay in the intense plastic flow imparted by the increased swaging temperature. The harder tungsten agglomerates featuring lower formability acted as "transferring agents" of the imposed strain during swaging. During processing, mutual distances between the individual agglomerates decrease, i.e., the volume of the ductile matrix between the hard agglomerates gradually decreases, by the effect of which the interactions of the individual agglomerates become more intense. Moreover, the occurring diffusion of tungsten to the NiCo matrix, which intensifies with increasing processing temperature, should also be considered. By the effect of this phenomenon, the matrix hardens, and residual stress develops, especially at the W/NiCo interfaces. In other words, the matrix then involves locations featuring varying chemical compositions, imparting local differences within the matrix, and consequently introducing strain gradients resulting in the presence of residual stress.

5. Conclusions

The study focused on the numerical prediction of the deformation behaviour of a 93W6Ni1Co tungsten heavy alloy rotary swaged at 20 °C and 900 °C. The study was supplemented with numerical investigations of the applied induction heating, and experimental observations of residual stress and substructure development within the swaged pieces.

The predictions showed the overall time for heating of the THA to 900 °C to be 24 s. The induction heating promoted shifting of the neutral zone the plastic flow vectors in which reverse primarily by the effect of flow stress gradient between the swaged and unswagged materials.

Warm swaging at 900 °C imparted a more homogeneous distribution of the imposed strain. However, for both the swaged pieces, the highest strain was observed in their surface regions, which were directly affected by the swaging dies. This finding corresponded to the predicted distribution of temperature throughout the swaged pieces; for both, the temperature increased in their surface regions primarily due to the development of deformation heat.

Both the swaged pieces exhibited the presence of residual stress in the peripheral areas of tungsten agglomerates, which could primarily be attributed to substructure development. However, the supported plastic flow of the warm-swaged piece also introduced the presence of residual stress, i.e., unrelaxed elastic deformation, within the warm-swaged piece matrix, and diffusion of W to the NiCo matrix. The UTS was the highest for the cold-swaged piece (1828 MPa), and slightly lower for the warm-swaged piece (1650 MPa); this phenomenon can primarily by attributed to the occurrence of softening of the matrix.

Author Contributions: The contributions of the individual authors were the following: methodology, R.K., L.K. (Lenka Kunčická), and A.M.; numerical prediction and evaluation, R.K., R.P., A.M., L.K. (Ludmila Krátká); experimental validation, R.K., L.K. (Lenka Kunčická), L.K. (Ludmila Krátká), and A.M.; microscopy investigation and evaluation, L.K. (Lenka Kunčická); writing—original draft preparation, L.K. (Ludmila Krátká), R.K., L.K. (Lenka Kunčická); writing—review and editing, L.K. (Lenka Kunčická), R.K.; project administration, A.M.; funding acquisition, R.K., R.P., and A.M.

Funding: This work was supported by Ministry of Education, Youth and Sports, Czech Republic in the framework of the projects SP 2019/74 and SP 2019/43 and by the project no. 19-15479S by the Grant Agency of the Czech Republic.

Acknowledgments: The experiments were performed using equipment available at the Technical University of Ostrava. We also acknowledge the help of IPMinfra research infrastructure and working team of Institute of Physics of Materials, CAS.

Conflicts of Interest: The authors declare no conflict of interest.

References

1. Pappu, S.; Kennedy, C.; Murr, L.E.; Magness, L.S.; Kapoor, D. Microstructure analysis and comparison of tungsten alloy rod and [001] oriented columnar-grained tungsten rod ballistic penetrators. *Mater. Sci. Eng. A* **1999**, *262*, 115–128. [CrossRef]
2. Gong, X.; Fan, J.; Ding, F. Tensile mechanical properties and fracture behavior of tungsten heavy alloys at 25–1100 °C. *Mater. Sci. Eng. A* **2015**, *646*, 315–321. [CrossRef]
3. Kunčická, L.; Kocich, R.; Hervoches, C.; Macháčková, A. Study of structure and residual stresses in cold rotary swaged tungsten heavy alloy. *Mater. Sci. Eng. A* **2017**, *704*, 25–31. [CrossRef]
4. Das, J.; Ravi Kiran, U.; Chakraborty, A.; Eswara Prasad, A. Hardness and tensile properties of tungsten based heavy alloys prepared by liquid phase sintering technique. *Int. J. Refract. Met. Hard Mater.* **2009**, *27*, 577–583. [CrossRef]
5. Bose, A.; Schuh, C.A.; Tobia, J.C.; Tuncer, N.; Mykulowycz, N.M.; Preston, A.; Barbati, A.C.; Kernan, B.; Gibson, M.A.; Krause, D.; et al. Traditional and additive manufacturing of a new Tungsten heavy alloy alternative. *Int. J. Refract. Met. Hard Mater.* **2018**, *73*, 22–28. [CrossRef]
6. Ravi Kiran, U.; Panchal, A.; Sankaranarayana, M.; Nageswara Rao, G.V.S.; Nandy, T.K. Effect of alloying addition and microstructural parameters on mechanical properties of 93% tungsten heavy alloys. *Mater. Sci. Eng. A* **2015**, *640*, 82–90. [CrossRef]

7. Upadhyaya, A. Processing strategy for consolidating tungsten heavy alloys for ordnance applications Mater. *Chem. Phys* **2001**, *67*, 101–110.
8. Ripoll, R.M.; Očenášek, J. Microstructure and texture evolution during the drawing of tungsten wires. *Eng. Fract. Mech.* **2009**, *76*, 1485–1499. [CrossRef]
9. Ravi Kiran, U.; Kumar, J.; Kumar, V.; Sankaranarayana, M.; Nageswara Rao, G.V.S.; Nandy, T.K. Effect of cyclic heat treatment and swaging on mechanical properties of the tungsten heavy alloys. *Mater. Sci. Eng. A* **2016**, *656*, 256–265. [CrossRef]
10. Kocich, R.; Szurman, I.; Kursa, M.; Fiala, J. Investigation of influence of preparation and heat treatment on deformation behaviour of the alloy NiTi after ECAE. *Mater. Sci. Eng. A* **2009**, *512*, 100–104. [CrossRef]
11. Kocich, R.; Greger, M.; Macháčková, A. Finite element investigation of influence of selected factors on ECAP process. In Proceedings of the Metal 2010, 19th International Conference on Metallurgy and Materials, Brno, Czech Republic, 18–20 May 2010; Tanger Ltd.: Ostrava, Czech Republic, 2010; pp. 166–171.
12. Tohidlou, E.; Bertram, A. Effect of strain hardening on subgrain formation during ECAP process. *Mech. Mater.* **2019**, *137*, 103077. [CrossRef]
13. Naizabekov, A.B.; Andreyachshenko, V.A.; Kocich, R. Study of deformation behavior, structure and mechanical properties of the AlSiMnFe alloy during ECAP-PBP. *Micron* **2013**, *44*, 210–217. [CrossRef] [PubMed]
14. Kocich, R.; Kunčická, L.; Král, P.; Macháčková, A. Sub-structure and mechanical properties of twist channel angular pressed aluminium. *Mater. Charact.* **2016**, *119*, 75–83. [CrossRef]
15. Kocich, R.; Kunčická, L.; Macháčková, A. Twist Channel Multi-Angular Pressing (TCMAP) as a method for increasing the efficiency of SPD. 6th International Conference on Nanomaterials by Severe Plastic Deformation (NANOSPD6), Metz, France, 30 Jun–4 Jul 2014. *IOP Conf. Ser. Mater. Sci.* **2014**, *63*, 012006. [CrossRef]
16. Kunčická, L.; Kocich, R. Comprehensive characterisation of a newly developed Mg–Dy–Al–Zn–Zr alloy structure. *Metals* **2018**, *8*, 73. [CrossRef]
17. Murashkin, M.; Medvedev, A.; Kazykhanov, V.; Krokhon, A.; Raab, G.; Enikeev, N.; Valiev, R.Z. Enhanced mechanical properties and electrical conductivity in ultrafine-grained Al 6101 alloy processed via ECAP-Conform. *Metals* **2015**, *5*, 2148–2164. [CrossRef]
18. Medvedev, A.; Arutyunyan, A.; Lomakin, I.; Bondarenko, A.; Kazykhanov, V.; Enikeev, N.; Raab, G.; Murashkin, M. Fatigue properties of ultra-fine grained Al-Mg-Si wires with enhanced mechanical strength and electrical conductivity. *Metals* **2018**, *8*, 1034. [CrossRef]
19. Kocich, R.; Macháčková, A.; Fojtík, F. Comparison of strain and stress conditions in conventional and ARB rolling processes. *Int. J. Mech. Sci.* **2012**, *64*, 54–61. [CrossRef]
20. Ghalehbandi, S.A.; Massoud, M.; Gupta, M. Accumulative Roll Bonding—A Review. *Appl. Sci.* **2019**, *9*, 3627. [CrossRef]
21. Kocich, R.; Kunčická, L.; Král, P.; Strunz, P. Characterization of innovative rotary swaged Cu-Al clad composite wire conductors. *Mater. Des.* **2018**, *160*, 828–835. [CrossRef]
22. Kuncicka, L.; Kocich, R. Deformation behaviour of Cu-Al clad composites produced by rotary swaging. 5th Global Conference on Polymer and Composite Materials (PCM), Kitakyushu, Japan, 10–13 April 2018. *IOP Conf. Ser. Mater. Sci.* **2018**, *369*, 012029.
23. Zhang, Q.; Jin, K.; Mu, D. Tube/tube joining technology by using rotary swaging forming method. *J. Mater. Process. Technol.* **2014**, *214*, 2085–2094. [CrossRef]
24. Moumi, E.; Ishkina, S.; Kuhfuss, B.; Hochrainera, T.; Struss, A.; Hunkel, M. 2D-simulation of material flow during infeed rotary swaging using finite element method. *Procedia Eng.* **2014**, *81*, 2342–2347. [CrossRef]
25. Lin, C.W.; Chen, K.J.; Hung, F.Y.; Lui, T.S.; Chen, H.P. Impact of solid-solution treatment on microstructural characteristics and formability of rotary-swaged 2024 alloy tubes. *J. Mater. Res. Technol.* **2019**, *8*, 3137–3148. [CrossRef]
26. Gan, W.M.; Huang, Y.D.; Wang, R.; Zhong, Z.Y.; Hort, N.; Kainer, K.U.; Schell, N.; Brokmeier, H.G.; Schreyer, A. Bulk and local textures of pure magnesium processed by rotary swaging. *J. Magnes. Alloy.* **2013**, *1*, 341–345. [CrossRef]
27. Yang, Y.; Nie, J.; Mao, Q.; Zhao, Y. Improving the combination of electrical conductivity and tensile strength of Al 1070 by rotary swaging deformation. *Results Phys.* **2019**, *13*, 102236. [CrossRef]

28. Kocich, R.; Kunčická, L.; Macháčková, A.; Šofer, M. Improvement of mechanical and electrical properties of rotary swaged Al-Cu clad composites. *Mater. Des.* **2017**, *123*, 137–146. [CrossRef]
29. Kunčická, L.; Kocich, R.; Dvořák, K.; Macháčková, A. Rotary swaged laminated Cu-Al composites: Effect of structure on residual stress and mechanical and electric properties. *Mater. Sci. Eng. A* **2019**, *742*, 743–750. [CrossRef]
30. Yang, X.R.; Zhang, W.Y.; Lui, X.Y.; Luo, L.; Feng, G.H.; Wang, X.H.; Zhao, X.C. Low Cycle Fatigue Behavior of Ultrafine Grained CP-Zr Processed by ECAP and RS. *Rare Met. Mater. Eng.* **2019**, *48*, 1202–1207.
31. Kocich, R.; Kunčická, L.; Dohnalík, D.; Macháčková, A.; Šofer, M. Cold rotary swaging of a tungsten heavy alloy: Numerical and experimental investigations. *Int. J. Refract. Met. Hard Mater.* **2016**, *61*, 264–272. [CrossRef]
32. Beausir, B.; Fundenberger, J.J. Analysis Tools for Electron and X-ray Diffraction, ATEX Software. Available online: www.atex-software.eu (accessed on 21 November 2018).
33. Yang, B.J.; Hattiangadi, A.; Li, W.Z.; Zhou, G.F.; McGreevy, T.E. Simulation of steel microstructure evolution during induction heating. *Mater. Sci. Eng. A* **2010**, *527*, 2978–2984. [CrossRef]
34. Wang, J.; Xiao, Y.; Song, W.; Chen, C.; Pan, P.; Zhang, D. Self-Healing Property of Ultra-Thin Wearing Courses by Induction Heating. *Materials* **2018**, *11*, 1392.
35. Kunčická, L.; Macháčková, A.; Lavery, N.P.; Kocich, R.; Cullen, J.C.T.; Hlaváč, L.M. Effect of thermomechanical processing via rotary swaging on properties and residual stress within tungsten heavy alloy. *Int. J. Refract. Met. Hard Mater.* **2019**, 105120, in press.
36. Humphreys, F.J.; Hetherly, M. *Recrystallization and Related Annealing Phenomena*, 2nd ed.; Elsevier Ltd.: Oxford, UK, 2004.

Article

Correlating Microstrain and Activated Slip Systems with Mechanical Properties within Rotary Swaged WNiCo Pseudoalloy

Pavel Strunz [1,*], Lenka Kunčická [2], Přemysl Beran [1,3], Radim Kocich [2] and Charles Hervoches [1]

1 Nuclear Physics Institute of the CAS, 250 68 Řež, Czech Republic; pberan@ujf.cas.cz (P.B.); hervoches@ujf.cas.cz (C.H.)

2 Faculty of Materials Science and Technology, VŠB-Technical University of Ostrava, 708 00 Ostrava-Poruba, Czech Republic; lenka.kuncicka@vsb.cz (L.K.); radim.kocich@vsb.cz (R.K.)

3 European Spallation Source ERIC, 225 92 Lund, Sweden; premysl.beran@esss.se

* Correspondence: strunz@ujf.cas.cz; Tel.: +420-266-173-553

Received: 27 September 2019; Accepted: 30 December 2019; Published: 3 January 2020

Abstract: Due to their superb mechanical properties and high specific mass, tungsten heavy alloys are used in demanding applications, such as kinetic penetrators, gyroscope rotors, or radiation shielding. However, their structure, consisting of hard tungsten particles embedded in a soft matrix, makes the deformation processing a challenging task. This study focused on the characterization of deformation behavior during thermomechanical processing of a WNiCo tungsten heavy alloy (THA) via the method of rotary swaging at various temperatures. Emphasis is given to microstrain development and determination of the activated slip systems and dislocation density via neutron diffraction. The analyses showed that the grains of the NiCo2W matrix refined significantly after the deformation treatments. The microstrain was higher in the cold swaged sample (44.2×10^{-4}). Both the samples swaged at 20 °C and 900 °C exhibited the activation of edge dislocations with <111> {110} or <110> {111} slip systems, and/or screw dislocations with <110> slip system in the NiCo2W matrix. Dislocation densities were determined and the results were correlated with the final mechanical properties of the swaged bars.

Keywords: tungsten; rotary swaging; neutron diffraction; dislocations; microstrain

---

1. Introduction

Given their excellent mechanical and physical properties, tungsten heavy alloys (THAs) are popular for demanding applications in the military, radiation shields, and highly demanding components such as aircraft counter-balances and gyroscope rotors. THAs usually consist of 90–97 wt.% of tungsten plus other elements, such as Co, Ni, Fe, and Cu [1]. THAs are generally two-phase composites consisting of spherical tungsten particles/agglomerates surrounded by a ductile matrix [2]. Nevertheless, the structure characteristics consequently impacting the performance of the final product can non-negligibly be altered by even the slightest modifications to the processing technology. By this reason, characterization of the occurring structural phenomena, such as the possible presence of adiabatic shear bands (ASBs), determination of the presence of microstrain and residual stress, and characterization of dislocations and active slip systems are of the utmost importance. The first mentioned phenomenon has been investigated quite thoroughly [3–5], but detailed works characterizing other structure phenomena during deformation processing of THAs are scarce [6,7].

Recently, several pieces of research have shown that imposing intensive shear strain into THAs during their production enhances their utility properties and ballistic performance [8,9]. Effective structure refinement can preferably be introduced via methods of severe plastic deformation (SPD), for example via the widely researched equal channel angular pressing (ECAP) and its modifications, which have been proven to effectively refine the grain size down to several hundreds of nanometres for various materials from aluminium to tungsten [10–16]. Nevertheless, the major drawback of prospective industrial use of SPD methods is their limited applicability for bulk volume samples. For example, the most effective high pressure torsion (HPT) method is only suitable to process coin-like samples [17]. By these reasons, THAs are mostly fabricated via various thermomechanical treatments and more conventional technologies, such as hot extrusion [18,19], cold rolling [20], and swaging [21–23].

Rotary swaging (RS) is an intensive plastic deformation method advantageously used in the industry to gradually reduce cross-sections and increase lengths of axisymmetric workpieces [24,25]. Given by its incremental character and dominant compressive stress state, the method can favourably be used to process sintered materials [26]. The dominant shear strain mechanism enables elimination of residual porosity and imparts significant structure refinement.

The primary aim of the presented study was to determine microstrain and characterize the dislocations and active slip systems in the original sintered THA, as well as in the rotary swaged bars, in order to characterize the effects of thermomechanical treatment on the microstructure and final mechanical properties. Throughout the paper, the term "microstrain" denotes the root mean square of the variations in the lattice parameters across the individual crystallites across microscopic distances (root mean square strain, RMSS). In contrast, the term "macrostrain" (not reported here) refers to the overall change in the lattice parameter caused, for example, by a residual stress distribution across the whole component. Generally, microstrain can be caused by a distribution of crystal defects such as vacancies, dislocations, stacking or twin faults.

Neutron powder diffraction method was used as the principal tool for the characterization of structure and microstructure. The strength of neutron diffraction lies in the possibility to provide information from the bulk of the sample, not only from its near-surface region. This fact is very important, especially for THA, where the material is composed mainly of tungsten which is highly absorbing other radiation (X-ray, electrons). When neutrons are used, the signal is averaged over a large volume and the effects of local variability, large grain size and possible local artefacts are minimized.

2. Materials and Methods

The W-Ni-Co (93-6-1) wt.% (80.9-16.4-2.7 at.%) pseudo-alloy was produced by powder metallurgy. The particle size of initial W, Ni and Co powders was in the range of 2–4 μm. The weighted mixture of the powders was homogeneously mixed and then sintered at 1500 °C under H_2 protective atmosphere, and subsequently quenched in water. The as-sintered material, i.e., bars with approximately 12×18 mm^2 elliptical cross sections, is denoted as W_0 throughout the following text.

The sintered bars were further processed by rotary swaging (RS) into circular swaged bars with a diameter of 10 mm. RS was performed in two different ways: at room temperature (sample W_A) and at 900 °C (sample W_B).

The neutron diffraction patterns for structure and microstrain determination were collected at ambient temperature on the MEREDIT diffractometer of CANAM infrastructure at NPI Řež near Prague [27]. A mosaic Cu monochromator (reflection 220) provided neutrons with a wavelength of $\lambda = 1.46$ Å. A small (0.4%) $\lambda/2$ (0.73 Å) contamination of the incoming beam was present and was taken into account during the analysis. The samples were fixed in the beam using a sample holder enabling a sample rotation along the vertical axis to average the texture and large-grain influence on the diffracted intensities within the diffraction plane. A neutron beam size was selected to submerge the sample fully in the beam. The diffraction patterns were collected from 4 to 144° of 2θ (where θ is the scattering angle) with a step size of 0.08° using a multi-detector bank (35 ^{3}He point counters with corresponding 10′ Soller collimators). In order to assess the grain size, an additional neutron diffraction measurement was performed using the TKSN-400 diffractometer [28] equipped with a 2D position-sensitive detector. This measurement was carried out with the neutron wavelength of $\lambda = 1.21$ Å.

Further sample analyses were performed using scanning and transmission electron microscopy (SEM and TEM) on ion-polished transversal samples taken from both the swaged pieces and from the original sintered material. SEM-EBSD (electron backscatter diffraction) analyses were performed using a TESCAN Lyra 3 device equipped with NordlysNano EBSD detector with the scan step of 0.1 μm. The substructures and grains analyses were performed using ATEX [29] and Channel 5 software. TEM images were acquired on ion-polished thin foils with a JEOL 2100F device.

The last step was characterization of mechanical properties via tensile tests performed to evaluate the mechanical behavior of the sintered and swaged material states and to determine their ultimate tensile strength (UTS) and maximum elongation. Tensile testing was performed with 100 mm long bars and a strain rate of 1.3×10^{-3} s^{-1} using a Zwick device. By the reason that determination of elastic properties on tungsten heavy alloys is complicated by tensile tests due to possible deflections of the stress-strain curves, the elastic moduli were additionally determined via ultrasound measurements by an Olympus 38DL Plus device which applies the Pulse Echo Overlap (PEO).

3. Results

3.1. Phase Identification and Grain Size

Prior to microstrain determination, phase identification was done using diffractograms measured at neutron diffractometer MEREDIT [27] for all the samples. An example of the measured and calculated neutron diffraction pattern for the sample W_B with recognized phases is shown in Figure 1. The other two samples exhibited similar phase composition and the diffractograms are similar, although they differ in details due to peak broadening, as will be discussed later in the text. The phase identification and analysis in all the samples was performed by full-pattern refinement using FullProf software [30].

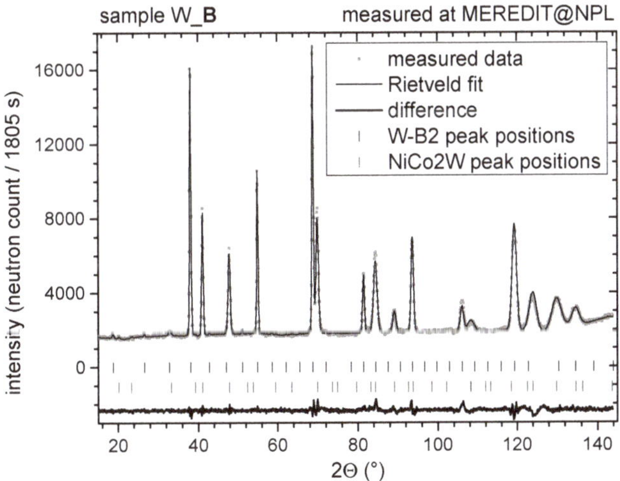

Figure 1. Measured and calculated neutron diffraction pattern for W_B sample used for phase determination as well as for microstrain characterization. The Bragg positions for individual recognized phases (from the top W-B2 and NiCo2W; it should be noted that also the peak positions for λ/2 contamination wavelength are shown) are below the intensity curve. The difference between the measured and calculated intensity is shown as well.

Two phases were identified in all samples. The main phase is α-W (B2 structure; within the text, it is referred to as W-B2 phase). The second phase with a weight fraction of 6%–7% has pure-Ni-like structure (fcc) with the lattice parameter of about 3.60 Å. The lattice parameter of this Ni-like phase is slightly larger than the one for pure nickel (3.55 Å), thus indicating alloying with larger W atoms (atomic radius of W, Ni and Co is 139, 124 and 125 pm, respectively). During the structural refinement, it was assumed that the second phase consists of Ni and Co in the same ratio as the initial composition (6:1) with a further addition of 2 at.% of W. This secondary phase is denoted NiCo2W in what follows.

It was found using data from 2D detector of TKSN-400 diffractometer [28] that the W_0 bar (i.e., the sample without rotary swaging forming) has a fine-grained microstructure for the W-B2 phase but very raw-grained microstructure of the NiCo2W phase. The NiCo2W phase produces spots on the 2D detector while the W grains of the W-B2 phase result in a smooth Debye–Scherrer diffraction conus, as can be seen in Figure 2a. Taking into account the gauge volume of 0.13 cm^3, detector characteristics, and the geometrical arrangement of the experiment, the grain size of the NiCo2W phase in the W_0 sample can be estimated to be in the range 0.2–1 mm.

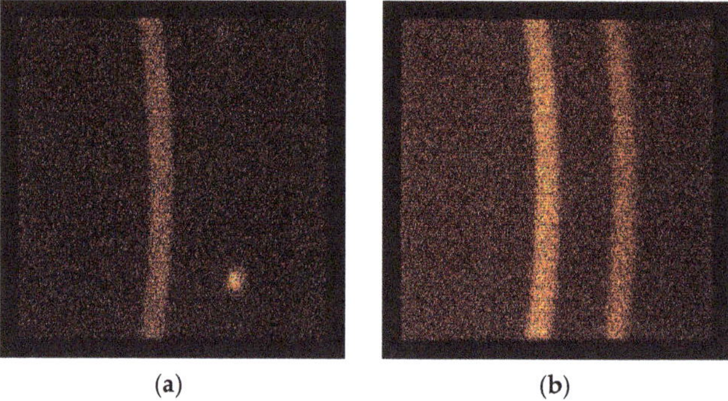

(**a**) (**b**)

Figure 2. Part of the diffractogram taken with 2D detector at TKSN-400 diffractometer (in the angular range 12.5–22°). (**a**) W_0 sample: Left strip, W-B2 110 reflection; right spot, NiCo2W 111 large-grain reflection; (**b**) W_B sample: The smooth strips of intensities from W-B2 110 (left) and NiCo2W 111 (right) reflections of fine-grained phases after rotary swaging.

The large-grain microstructure is refined by rotary swaging. W_A and W_B samples already exhibit the fine-grained NiCo2W phase, as can be seen in Figure 2b, taken for the same angular range as Figure 2a. The spotty pattern of the NiCo2W phase changed here to a smooth pattern of fine-grained NiCo2W 111 reflection on the right side of the angular range while the character of the W-B2 110 reflection (on the left) remained unchanged after rotary swaging.

The finding of the large-grain microstructure of the NiCo2W phase in the W_0 sample bar stressed the necessity to rotate the samples around the vertical axis in order to minimize the influence of the large grains on the resulting diffractogram and consequently on the microstrain determination.

The results of neutron diffraction analyses were further supported via electron backscattering observations. Figure 3a,b shows EBSD scans depicting the orientations of the NiCo2W grains in W_A and W_B samples, respectively. The depicted colours in the unit triangle indicate the orientation of the axis normal to the investigated sample surface in the crystal reference frame. With respect to the W_0 sample (not shown here), the size of NiCo2W grains was significantly refined and substructure developed. The average NiCo2W grain sizes were 1.35 μm for the W_A sample and 1.0 μm for the W_B sample. Regarding the W-B2 grains, the original W powder agglomerated during sintering and formed particles with the sizes of several dozens of micrometres, as can be seen in Figure 3 and as was also reported previously [6].

It comes from the concurrent texture investigation by neutron scattering, which will be published elsewhere, that there was no preferential orientation of the NiCo2W phase in the W_0 bar. On the other hand, the NiCo2W phase was textured with <111> crystallographic directions preferentially oriented along the sample bar axis after rotary swaging. The texture was significantly stronger for the cold swaged sample W_A than for the warm swaged sample (W_B). There is a certain relationship between the texture and mechanical properties. Nevertheless, the relationship is not described here purposely, as it will be a topic of a detailed study published in a future paper.

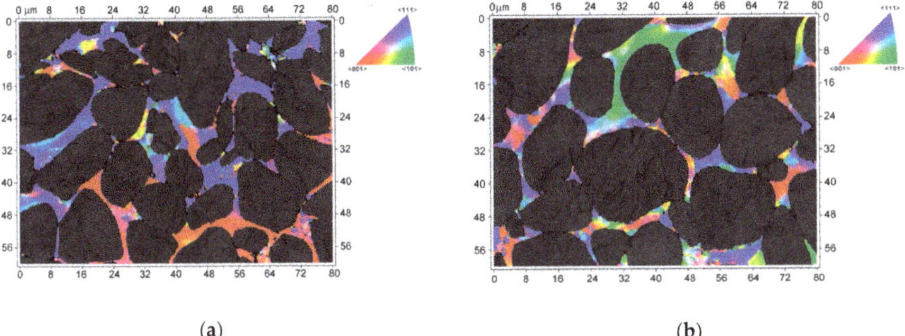

Figure 3. Electron backscatter diffraction (EBSD) scan of the surface perpendicular to the sample-bar axis depicting NiCo2W phase for sample: (**a**) W_A; (**b**) W_B. The depicted colors in the unit triangle indicate the orientation of the axis normal to the investigated sample surface in the crystal reference frame. The gray areas are W-B2 agglomerates.

3.2. Data for Microstrain Determination

Figure 4a–c display the zoomed selected angular range of the measured and calculated diffractograms for all three samples with indexed reflections for both W-B2 and NiCo2W phases. In order to display the extent of sample peak broadening, Figure 4d additionally shows W_B sample data together with a pattern calculated without any sample broadening effect.

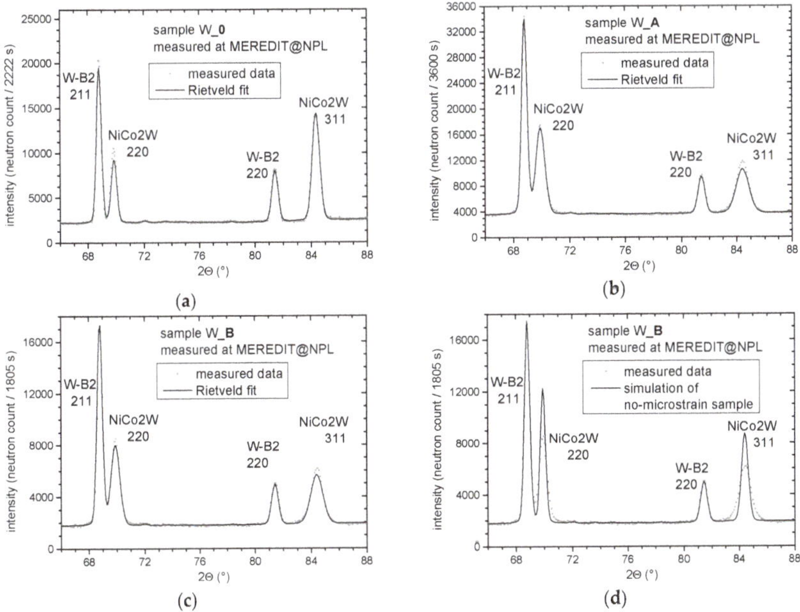

Figure 4. Detail of measured and calculated diffractograms showing the same indexed reflections from both W-B2 and NiCo2W phases in (**a**) W_0, (**b**) W_A and (**c**) W_B samples. (**d**) Additional W_B sample data together with theoretical simulation of a sample profile without any microstrain present in the NiCo2W phase (i.e., only the instrumental broadening effect is present).

By comparing the measured peaks (see Figure 4) of the W-B2 phase for the sample without deformation and samples after the RS process, no significant changes are recognized. On the other hand, NiCo2W reflections broadened significantly with respect to pure instrumental broadening (see Figure 4d). This indicates an increase in microstrain and dislocation density in NiCo2W phase while the W-B2 phase was not affected significantly.

The microstructural fit to the measured data is discussed in Section 4.

3.3. Electron Microscopy

Figure 5 shows a TEM scan of the W_A sample taken in a location near the NiCo2W/W-B2 interface. The substantial presence of dislocations within the NiCo2W matrix was observed.

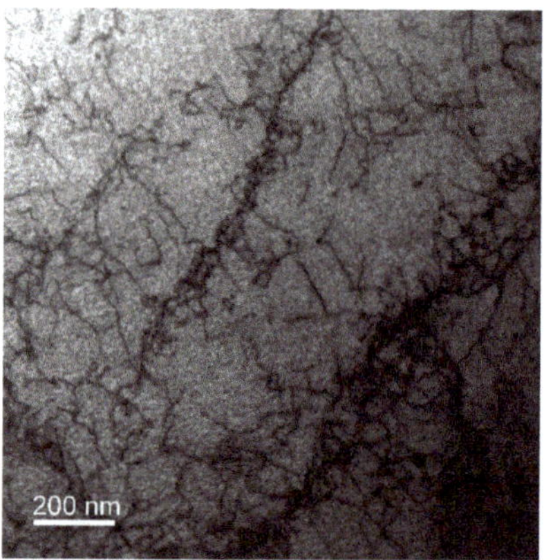

Figure 5. Transmission electron microscope image of W_A sample NiCo2W phase.

3.4. Material Properties

The stress-strain curves for W_0, W_A, and W_B samples are depicted in Figure 6. The sintered W_0 sample exhibited the lowest UTS of approximately 860 MPa. On the other hand, the sample featured a relatively high maximum elongation of more than 18%. As a result of the intensive imposed shear strain, the strength increased substantially; however, plasticity (maximum elongation) decreased after both swaging regimes. The total strengthening, i.e., UTS, was higher for the W_A sample, whereas the W_B sample featured higher plasticity.

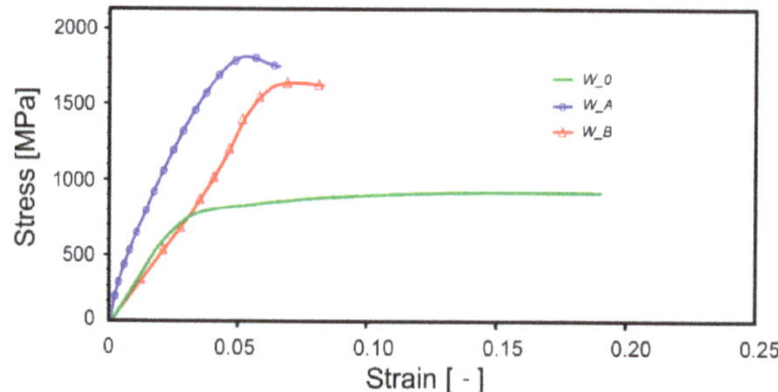

Figure 6. Experimental stress-strain curves for W_0, W_A, and W_B samples.

The physical properties measured via ultrasound are depicted in Table 1 (the average value from 5 independent measurements taken per sample).

Table 1. Physical properties resulting from ultrasound measurements.

Sample	Young's Modulus (GPa)	Shear Modulus (GPa)	Poisson's Ratio (-)
W_0	340	130	0.280
W_A	350	137	0.278
W_B	359	141	0.270

4. Data Analysis and Discussion

Detailed analysis of the measured neutron diffraction data was carried out with the intention to determine microstrain, dislocation type and dislocation density. With the support of the data obtained by other techniques, these microstructural parameters are related to the determined mechanical properties.

4.1. Microstrain Determination

First, a phenomenological approach was used to determine microstrain and the line integral breaths of all the measured reflections. FullProf software [30] enables the fitting of a phenomenological model of peak broadening caused by microstructural features, particularly microstrain and grain size, provided that the instrumental broadening is known across the whole measured 2θ range. The instrument profile dependency on scattering angle was obtained by measuring and fitting the standard SiO_2 powder sample in the identical instrument setup. The profile function parameters were extracted and stored in the instrument resolution file and this file was used during structural and microstructural full-pattern refinement. Then, the FullProf refinement results in the sample contribution to the reflection broadening. This sample broadening of the diffraction peaks is reported in the further text.

The reflection profiles of the tungsten grains (W-B2 phase) in the W_0 sample exhibited no sample broadening. Therefore, no measurable microstrain is present for this phase in the sintered sample. The small broadening of diffraction peaks in the W-B2 phase after rotary swaging was observed. As the W-B2 phase grains were sufficiently large, no grain-size broadening was present. The sample broadening of the W-B2 peaks was then satisfactorily fit using isotropic microstrain. The output was the maximum (upper limit) strain e [31] which in fact represents the microstrain as it is connected with the root mean square strain (RMSS, $\langle \varepsilon_0^2 \rangle^{1/2}$) through a constant scaling factor, $\langle \varepsilon_0^2 \rangle^{1/2} = \sqrt{2/\pi}e$. The determined W-B2 phase upper limit strain for the W_A and W_B samples was $e_{W\_A} = 12.4 \times 10^{-4}$ and $e_{W\_B} = 10.8 \times 10^{-4}$, respectively. It can be seen that there is slightly higher microstrain in the W_A sample than in the W_B sample.

In the case of the NiCo2W phase, peak broadening was already present in the W_0 sample. After rotary swaging, the sample broadening effect still significantly increased. Although RS procedure refined the NiCo2W grains, they still remained sufficiently large in the measured samples (see Figure 3). The grain-size broadening was thus not expected. As the reflections from (200) family were visibly more broadened, the anisotropic strain broadening using Stephens formalism [32,33] was used to determine the microstrain contribution. The fit was successfully carried out assuming only microstrain broadening (i.e., no size broadening). Gaussian profile, as usually done for microstrain broadening, was used to satisfactorily describe the sample broadening effect.

The outputs from the refinement using the anisotropic strain broadening are values of integral breadth for the individual reflections of the NiCo2W phase in reciprocal space, $\beta^* = (\beta \cos\theta)/\lambda$, where β is the integral breath in 2θ scale, and also microstrain values for the measured diffraction peaks of the NiCo2W phase in all three samples. The sample integral breadth β^* of each NiCo2W reflection is plotted as a function of reciprocal lattice spacing of the particular reflection d^* in Figure 7 (classical Williamson-Hall plot [31]), for W_0 and W_B samples.

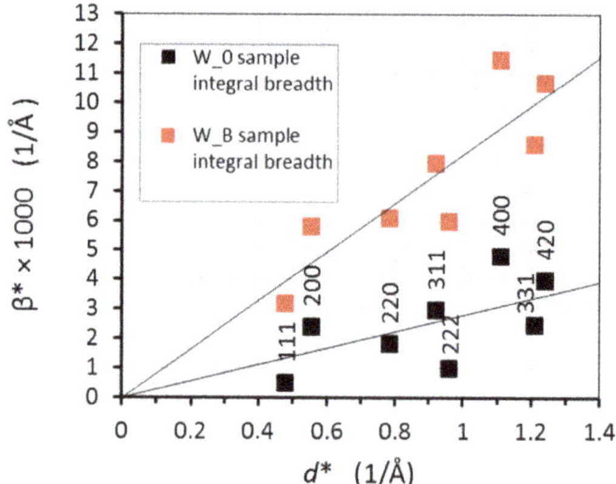

Figure 7. Williamson-Hall plot of integral breadths in reciprocal space for W_0 and W_B samples shown with linear fits through the points.

When considering only the strain broadening component as mentioned above, the upper limit strain e is connected with the integral breath by the formula $\beta^* = 2\,ed^*$, i.e., its average value can be calculated from the slope of the linear dependence of β^* on d^* [31]. The linear fits through the points are shown in Figure 7. The determined average sample microstrain for the individual W_0, W_A, and W_B samples is then $e_{NiCo\_0} = 14.2 \times 10^{-4}$, $e_{NiCo\_A} = 44.2 \times 10^{-4}$ and $e_{NiCo\_B} = 41.2 \times 10^{-4}$, respectively.

It can be seen that the microstrain in NiCo2W phase very significantly (approximately 3 times) increased after rotary swaging. Further, there is a slightly higher NiCo2W-phase microstrain in the W_A sample than in the W_B sample. Most probably, swaging at the temperature of 900 °C (W_B sample) enabled a partial dynamic recrystallization of NiCo2W phase.

The anisotropic character of the NiCo2W phase integral breadths, visible in the Williamson-Hall plot (Figure 7), indicates that microstrain is caused by dislocations. Further, the substantial presence of dislocations in the NiCo2W matrix was confirmed by TEM observations (Figure 5). Therefore, the data (i.e., acquired integral breaths for the NiCo2W phase individual reflections) were further analyzed in order to determine the active slip systems and to estimate the dislocation density after the thermomechanical processing.

4.2. Dislocation and Slip System Type

The fact that dislocation line broadening is usually anisotropic, i.e., depends on *hkl* reflection, is well known (see [34] and references therein). It is given by the anisotropic characters of the displacement fields of dislocations (line defects), which, moreover, are different for different types of dislocations and slip systems. Then, the anisotropy analysis can be in principle used to characterize the particular types of occurring dislocations and activated slip system [35]. Therefore, we tested this possibility also in the NiCo2W phase for the sintered and rotary swaged samples.

The anisotropy is characterized by the dislocation average contrast factors $C_{\alpha-hkl}$, which can be calculated with help of the ANIZC program [36] for various types of dislocations of given characters and slip system α and for all the measured *hkl* reflections. The calculations (i.e., determination of the average contrast factor for the possible types of dislocations) were performed for the measured NiCo2W reflections. To perform the calculations, elastic constants of the material were needed [37]. In order to use as precise values of the elastic constants for the NiCo2W solid solution as possible, the assumption of a linear combination of the individual elastic constants of the original constituents, i.e., Ni and Co, present in the ratio of 6:1, was made. The considered values found in the literature were the following: nickel [38]—C_{11} = 256.5 GPa, C_{12} = 151.5 GPa, C_{44} = 123.9 GPa; β-cobalt (fcc) [39]—C_{11}

= 239.8 GPa, C_{12} = 163.4 GPa, C_{44} = 133.4 GPa. Their linear combinations used to estimate the elastic constants of the alloy were then: C_{11} = 246.9 GPa, C_{12} = 156.1 GPa, C_{44} = 125.5 GPa.

After calculation of the average contrast factors $\overline{C^{\alpha}_{hkl}}$, a modified Williamson-Hall plot [37], i.e., the dependence of β* on $d^*C^{1/2}$, can be drawn. The results for the W_0, W_A, and W_B samples are depicted in Figure 8a–o. As can be seen, the best result (concerning fitting the linear dependence) for the W_0 sample was acquired for screw dislocations and <111> slip system (Figure 8e; the corresponding modified Williamson-Hall plot is marked by a red frame). All the other tested dislocation types and slip systems were far worse.

On the other hand, the slip system of <111> {110} edge dislocations fits the best to the measured integral breadths of W_A and W_B samples (Figure 8g,l, respectively, marked by a red frame). Nevertheless, the edge dislocations with <110> {111} slip system (Figure 8f,k), as well as the screw dislocations with <110> slip system (Figure 8i,n), exhibited very good linear fits for the rotary swaged samples, too. The two remaining tested dislocation types and slip systems (screw dislocations and <111> slip system, and edge dislocations with <111> {211} slip system) did not exhibit a sufficiently good fit for any of the investigated samples.

Obviously, certain microstrain was present within the structure of the NiCo2W phase already after sintering and subsequent cooling. The dislocations produced during the sintering/quenching process were predominantly of a screw type with <111> slip system. After rotary swaging (W_A and W_B samples), the microstrain increased significantly (~3 times) and the deformation mechanism changed either to edge dislocations with <111> {110} or <110> {111} slip system (the latter occurs typically in fcc structures [40]), or to screw dislocations with <110> slip system.

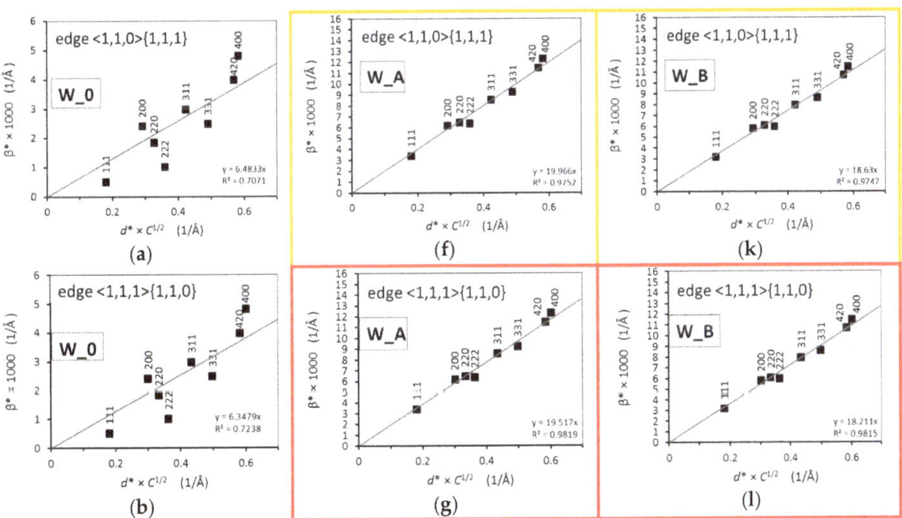

Figure 8. *Cont.*

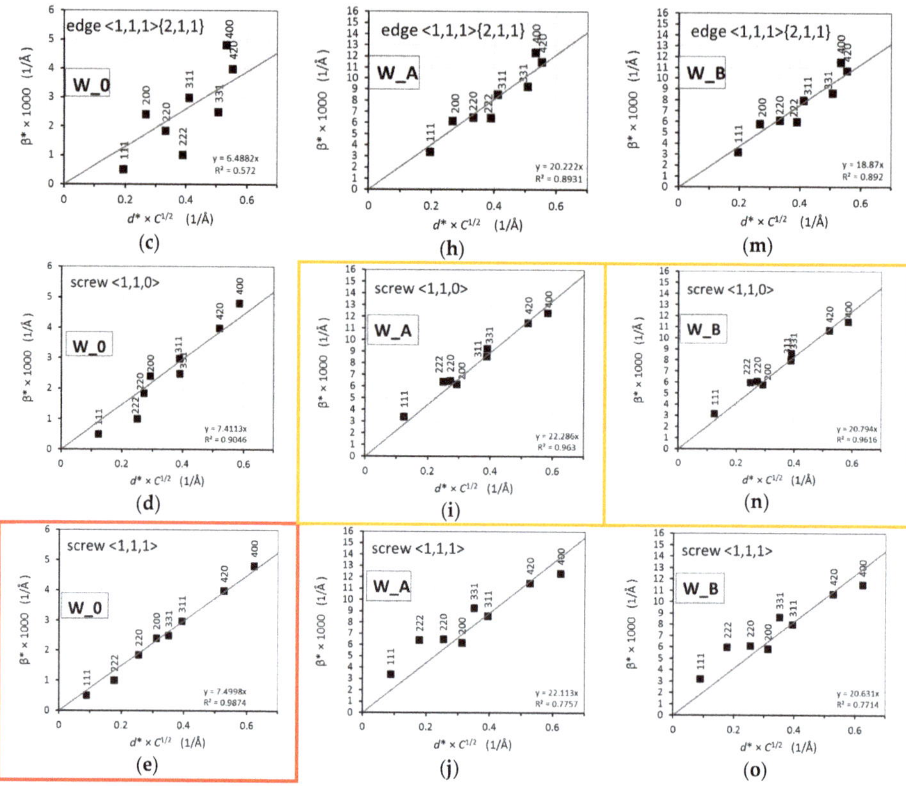

Figure 8. Modified Williamson-Hall plots considering various dislocation types and slip systems for samples: (**a–e**) W_0; (**f–j**) W_A; and (**k–o**) W_B. The red color framed plots represent the best-fitting model, the orange ones represent sufficiently good fits. These dislocation types and slip systems are considered as the most probable for the NiCo2W phase of the pseudoalloy and are used for a further evaluation (see text).

Figure 9a then summarizes the modified Williamson-Hall plots of integral breadths for the best fitting slip systems for W_0, W_A, and W_B samples (edge dislocations with <111> {110} slip system for W_A and W_B samples). It should be stressed that the quality of fitting for the edge dislocations with <110> {111} slip system and screw dislocations with <110> slip system was almost equal (see Figure 8). A combination of edge dislocation with screw dislocation system was tested as well for W_A and W_B samples. The result for the combination of edge dislocations with <110> {111} slip system and screw dislocations with <110> slip system (the ratio of influence on integral breadth was assumed to be 50%:50%) is shown in Figure 9b. A very similarly good result (not shown here) was obtained for the 50%:50% combination of the edge dislocation with <1,1,1> {1,1,0} slip system and the screw dislocation with <1,1,0> slip system.

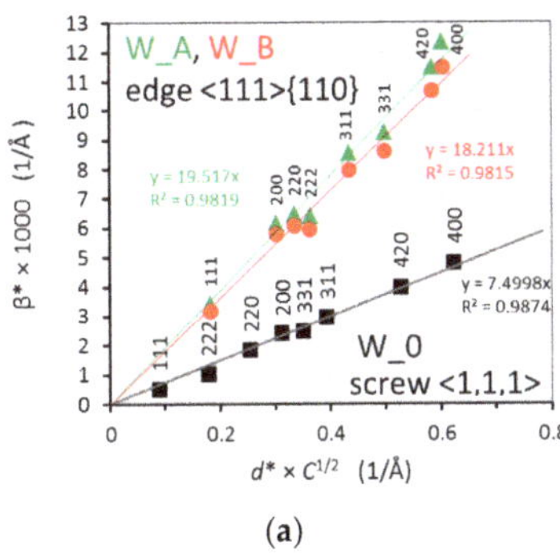

(a)

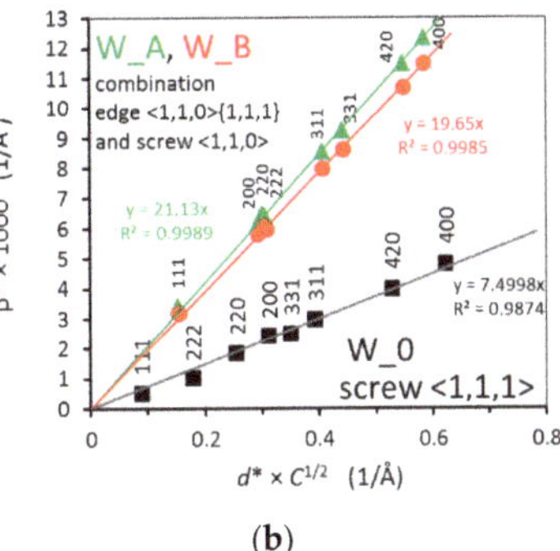

(b)

Figure 9. Summarized modified Williamson-Hall plots of integral breadths for W_0, W_A, and W_B samples shown with linear fits through the points: the edge dislocation with <1,1,1> {1,1,0} slip system for W_A and W_B samples (**a**); 50%:50% combination of edge dislocations with <110> {111} slip system and screw dislocations with <110> slip system (**b**).

4.3. Dislocation Density

Information on dislocation types, slip systems and also on dislocation densities may be very useful for modelling the material behavior using crystal plasticity [41]. Based on the results of line broadening of the NiCo2W phase, dislocation densities ρ can in principle be estimated using various numerical procedures [34,35,42–46].

Powder diffraction techniques, in general, have limited accuracy in determining dislocation densities. Precise descriptions of the instrument profile function and the size of the instrumental broadening are essential for this purpose. However, even for precisely determined instrument parameters, another important aspect is the source intensity and the peak-to-background ratio, which are usually limiting in the case of neutron diffraction. Thus, only a rather limited accuracy on the absolute scale of the estimated dislocation densities can be achieved using the above-mentioned analysis of the neutron diffraction line broadening for the NiCo2W phase. Nevertheless, such information has an important indicative value when comparing relatively the obtained dislocation densities for the individual samples.

For the estimation of the density, we used one of the approaches presented in [35]. The integral breath of each reflection can be approximated by,

$$\beta^*_{hkl} = \frac{2\sqrt{2}\,(\ln P)^{\frac{3}{2}}}{4\ln P - \ln(\ln P)}\, d^*_{hkl} C^{\frac{1}{2}}_{hkl}\, b\, \rho^{\frac{1}{2}} \tag{1}$$

where b is Burgers vector, and factor P is related to the correlation in the dislocation arrangement.

Assuming that the dislocation correlation factor P [35] can be reasonably estimated, dislocation density ρ can be determined based on Equation (1). Due to the moderate resolution of the diffraction data, P could not be determined directly from the diffraction profile analysis. For further calculations, the value 10 for the P factor was used which is a reasonable value for Gaussian profiles [35] employed in our analysis of the strain broadening. The estimated error in absolute value for ρ was −20% (in this case, P would be equal to 15) and +50% (in this case, P would be equal to 5).

Considering the above-mentioned assumptions, the dislocation densities for the most favorable slip system in NiCo2W for all the investigated samples were calculated and they are listed below:

W_0: 1.7×10^{11} cm/cm^3 (screw dislocations with <111> slip system);
W_A: 8.7×10^{11} cm/cm^3 (either edge dislocations with <110> {111} slip system or <110> screw dislocations);

W_B: 7.6 × 10^{11} cm/cm^3 (either edge dislocations with <110> {111} slip system or <110> screw dislocations).

As explained above, absolute values of dislocation densities are burdened by a large error. Nevertheless, a relative comparison of the obtained values of dislocation densities between the individual samples brings significantly more precise information as the relative error of the dislocation density values is expected to be similar for all the samples. Then, it can be seen that the dislocation densities increased approximately 5 times after rotary swaging, and that the dislocation density is 15% higher for the sample swaged at room temperature than for the sample deformed at 900 °C.

4.4. Mechanical Properties

The measured stress-strain curves (Figure 6) are in accord with the results of the microstructural study. Relatively low UTS and high maximum elongation for the sintered W_0 sample correspond to the coarse grain size of the NiCo2W phase, which, together with the low dislocation density, did not ensure sufficient strengthening.

The plasticity (maximum elongation) decreased after both swaging regimes. Nevertheless, the intensive imposed shear strain imparted significant accumulation of dislocations in the NiCo2W phase and the strength of the material increased substantially. The accumulated dislocations after RS provide hardening to the NiCo2W phase and thus enable effective transfer of the imposed strain to the W-B2 phase. Although the imposed deformation is predominantly accumulated in the NiCo2W phase, it does not mean that the W-grains are not strained at all. Tungsten grains deformation was already observed in the past [19,20,47]. Also in this study, a small level of microstrain within the W-B2 phase in the rotary swaged samples was detected (see Section 4.1).

The W_A sample, featuring the highest dislocation density and notable presence of microstrain in the NiCo2W phase, exhibited higher total strengthening (i.e., UTS), whereas the W_B sample featured higher plasticity. Mutual comparison of the stress-strain curves for the W_A and W_B samples also revealed a more gradual strengthening (i.e., smoother curve shape) of the W_B sample. Swaging at the temperature of 900 °C enabled the NiCo2W phase to dynamically recrystallize, which was documented by the very small average grain size of 1.0 μm (Section 3.1) and decreased microstrain values (Section 3.2), and provided it with the ability to consume a greater amount of the imposed energy. This finding is in accordance with the results documented by Katavič et al. [48] who reported the hardening rate of the matrix during intensive shear deformation of tungsten pseudoalloys to be more than two times higher than the hardening rate of W particles up to approximately 15% deformation.

5. Conclusions

This study focused on characterization of the effects of rotary swaging at various temperatures on a WNiCo tungsten heavy alloy via determination of microstrain and characterization of dislocation types and activated slip systems.

The results showed that the original sintered sample consisted of fine-grained spherical W-B2 type agglomerates surrounded by a coarse-grained NiCo2W matrix. The W-B2 agglomerates did not feature any significant microstrain. However, the NiCo2W matrix exhibited microstrain (magnitude 14.2 × 10^{-4}) resulting mainly from the presence of screw dislocations with <111> slip system.

Both cold and warm rotary swaging then imparted grains fragmentation for the NiCo2W matrix and resulted in formation of fine-grained structures within the NiCo2W phase. From results presented elsewhere [49,50], it is also clear that the W-B2 phase is refined by rotary swaging. Further, neutron diffraction revealed that the microstrain increased three times in the NiCo2W phase (44.2 × 10^{-4} and 41.2 × 10^{-4} for W_A and W_B samples, respectively). On the other hand, a rather low level of microstrain was detected in the tungsten W-B2 phase of the composite after the rotary swaging.

The measured mechanical parameters correspond to the results of the microstructural characterization. The swaged samples exhibited substantial strengthening which was primarily caused by the increase in dislocation density (~5× for the 900 °C sample, and even approximately 10% more for the cold swaged one) in the NiCo2W phase. The 20 °C swaged bar featured the ultimate tensile strength of almost 1900 MPa. It can be concluded from the neutron diffraction that the dominant deformation mechanisms for both the 20 °C and 900 °C rotary swaged samples were edge dislocations with <111> {110} or <110> {111} slip system, or screw dislocations with <1,1,0> slip system in NiCo2W phase. A combination of the above-mentioned systems is most probable for the NiCo2W phase as these combinations lead to the best modified Williamson-Hall plot.

Author Contributions: Conceptualization, P.S., L.K. and R.K.; Data curation, P.S., L.K., P.B. and R.K.; Formal analysis, P.S. and L.K.; Funding acquisition, P.S. and R.K.; Investigation, P.S., L.K., P.B., R.K. and C.H.; Methodology, P.S., P.B. and C.H.; Project administration, P.S. and R.K.; Resources, P.S., L.K., R.K. and C.H.; Software, P.B.; Supervision, P.S. and R.K.; Validation, P.S., L.K., P.B. and R.K.; Visualization, P.S. and L.K.; Writin—original draft, P.S.; Writing—review & editing, L.K., P.B. and R.K. All authors have read and agreed to the published version of the manuscript.

Funding: This research was funded by Czech Science Foundation, grant number 19-15479S. Measurements were carried out at the CANAM infrastructure of the NPI CAS Řež, supported through MŠMT project No. LM2015056, using the neutron source included in the infrastructure Reactors LVR-15 and LR-0, which is financially supported by the Ministry of Education, Youth and Sports-project LM2015074. In the study, equipment acquired within OPCZ.02.1.01/0.0/0.0/16_013/0001812 project supported by the European Structural and Investment Funds was used.

Acknowledgments: The authors are thankful for the technical support of Pravdomil Toman, Božena Michalcová and Alex Pylypchuk.

Conflicts of Interest: The authors declare no conflict of interest.

References

1. Kocich, R.; Kunčická, L.; Dohnalík, D.; Macháčková, A.; Šofer, M. Cold rotary swaging of a tungsten heavy alloy: Numerical and experimental investigations. *Int. J. Refract. Met. Hard Mater.* **2016**, *61*, 264–272. [CrossRef]
2. Durlu, N.; Caliskan, N.K.; Sakir, B. Effect of swaging on microstructure and tensile properties of W-Ni-Fe alloys. *Int. J. Refract. Met. Hard Mater.* **2014**, *42*, 126–131. [CrossRef]
3. Chen, B.; Cao, S.; Xu, H.; Jin, Y.; Li, S.; Xiao, B. Effect of processing parameters on microstructure and mechanical properties of 90W–6Ni–4Mn heavy alloy. *Int. J. Refract. Met. Hard Mater.* **2015**, *48*, 293–300. [CrossRef]
4. Liu, J.; Li, S.; Fan, A.; Sun, H. Effect of fibrous orientation on dynamic mechanical properties and susceptibility to adiabatic shear band of tungsten heavy alloy fabricated through hot-hydrostatic extrusion. *Mater. Sci. Eng. A* **2008**, *487*, 235–242. [CrossRef]
5. Liu, J.; Li, S.; Zhou, X.; Wang, Y.; Yang, J. Dynamic Recrystallization in the Shear Bands of Tungsten Heavy Alloy Processed by Hot-Hydrostatic Extrusion and Hot Torsion. *Rare Met. Mater. Eng.* **2011**, *40*, 957–960. [CrossRef]
6. Kunčická, L.; Kocich, R.; Hervoches, C.; Macháčková, A. Study of structure and residual stresses in cold rotary swaged tungsten heavy alloy. *Mater. Sci. Eng. A* **2017**, *704*, 25–31. [CrossRef]
7. Jinglian, F.; Xing, G.; Meigui, Q.; Tao, L.; Shukui, L.; Jiamin, T. Dynamic Behavior and Adiabatic Shear Bands in Fine-Grained W-Ni-Fe Alloy under High Strain Rate Compression. *Rare Met. Mater. Eng.* **2009**, *38*, 2069–2074. [CrossRef]
8. Zhigang, W.; Shisheng, H.; Yongchi, L.; Cunshan, F. Dynamic Properties and Ballistic Performance of Pre-Torqued Tungsten Heavy Alloys. In Proceedings of the 17th International Symposium on Ballistics, Midrand, South Africa, 23–27 March 1998; pp. 391–398.
9. Xiaoqing, Z.; Shukui, L.; Jinxu, L.; Yingchun, W.; Xing, W. Self-sharpening behavior during ballistic impact of the tungsten heavy alloy rod penetrators processed by hot-hydrostatic extrusion and hot torsion. *Mater. Sci. Eng. A* **2010**, *527*, 4881–4886. [CrossRef]
10. Kocich, R.; Kunčická, L.; Král, P.; Macháčková, A. Sub-structure and mechanical properties of twist channel angular pressed aluminium. *Mater. Charact.* **2016**, *119*, 75–83. [CrossRef]
11. Kunčická, L.; Kocich, R.; Král, P.; Pohludka, M.; Marek, M. Effect of strain path on severely deformed aluminium. *Mater. Lett.* **2016**, *180*, 280–283. [CrossRef]

12. Yuan, Y.; Ma, A.; Gou, X.; Jiang, J.; Arhin, G.; Song, D.; Liu, H. Effect of heat treatment and deformation temperature on the mechanical properties of ECAP processed ZK60 magnesium alloy. *Mater. Sci. Eng. A* **2016**, *677*, 125–132. [CrossRef]
13. Lukac, P.; Kocich, R.; Greger, M.; Padalka, O.; Szaraz, Z. Microstructure of AZ31 and AZ61 Mg alloys prepared by rolling and ECAP. *Kov. Mater.* **2007**, *45*, 115–120.
14. Hlaváč, L.M.; Kocich, R.; Gembalová, L.; Jonšta, P.; Hlaváčová, I.M. AWJ cutting of copper processed by ECAP. *Int. J. Adv. Manuf. Technol.* **2016**, *86*, 885–894. [CrossRef]
15. Levin, Z.S.; Srivastava, A.; Foley, D.C.; Hartwig, K.T. Fracture in annealed and severely deformed tungsten. *Mater. Sci. Eng. A* **2018**, *734*, 244–254. [CrossRef]
16. Wu, Y.; Hou, Q.; Luo, L.; Zan, X.; Zhu, X.; Li, P.; Xu, Q.; Cheng, J.; Luo, G.; Chen, J. Preparation of ultrafine-grained/nanostructured tungsten materials: An overview. *J. Alloys Compd.* **2019**, *779*, 926–941. [CrossRef]
17. Kocich, R.; Kunčická, L.; Král, P.; Lowe, T.C. Texture, deformation twinning and hardening in a newly developed Mg–Dy–Al–Zn–Zr alloy processed with high pressure torsion. *Mater. Des.* **2016**, *90*, 1092–1099. [CrossRef]
18. Ma, Y.; Zhang, J.; Liu, W.; Yue, P.; Huang, B. Microstructure and dynamic mechanical properties of tungsten-based alloys in the form of extruded rods via microwave heating. *Int. J. Refract. Met. Hard Mater.* **2014**, *42*, 71–76. [CrossRef]
19. Gong, X.; Fan, J.L.; Ding, F.; Song, M.; Huang, B.Y.; Tian, J.M. Microstructure and highly enhanced mechanical properties of fine-grained tungsten heavy alloy after one-pass rapid hot extrusion. *Mater. Sci. Eng. A* **2011**, *528*, 3646–3652. [CrossRef]
20. Fortuna, E.; Zielinski, W.; Sikorski, K.; Kurzydlowski, K.J. TEM characterization of the microstructure of a tungsten heavy alloy. *Mater. Chem. Phys.* **2003**, *81*, 469–471. [CrossRef]
21. Yu, Y.; Zhang, W.; Chen, Y.; Wang, E. Effect of swaging on microstructure and mechanical properties of liquid-phase sintered 93W-4.9(Ni, Co)-2.1Fe alloy. *Int. J. Refract. Met. Hard Mater.* **2014**, *44*, 103–108. [CrossRef]
22. Kocich, R.; Kunčická, L.; Král, P.; Strunz, P. Characterization of innovative rotary swaged Cu-Al clad composite wire conductors. *Mater. Des.* **2018**, *160*, 828–835. [CrossRef]
23. Kiran, U.R.; Panchal, A.; Sankaranarayana, M.; Nandy, T.K. Tensile and impact behavior of swaged tungsten heavy alloys processed by liquid phase sintering. *Int. J. Refract. Met. Hard Mater.* **2013**, *37*, 1–11. [CrossRef]
24. Kunčická, L.; Kocich, R. Deformation behaviour of Cu-Al clad composites produced by rotary swaging. *IOP Conf. Ser. Mater. Sci. Eng.* **2018**, *369*, 012029.
25. Kunčická, L.; Kocich, R.; Strunz, P.; Macháčková, A. Texture and residual stress within rotary swaged Cu/Al clad composites. *Mater. Lett.* **2018**, *230*, 88–91. [CrossRef]
26. Kunčická, L.; Lowe, T.C.; Davis, C.F.; Kocich, R.; Pohludka, M. Synthesis of an Al/Al2O3 composite by severe plastic deformation. *Mater. Sci. Eng. A* **2015**, *646*, 234–241. [CrossRef]
27. Beran, P.; Ivanov, S.A.; Nordblad, P.; Middey, S.; Nag, A.; Sarma, D.D.; Ray, S.; Mathieu, R. Neutron powder diffraction study of Ba3ZnRu2-xIrxO9 (x = 0, 1, 2) with 6H-type perovskite structure. *Solid State Sci.* **2015**, *50*, 58–64. [CrossRef]
28. Davydov, V.; Lukáš, P.; Strunz, P.; Kužel, R. Single-Line Diffraction Profile Analysis Method Used for Evaluation of Microstructural Parameters in the Plain Ferritic Steel upon Tensile Straining. *Mater. Sci. Forum* **2008**, *571*, 181–188. [CrossRef]
29. Beausir, B.; Fundenberger, J.J. Analysis Tools for Electron and X-ray Diffraction, ATEX–Software. Available online: www.atex-software.eu (accessed on 22 May 2019).
30. Rodríguez-Carvajal, J. Recent advances in magnetic structure determination by neutron powder diffraction. *Phys. B Condens. Matter* **1993**, *192*, 55–69. [CrossRef]
31. Williamson, G.; Hall, W. X-ray line broadening from filed aluminium and wolfram. *Acta Metall.* **1953**, *1*, 22–31. [CrossRef]
32. Stephens, P.W. Phenomenological model of anisotropic peak broadening in powder diffraction. *J. Appl. Crystallogr.* **1999**, *32*, 281–289. [CrossRef]
33. Rodriguez-Carvajal, J.; Fernandez-Diaz, M.T.; Martinez, J.L. Neutron diffraction study on structural and magnetic properties of La_2NiO_4. *J. Phys. Condens. Matter* **1991**, *3*, 3215–3234. [CrossRef]

34. Mittemeijer, E.J.; Welzel, U. The "state of the art" of the diffraction analysis of crystallite size and lattice strain. *Z. Für Krist.* **2008**, *223*, 552–560.
35. Kužel, R. Kinematical diffraction by distorted crystals–dislocation X-ray line broadening. *Z. Für Krist. Cryst. Mater.* **2007**, *222*, 136–149. [CrossRef]
36. Borbély, A.; Dragomir-Cernatescu, J.; Ribárik, G.; Ungár, T. Computer program ANIZC for the calculation of diffraction contrast factors of dislocations in elastically anisotropic cubic, hexagonal and trigonal crystals. *J. Appl. Crystallogr.* **2003**, *36*, 160–162. [CrossRef]
37. Ungár, T.; Borbély, A. The effect of dislocation contrast on x-ray line broadening: A new approach to line profile analysis. *Appl. Phys. Lett.* **1996**, *69*, 3173–3175. [CrossRef]
38. Ledbetter, H.; Kim, S. *Monocrystal Elastic Constants and Derived Properties of the Cubic and the Hexagonal Elements*; Handbook of Elastic Properties of Solids, Fluids, and Gases; Academic Press: Cambridge, MA, USA, 2001; pp. 97–106. [CrossRef]
39. Leamy, H.J.; Warlimont, H. The Elastic Behaviour of Ni–Co Alloys. *Phys. Status Solidi B* **1970**, *37*, 523–534. [CrossRef]
40. Russell, A.; Lee, K.L. *Structure-Property Relations in Nonferrous Metals*, 1st ed.; John Wiley & Sons, Inc.: New York, NY, USA, 2005.
41. Rodney, D.; Ventelon, L.; Clouet, E.; Pizzagalli, L.; Willaime, F. Ab initio modeling of dislocation core properties in metals and semiconductors. *Acta Mater.* **2017**, *124*, 633–659. [CrossRef]
42. Wilkens, M. The determination of density and distribution of dislocations in deformed single crystals from broadened X-ray diffraction profiles. *Phys. Status Solidi* **1970**, *2*, 359–370. [CrossRef]
43. Krivoglaz, M.A.; Martynenko, O.V.; Ryaboshapka, K. Influence of correlation in dislocation arrangement on the X-ray diffraction by deformed crystals. *Fiz. Met. Met.* **1983**, *55*, 5–17.
44. Scardi, P.; Leoni, M.; Dong, Y.H. Whole diffraction pattern-fitting of polycrystalline fcc materials based on microstructure. *Eur. Phys. J. B* **2000**, *18*, 23–30. [CrossRef]
45. Ungár, T.; Gubicza, J.; Ribárik, G.; Borbély, A. Crystallite size distribution and dislocation structure determined by diffraction profile analysis: Principles and practical application to cubic and hexagonal crystals. *J. Appl. Cryst.* **2001**, *34*, 298–310. [CrossRef]
46. Vermeulen, A.C.; Delhez, R.; de Keijser, T.H.; Mittemeijer, E.J. Changes in the densities of dislocations on distinct slip systems during stress relaxation in thin aluminium layers: The interpretation of x-ray diffraction line broadening and line shift. *J. Appl. Phys.* **1995**, *77*, 5026–5049. [CrossRef]
47. Gong, X.; Fan, J.L.; Ding, F.; Song, M.; Huang, B.Y. Effect of tungsten content on microstructure and quasi-static tensile fracture characteristics of rapidly hot-extruded W–Ni–Fe alloys. *Int. J. Refract. Met. Hard Mater.* **2012**, *30*, 71–77. [CrossRef]
48. Katavić, B.; Odanović, Z.; Burzić, M. Investigation of the rotary swaging and heat treatment on the behavior of W- and γ-phases in PM 92.5W-5Ni-2.5Fe-0.26Co heavy alloy. *Mater. Sci. Eng. A* **2008**, *492*, 337–345. [CrossRef]
49. Kiran, U.R.; Khaple, S.K.; Sankaranarayana, M.; Nageswara Rao, G.V.S.; Nandy, T.K. Effect of swaging and aging heat treatment on microstructure and mechanical properties of tungsten heavy alloy. *Mater. Today Proc.* **2018**, *5*, 3914–3918. [CrossRef]
50. Kumari, A.; Sankaranarayana, M.; Nandy, T.K. On structure property correlation in high strength tungsten heavy alloys. *Int. J. Refract. Met. Hard Mater.* **2017**, *67*, 18–31. [CrossRef]

Article

Fatigue Life of 7475-T7351 Aluminum After Local Severe Plastic Deformation Caused by Machining

Petra Ohnistova [1,*], Miroslav Piska [1], Martin Petrenec [1], Jiri Dluhos [2], Jana Hornikova [3] and Pavel Sandera [3]

1 Department of Manufacturing Technology, Faculty of Mechanical Engineering, Brno University of Technology, 616 69 Brno, Czech Republic; piska@fme.vutbr.cz; mpetrenec@gmail.com
2 TESCAN ORSAY HOLDING a.s., 623 00 Brno, Czech Republic; jiri.dluhos@tescan.com
3 Central European Institute of Technology, Brno University of Technology, Brno 612 00, Czech Republic; hornikova@fme.vutbr.cz (J.H.); sandera@fme.vutbr.cz (P.S.)
* Correspondence: petra.ohnistova@vutbr.cz; Tel.: +602-578-656

Received: 7 October 2019; Accepted: 31 October 2019; Published: 3 November 2019

Abstract: The fatigue properties of thermo-mechanically treated and machined aluminum alloy 7475-T7351 have been studied. The applied advanced machining strategy induced intensive plastic deformation on the machined surface under defined cutting conditions. Therefore, a detailed study of 3D surface topography was performed. Advanced characterization of the material structure and electron back scattered diffraction mapping of selected chemical phases were performed, as well as energy dispersive X-ray analysis of the surface. Advanced mechanical properties of the material were investigated in situ with a scanning electron microscope that was equipped with a unique tensile fixture. The fatigue results confirmed an evident dispersion of the data, but the mechanism of crack nucleation was established. Fracture surface analysis showed that the cracks nucleated at the brittle secondary particles dispersed in the material matrix. The surface topography of samples that had been machined in wide range of cutting/deformation conditions by milling has not proved to be a decisive factor in terms of the fatigue behavior. The incoherent interface and decohesion between the alumina matrix and the brittle secondary phases proved to significantly affect the ultimate strength of the material. Tool engagement also affected the fatigue resistance of the material.

Keywords: crack nucleation; fatigue; plastic deformation; surface topography

---

1. Introduction

Aluminum and its alloys are used in a wide range of industrial applications. Duralumins, are employed in the aerospace industry. Due to an ideal combination of low density, high strength, good corrosion resistance, and high resistance to fatigue crack propagation, these special alloys take precedence over other structural materials [1]. Aluminum alloys of the 2000 and 7000 series are widely used for primary and secondary aircraft structures, such as frames, spares, and ribs, where any damage has a crucial impact on safety.

The high strength Al-Zn-Mg-Cu 7475 is an alloy with controlled toughness made in the form of sheets and plates, that has an ideal combination of high strength, good fracture toughness, and resistance to fatigue crack propagation. The 7475 alloy has almost 40% greater fracture toughness than the previous version, 7075 [2]. This progress in mechanical properties is a result of the reduction of the content of iron, silicon, and magnesium, and application of thermo-mechanical and heat treatments which achieve a refined grain size [3]. The 7475 alloy, in the form of plates, is usually available in different tempered conditions such as T651, T7351, and T7651 [4]. However, among modifying the chemical composition of the alloys, the mechanical properties of Al-based alloys can also be enhanced via severe plastic deformation (SPD) methods, including high pressure torsion (HPT) [5], equal channel angular pressing (ECAP) and its modifications [6–8], or rotary swaging (RS) [9].

As the significant portion of components made of alloy 7475-T7351 are processed by different machining strategies, it is difficult to determine the mechanisms of the severe plastic deformation imposed by the each machining process.

The machining process is generally characterized by chip formation as the direct result of the force interactions between the tool and the workpiece. To achieve chip separation, the stress applied between the tool and the workpiece in the chip formation zone must exceed the ultimate strength of the workpiece material. This force interaction is highly dependent on the studied material, the geometry and material of the tool, the cutting environment, and on the defined cutting conditions. However, during chip formation, three zones of plastic deformation can be always observed: (1) a primary deformation zone in front of the cutting tool edge with extreme shear deformation (γ = 2–5) and deformation rates (10^3–10^8 s^{-1}), (2) a secondary deformation zone between the chip and the tool rake face, and (3) a tertiary deformation zone between the machined surface and the tool flank face. In the zone of primary plastic deformation, the chip reaches a high temperature over its entire cross-section (in some cases near to the melting point). As a result of this elevated temperature, the metallurgical and mechanical properties change, the chip softens, the frictional force and the cutting resistance between tool and workpiece decreases, the shear plane angle increases, the chip cross section becomes thinner, and the chip speed increases. In the zone of tertiary plastic deformation, the cutting edge of the tool initiates a stress concentration at the contact zone between the tool and the workpiece [10–12]. The machined surface layer is subjected to elastic and plastic deformation at a temperature lower than the temperature of recrystallization. Therefore, material is not melted and any material texture change is permanent and the surface layer is hardened. The imposed force interaction and the thermal load result in residual stress concentration at different depths of the surface layer. These residual stresses can be compressive, increasing the fatigue limit, or tensile, reducing the fatigue limit [13]; they can also alter other mechanical and utility properties [14,15].

As it is well known that nuclei of fatigue cracks are mostly observed on the free surface of the loaded component, all previously described factors have to be taken into consideration while choosing an appropriate machining strategy and machining conditions. Initiation of fatigue cracks on the machined surface is usually closely associated with the combination of severe plastic deformation and the presence of material inclusions or severe plastic deformation and significantly deteriorated surface topography [16].

Ojolo at al. [17] presented results of the four-point bending fatigue testing of a end-milled specimen of alloy 2024. The important fatigue life increase (from 2.67×10^3 to 3.6×10^3 cycles to failure) was observed for samples machined with use of higher cutting speeds (from 3.77 m/min to 48.25 m/min). A decrease of the surface roughness was observed with an increase of the cutting speed, which may be the result of the thermal softening effect due to accumulated heat which caused a temperature rise in the machining zone. The feed speed was described as the most influential factor affecting the fatigue life. The increase of fatigue life was observed with a decrease of the feed speed (from 60 mm/min to 7 mm/min). Surface topography, on the other hand, deteriorated with increase of the feed speed. Ojolo therefore supposed that while using higher feed speeds, the teeth of the end-mill cutter do not perform perfect swiping of the entire surface of the machined zone to make a perfectly smooth surface. It was also observed that an increase in rake angle from 30° to 45° resulted in a better surface finish and increased the fatigue life of the specimens (from 2.53×10^3 to 3.49×10^3 cycles to failure).

Many other studies have been devoted to determining the effect of machining strategies on the fatigue life. Some have demonstrated the influence of the surface topography. The effect of machining and surface integrity on the fatigue life has been summarized by Novovic et al. [18]. They reported that a large dispersion of results has been found in the literature; however, fatigue life increase was mostly observed with decreasing surface roughness. Koster [19] found that for a roughness parameter (*Ra*) between 2.5 and 5 µm, the residual stress imposed by the machining process is the most important factor affecting the fatigue life of structural alloys. Koster [19] also reported that this effect was suppressed by elevated temperature, which allowed relaxation of imposed residual stresses. However, a study of progressive milling technology on the surface topography and fatigue life of aluminum alloy 7475 of Piska et al. [20] showed that in the case of the presence of material inclusions or secondary phases larger than standard topography parameters (such as average roughness of the profile, maximum depth of the valley of the roughness profile, and others), the effect of the surface topography is usually suppressed.

Regarding ever-increasing safety requirements, it is necessary to carefully analyze the effect of the surface quality, including material structure and surface topography together with residual stresses and severe plastic deformation, imposed by the machining process before releasing components into operation.

This study is, therefore, focused on the influence of the different cutting conditions and tool inclination of the face milling strategy applied on the bottom wing panel made of alloy 7475-T7351 with regard to its fatigue life during operational use. The main goal of this study is therefore to define the milling condition range that allows maintaining the optimum balance between the productivity of the production process, the quality of machined surface, and the required fatigue properties.

2. Experimental Materials and Methods

2.1. Material

The aluminum alloy 7475-T7351 in the form of 70 mm thick plates was used in this study. The heat treatment with designation T7351 denotes solution heat treatment at 470 °C, water quenching, controlled stretching, and artificial ageing (over-aged in two stages: first at 121 °C for 25 h, second at 163 °C for a period of 24–30 h). The average chemical composition of the alloy is presented in Table 1 and its basic mechanical properties are shown in Table 2.

Table 1. Chemical composition of the 7475-T7351 alloy (in wt.%) [4].

Si	Fe	Cu	Mn	Mg	Cr	Zn	Ti	Other, Each	Al
0.10 max	0.12 max	1.20–1.90	0.06 max	1.90–2.60	0.18–0.25	5.20–6.20	0.06 max	<0.05	Balance

Table 2. Mechanical properties of the 7475-T7351 alloy [4].

Thickness of the Blank Sheet (mm)	25–38	50–63	75–89
Tensile strength (MPa)	490	476	448
Yield strength (MPa]	414	393	365
Elongation (%]	9	8	8

The electron back scattered diffraction mapping (EBSD) study showed a heavily deformed structure with high anisotropy and texture of the grains (and very fine subgrains), as shown in Figure 1.

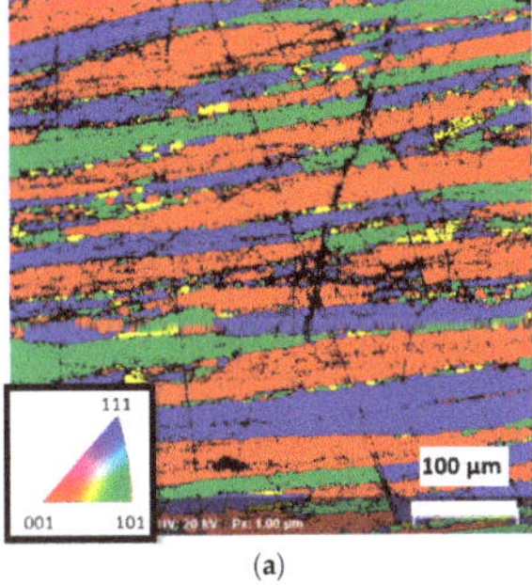

(a)

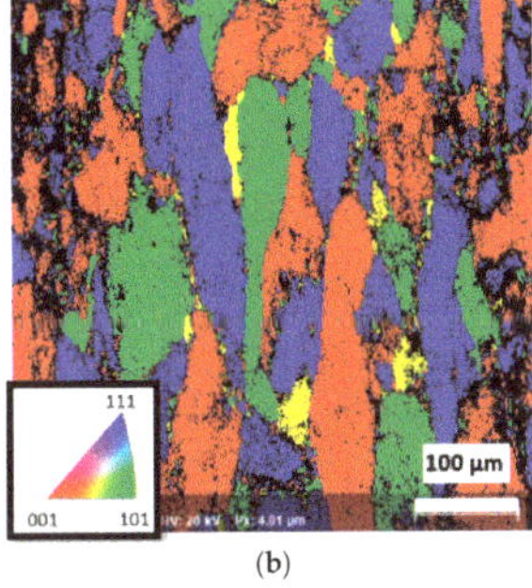

(b)

Figure 1. Structure of aluminum alloy 7475-T7351 determined by electron back scattered diffraction mapping (EBSD): (**a**) Longitudinal direction; (**b**) Transversal direction.

During STEM (Scanning Transmission Electron Microscopy) lamella analysis, three different secondary phases have been observed in the material matrix, as indicated in Figure 2. Coarse intermetallic particles Al-Cu-Fe (possibly Al_7Cu_2Fe) [21,22] and Al-Cr-Fe-Cu-Si in the range from 2 μm up to 20 μm were formed during solidification phase. Precipitated Al-Fe-Si and Al-Mg-Cr dispersoids (possibly $Al_{12}Fe_3Si$; $Al_{12}Mg_2Cr$) were formed by solid state precipitation in the grain boundaries. Third, observed secondary phases can be described as fine metastable precipitates in the material matrix (sizes from 2 nm up to 0.6 μm) and these are responsible for strengthening of the alloy (via *GP*, $\acute{\eta}$ or η) [22,23].

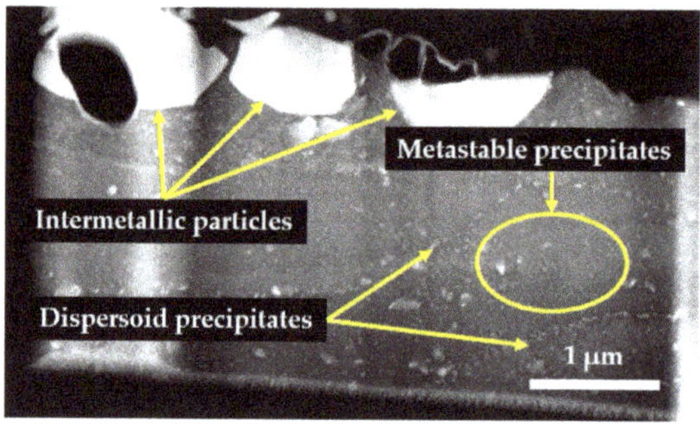

Figure 2. STEM lamella of aluminum alloy 7475-T7351—occurrence of secondary phases.

Energy dispersive X-ray spectroscopy (EDX) was used for elemental analysis of the large intermetallic particles, as shown in Figure 3.

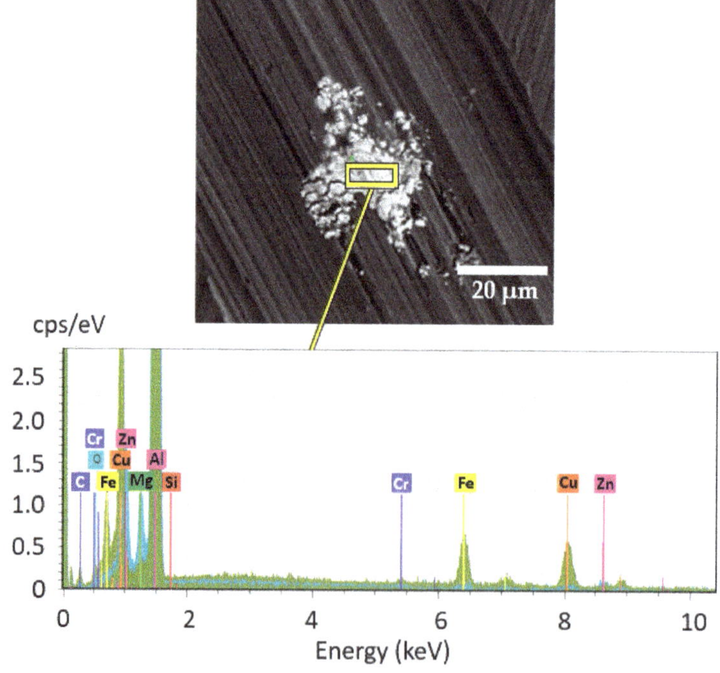

Figure 3. EDX of a selected intermetallic particle in alloy 7475-T7351.

2.2. Tool Geometry

The end-milling whole carbide tool ⌀16×55-115 mm JHF 980 Special provided by SECO Tools company, with (Ti, Al)N coating was used for advanced high feed face milling of the specimens. To exclude any potential impact of inaccurate tool geometry, optical 3D tool geometry analysis was performed using a special software subprogram of the ALICONA-IF G5 optical microscope called "Alicona Edge Master". The principle of this analysis was a gradual positioning of the reference plane perpendicular to the cutting edges of the tool. The measured results of the intersection of the reference plane with the cutting edges were statistically processed to obtain final results of the mean radius of the mean cutting edges and information about the mean cutting angles.

2.3. Surface Topography Analysis

Complex measurement of the surface topography was performed on a set of samples machined by different cutting parameters of face milling, as presented in Table 3, with the tool positioned perpendicular to the machined surface. Measurement of the surface topography for a set of samples with tool inclination of 1° was also performed. The high-resolution optical microscope ALICONA IF-G5 was used for analysis of roughness parameters ("*R*"), waviness parameters ("*W*"), Firestone–Abbott parameters, and other advanced 3D surface texture parameters ("*S*"). The measurement methodology was based on the combination of the small depth of focus of the optical system with vertical scanning. In order to perform complex detection of the surface, the high-precision optics moved vertically along the optical axis and continuously captured data from the surface. A corresponding algorithm converted the acquired sensor data into 3D information and true colour images with full depth of field [24]. Nonmeasured points in the datasets were not taken into consideration for further processing or for calculation of corresponding parameters due their low ratio (flat surfaces of samples, good fits of data with the Gaussian distribution, very low occurrence of nonmeasured points in the whole dataset). The measurement methodology was in accordance with the standard EN ISO 25178-606 [25].

Table 3. Combinations of cutting parameters of high feed face milling.

a_p = 1.50 mm; a_e = 8.00 mm		Feed per tooth, f_z (mm)		
Cutting speed, v_c (m.min^{-1})	-	**0.05**	**0.50**	**0.90**
	90	Combination 1	Combination 2	Combination 3
	200	Combination 4	Combination 5	Combination 6
-	**400**	Combination 7	N/A	N/A

a_p— axial depth of cut, a_e — radial depth of cut, f_z — feed per tooth

2.4. Force Loading Analysis of High Feed Face Milling and Induced Severe Plastic Deformation

Cutting experiments for various cutting speeds and cutting feeds were carried out with a five-axis milling center MCV 1210/Sinumerik 840D. A stationary KISTLER 957B/SW dynamometer was used for measurement of the force loading during high feed face milling for the different cutting conditions, as presented in Table 4. The results have been analyzed with DynoWare software (type 2825A, Kistler, Winthherthur, Switzerland), where mean values of the maximal instantaneous force loading in the X, Y, and Z directions were used for graphical determination of the resultant force F_{1M} and its vector decomposition to the cutting force F_C and the force perpendicular F_{CN}. The cutting force and non-deformed chip cross section A_D was used for calculation of the specific cutting energy k_c for a given material characterized by the constants c_o, the axial depth of cut a_p, the radial with of cut a_e, a angular tooth engagement ϕ and the effect of chip thickness on specific force loading expressed with the parameter mc:

$$k_c = \frac{F_c}{A_D} = \left(\frac{c_o}{a_p \times a_e}\right) \times \int_{\varphi 1}^{\varphi 2} sin^{1-mc}\varphi \times d\varphi \quad (1)$$

Table 4. Set of specimens used for force loading analysis.

a_p = 1.50 mm; a_e = 8.00 mm		Feed per tooth, (f_z mm)				
	-	**0.05**	**0.25**	**0.50**	**0.75**	**0.90**
Cutting Speed, v_c (m.min$^{-1)}$	**90**	Combination 1	Combination 2	Combination 3	Combination 4	Combination 5
	200	Combination 6	Combination 7	Combination 8	Combination 9	Combination 10
	300	Combination 11	Combination 12	Combination 13	Combination 14	Combination 15
	400	Combination 16	Combination 17	Combination 18	Combination 19	Combination 20

The basic model of continuous chip formation and the individual parameters for shear deformation and rate of the deformation can be seen in Figure 4.

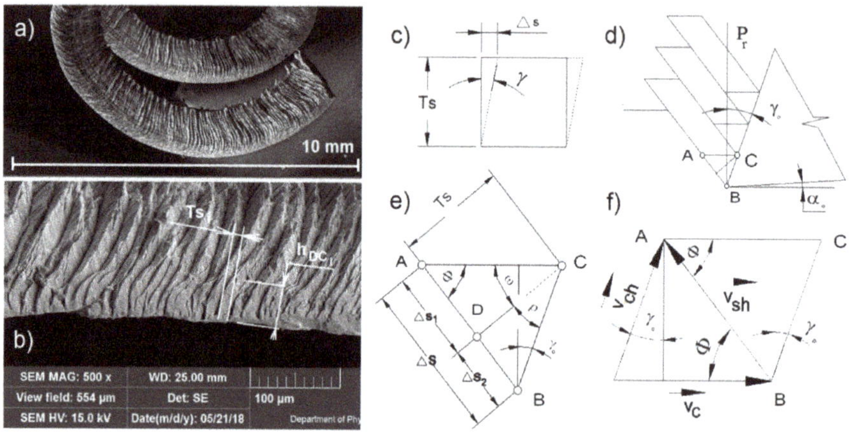

Figure 4. Analysis of the chip formation model when cutting: (**a**) Continuous chip; (**b**) Position of the measured lamella; (**c**) The shear deformation of the lamella; (**d**) Geometry of the cutting tool; (**e**) Model of the shear deformation in the primary zone; (**f**) Speed vector diagram.

The plastic flow of the material is defined with the condition of a constant volume of machined material V passing through the first deformation zone and converted to the chip:

$$Div \times V = 0 \tag{2}$$

$$V = A_D.\times v_c = A_{Dc} \times v_{ch} \tag{3}$$

where A_D is the cross section of the undeformed material entering with cutting speed v_c and A_{DC} is the cross section of the material converted to a chip, leaving with speed of chip v_{ch}. The other variables can be understood according to Figure 4a–f.

The angle of the shear plane φ is defined as

$$\phi = arctg\frac{sin\delta_o}{\Lambda - \cos\delta_o} \tag{4}$$

where the orthogonal cutting angle δ_o is sum of the orthogonal flank angle and orthogonal cutting edge angle β_o, i.e., $\delta_o = \alpha_o + \beta_o$, and Λ means the chip thickness coefficient, which is expressed as

$$\Lambda = \frac{h_{DC}}{h_D} \tag{5}$$

The shear deformation γ in the primary zone can be derived as function of shear angle φ and orthogonal rake angle γ_o

$$\gamma = \frac{cos_{\gamma_o}}{sin\phi cos(\phi - \gamma_o)} \tag{6}$$

and the rate of shear deformation sequentially as

$$\dot{\gamma} = \frac{cos\gamma_o}{cos(\phi - \gamma_o)} \frac{v_c}{T_s} \tag{7}$$

The average thickness T_s and thickness h_{DC} of the material lamella can be measured and calculated statistically by electron microscopy, as seen in Figure 4, and the orthogonal rake angle γ_o can be measured with the Alicona G5 microscope. The parameter h_D corresponds to the feed per tooth.

2.5. Fatigue Testing and Frature Surface Analysis

The objective of the fatigue testing was to examine the influence of the defined cutting conditions of the face milling on the fatigue life, as summarized in Table 5. The effect of tool inclination of 1° has also been examined.

Table 5. Combinations of cutting parameters of high feed face milling.

a_p = 1.50 mm; a_e = 8.00 mm		Feed per tooth, f_z (mm)		
Cutting Speed, v_c (m.min^{-1})	-	**0.05**	**0.50**	**0.90**
	90	Combination 1	Combination 2	Combination 3
	200	Combination 4	Combination 5	Combination 6

The geometry of the fatigue specimens was chosen to achieve the best match with final operational use of the bottom wing panel. The main criteria for appropriate specimen geometry are defined as follows:

- Specimen must allow performance analysis of the effect of the high feed face milling on the fatigue life. Therefore, the flat specimen with largest possible functional area must be chosen.
- Specimen must allow simulation of the tensile cyclic loading during operation.
- Specimen must comply with ASTM E466-15 [26] and EN 6072 [27] aviation standards.

All specimens were oriented in the L-T direction of the rolled plate. Specimens were machined at the five-axis MCV 1210/Sinumerik 840D milling center (TAJMAC ZPS, share company, Zlin, Czech Republic/Siemens AG, Erlangen, Germany). Specimens were specially protected against bending or torsion during machining, and excess heating was limited by use of CIMSTAR 597 coolant (Cimcool Industrial Products B.V., Vlaardingen, the Netherlands) of 10% volume concentration, 20 bar pressure, and 20 L/min flow rate. All functional areas of the specimens have been protected against scratches, all sharp edges rounded to 0.3 mm radius, and all other stress concentrators have been removed.

Fatigue testing has been performed at special axial testing machine BISS and parameters of the testing are defined as follows:

- Fluctuating tensile cycle with stress ratio, $R = 0.1$.
- Frequency, $f = 10$ Hz.
- Stress levels: 180 MPa, 220 MPa, 250 MPa, and 300 MPa.

The source of fatigue crack nucleation was examined with the scanning electron microscope (TESCAN ORSAY HOLDING share company, Brno, Czech Republic) TESCAN MIRA 3 operating in both secondary and backscattered electron mode. The fatigue crack initiation and propagation mechanism and the integrity of the adjacent surfaces were investigated.

2.6. In Situ Testing

A specimen with special geometry was designed for in situ tensile mechanical testing. Profiles of the specimen were cut by EDM wire cutting (Electrical Discharge Machining) and polished. Flat functional surface areas were face milled with a special high feed end-mill (SECO tool JHF 980 Special, $f_z = 0.05$ mm, $v_c = 200$ m/min, $a_p = 0.2$ mm), and all sharp edges were rounded and polished.

Testing was performed on a special in situ tensile stage MT1000 made by NewTec (10 kN, a tensile stage) and all analyses were carried out with the SEM TESCAN MIRA 3, equipped with the NewTec SoftStrain software (version 1, NEWTEC, Nîmes, France), as shown in Figure 5.

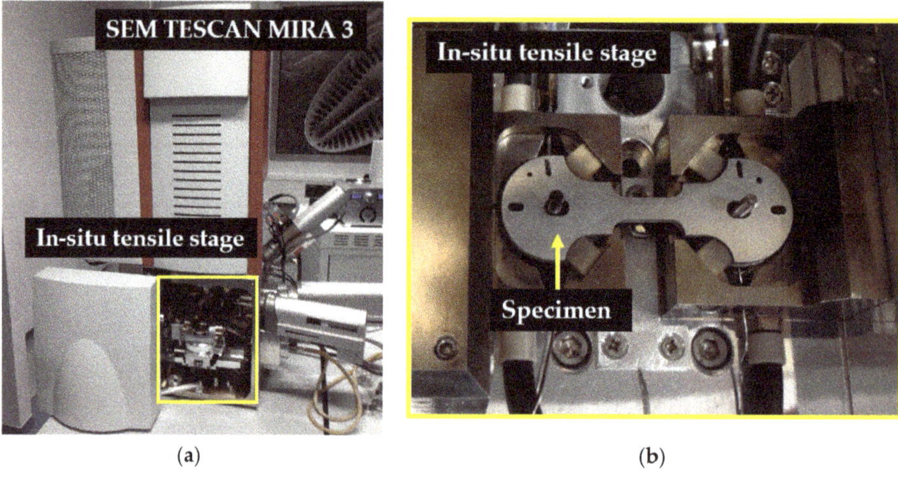

(**a**) (**b**)

Figure 5. (**a**) SEM TESCAN MIRA 3 and in situ tensile stage MT1000 made by NewTec; (**b**) Specimen clamping in the in situ tensile stage.

2.6.1. In Situ Tensile Testing

The main scope of the in situ tensile testing was to observe the crack initiation and propagation mechanism in the 7475-T7351 alloy. Two main analyses were performed: (a) Observation of the entire functional area of the specimen; and (b) detailed observation of selected intermetallic particles located at the free machined surface of the specimen. Engineering strain distribution under tensile loading in selected particle was analyzed by DIC (Digital Image Correlation) in MERCURY real-time tracking software.

2.6.2. In Situ Cyclic Testing

The aim of the in situ tensile cyclic loading was to simulate cyclic loading under the operation mode and to observe crack nucleation and short crack propagation. Detailed observation of twenty selected large intermetallic particles was performed in parallel. Parameters of the fatigue loading were defined as follow:

- Stress control in the range from 35 MPa to 350 MPa (corresponding to force loading from 245 N to 2450 N).
- Stress ratio, $R = 0.1$.
- Speed of loading: 100 N/s.

3. Results

3.1. Tool Geometry

The geometry of the tool and profile roughness of the cutting edges were analyzed and the results are presented in Tables 6 and 7. All parameters complied with manufacturer's specifications. The standard deviations were in the range of 3–4% of the mean values.

Table 6. SECO end-milling tool ⌀16×55-115 mm JHF 980 Special geometry.

Cutting Edge	Cutting Edge Radius r_n (μm)	Orthogonal Clearance Angle α_o (°)	Orthogonal Edge Angle β_o (°)	Orthogonal Rake Angle γ_o (°)
1.	7.58	9.00	65.57	15.43
2.	7.46	9.31	65.42	15.28
3.	7.48	9.28	65.18	15.54
4.	7.45	9.09	65.19	15.72
5.	7.46	9.07	65.29	15.61

Table 7. SECO end-milling tool ⌀16×55-115 mm JHF 980 Special surface profile roughness parameters.

Cutting Edge	*Ra* (μm)	*Rq* (μm)	*Rz* (μm)	*Rp* (μm)	*Rv* (μm)
1.	0.29	0.37	0.67	0.37	0.63
2.	0.27	0.34	0.86	0.34	0.69
3.	0.24	0.31	0.75	0.31	0.52
4.	0.21	0.27	0.44	0.27	0.52
5.	0.25	0.30	0.59	0.32	0.54

Ra—average roughness of the profile, *Rq*—root-mean-square roughness of the profile, *Rz*—mean peak to valley height of the roughness profile, *Rp*—maximum peak height of the roughness profile, *Rv*—maximum valley depth of the roughness profile.

3.2. Surface Topography Analysis

The effect of the defined cutting conditions and tool inclination on *R*-parameters of the surface topography was evident. For the set of samples machined with tool positioning perpendicular to the machined surface, a digression of the average values of the profile roughness parameter (measured perpendicularly to the cutting speed, along the feed speed and longitudinal axis of the samples) was observed:

- The average roughness (*Ra*) and root-mean-square roughness (*Rq*) dropped by 35% for the highest cutting speed (v_c = 400 m min^{-1})
- No linear function was observed between the cutting speed and the profile roughness parameters, as indicated in Figure 6.
- Standard deviations of the repeated measurements varied between 5% and 8% of the average values for all measurements and conditions.

On the other hand, the increase of the profile roughness parameters was observed with the increase of the feed speed (increase of the feed per tooth, f_z, from 0.05 to 0.90 mm). Some examples of the profile roughness values increasing while increasing the feed speed are mentioned below:

- The average roughness (*Ra*) increased from 2.71 to 4.30 μm (an increase of 37%), as indicated in Figure 7.
- The root-mean-square roughness (*Rq*) increased from 3.49 to 5.27 μm (an increase of 34%), as indicated in Figure 7.
- The maximum peak to valley height of roughness profile (*Rt*) increased from 28.59 to 33.78 μm (an increase of 15%), as indicated in Figure 7.

- The maximum valley depth of roughness profile (*Rv*) increased from 13.23 to 16.92 μm (an increase of 22%), as indicated in Figure 7.

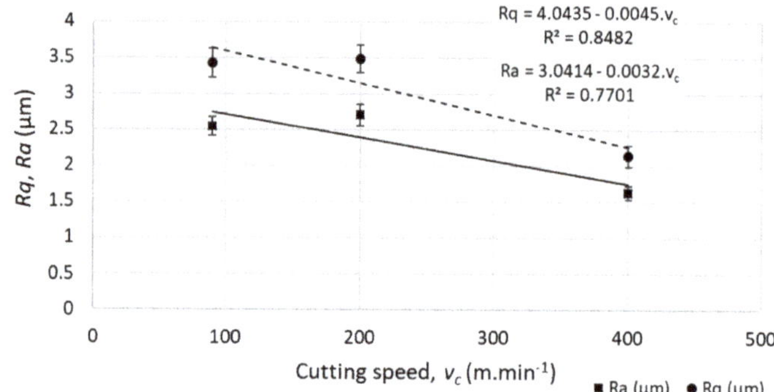

Figure 6. Example of profile roughness parameters (*Ra*—average roughness of the profile, *Rq*—root-mean-square roughness of the profile) for different cutting speed (at feed per tooth f_z = 0.05 mm). The parameter R^2 reflects the level of statistical correlation.

No linear function was observed between the cutting speed and the waviness parameters, or between the feed speed and the waviness parameters (*R*-squared parameter varied in the range from 0.1 to 0.7 for different waviness parameters), as shown in Figures 8 and 9.

The surface parameters under different cutting parameters of the face milling were examined (with tool positioning perpendicular to the machined surface). The results of the measurement can be summarized as follows:

- The increase of the feed speed (increase of the feed per tooth from 0.05 to 0.90 mm) caused an increase of surface topography parameters, as indicated in Figure 10.
- The average height of the selected area (*Sa*) increased from 2.60 to 5.30 μm.
- The root-mean-square height of the selected area (*Sq*) increased from 3.40 to 6.43 μm.
- The maximum valley depth of the selected area (*Sv*) increased from 26.77 to 35.35 μm.
- No statistically significant linear function (probability 95%) was found between the cutting speed and the 3D surface topography parameters.

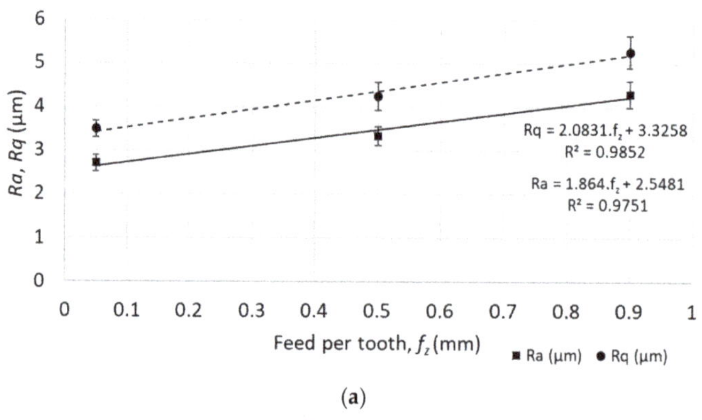

(**a**)

Figure 7. *Cont.*

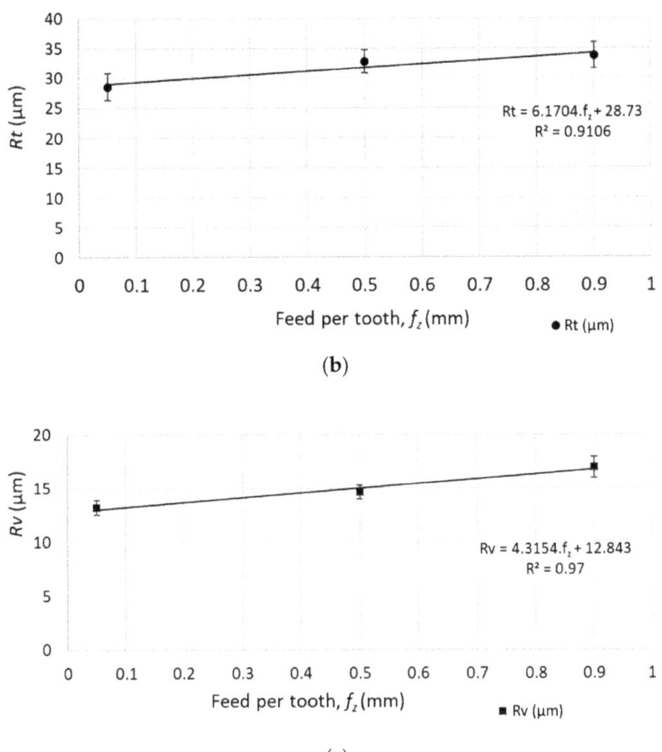

(c)

Figure 7. Selected profile roughness parameters for different feed speeds (at cutting speed v_c = 200 m.min^{-1}): (**a**) *Ra* – Average roughness, *Rq* – root-mean-square roughness; (**b**) *Rt* - maximum peak to valley height of roughness profile; (**c**) *Rv* - maximum valley depth of roughness profile.

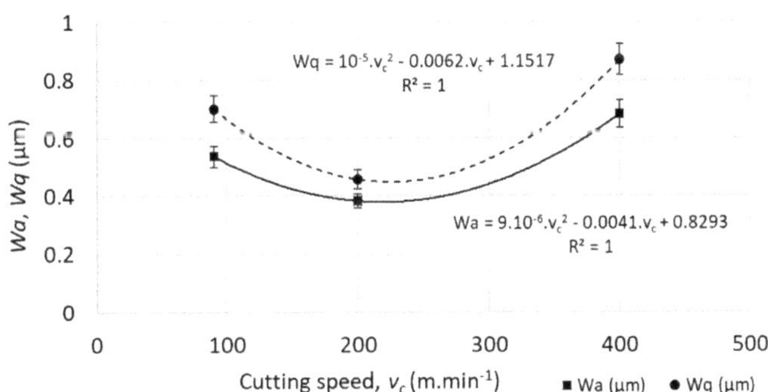

Figure 8. Example of profile waviness parameters (*Wa*—average waviness of the profile, *Wq*—root-mean-square waviness of the profile) for different cutting speeds (at feed per tooth f_z = 0.05 mm). The parameter R^2 reflects the level of statistical correlation.

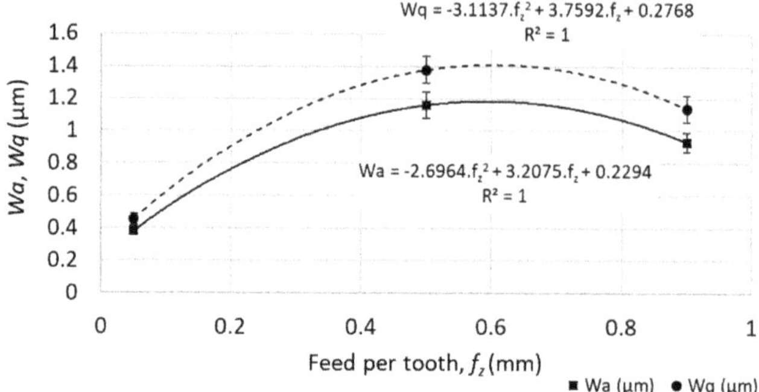

Figure 9. Example of profile waviness parameters (*Wa*—average waviness of the profile, *Wq*—root-mean-square waviness of the profile) for different feed speeds (at feed per tooth v_c = 200 m.min^{-1}). The parameter R^2 reflects the level of statistical correlation.

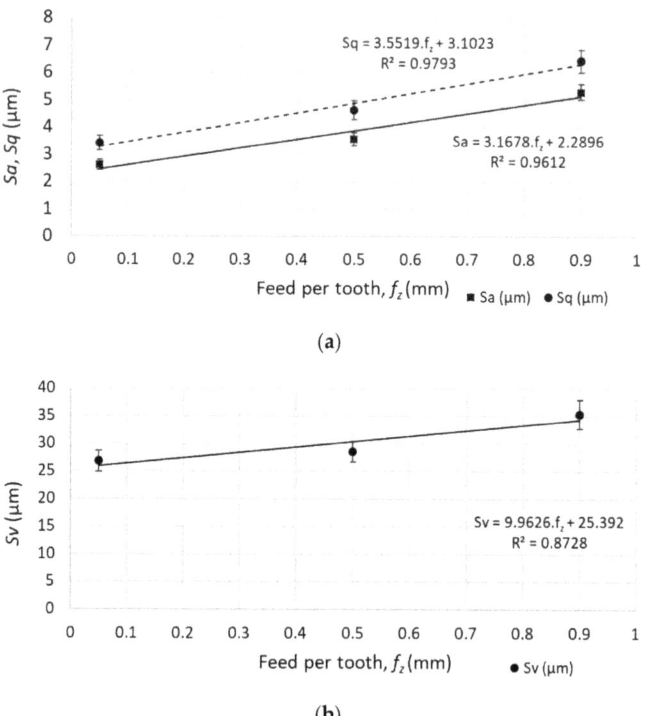

Figure 10. Surface parameters of the selected area for different feed speeds (at cutting speed v_c = 200 m.min^{-1}): (**a**) *Sa* - average height of the selected area, *Sq* - root-mean-square height of the selected area; (**b**) *Sv* - maximum valley depth of the selected area.

The machined surface can by described by the Firestone–Abbott curve, which indicates the percentage of the material of the profile elements at a defined height relative to the evaluation profile length (*R*) or surface area (*S*). This specific surface criterion is characterized by several parameters. The parameters of the core roughness depth (*Rk* and *Sk*) indicate the volume of the material above the core material which can be worn during operational use. The parameters of reduced peak height

(*Rpk* and *Spk*) describe the mean height of peaks above the core material. Furthermore, the parameters of reduced peak height (*Rpk* and *Spk*) express the amount of the material that will be removed during the initial operational wearing process. The parameters of reduced valley height (*Rvk* and *Svk*) describe the mean depth of the valleys below the core material. Therefore, *Rvk* and *Svk* parameters indicate the ability of the machined surface to retain liquids. The parameter *Rmr1* indicates the fraction of the surface which consists of peaks above the core material, and the parameter *Rmr2* indicates the fraction of the surface which will carry the load [28]. An examples of the Firestone–Abbott curve are presented in Figures 11 and 12.

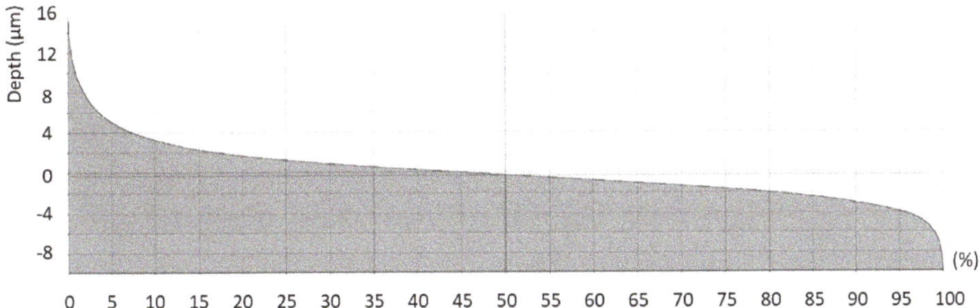

Figure 11. Example of Firestone–Abbott curve of the roughness profile (f_z = 0.05 mm and v_c = 90 m.min^{-1}).

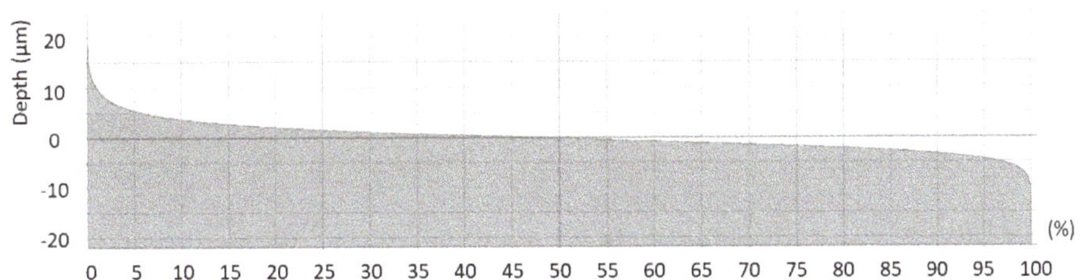

Figure 12. Example of Firestone–Abbott curve of the selected area (f_z = 0.05 mm and v_c = 90 m.min^{-1}).

The machined surface under the defined cutting conditions of the face milling showed the following results:

- *Rpk* and *Spk* decrease linearly with the increase of the cutting speed (while increasing the cutting speed from 90 m.min^{-1} to 400 m.min^{-1}, *Rpk* dropped by approximately 58%, and *Spk* dropped by 25%), as shown in Figure 13. Therefore, less material will be removed during the initial wearing process of operational use while implementing higher cutting speeds.
- *Rk*, *Sk* and *Rpk*, *Spk* increase linearly with increasing feed speed (while increasing feed per tooth from 0.05 mm to 0.90 mm, *Rpk* increased by 16% and *Spk* increased by 26%), as shown in Figure 14. Therefore, a higher volume of the material will be removed during the initial wearing process of operational use when implementing higher feed speeds.

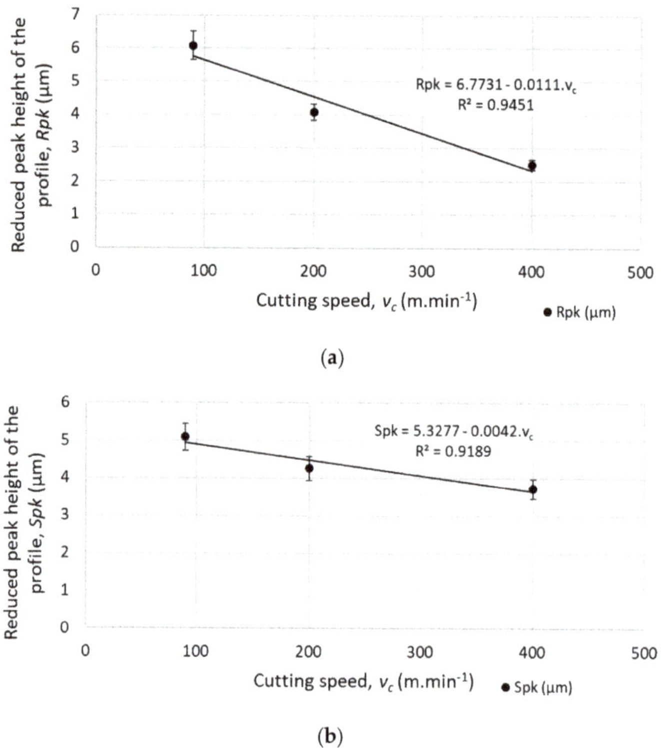

Figure 13. Mean height of peaks above the core material for different cutting speeds (at feed per tooth f_z = 0.05 mm) (**a**) *Rpk* - mean height of peaks above the core material for a profile; (**b**) *Spk* - mean height of peaks above the core material for a selected area.

Analysis of the effect of the tool inclination on the surface parameters can be summarized as follows:

- Roughness parameters are similar for both types of strategies; however, a greater increase of roughness parameters while using higher feed speeds was observed for samples machined with tool inclined by 1°.
- Roughness parameters were affected by the feed per tooth, not by the tested cutting speeds.
- The effect of the tool inclination cannot be compared properly if no 3D surface topography parameters are used (see example of two machined samples with different tool positioning in Figure 15).
- The average height of the selected area (*Sa*) showed higher values for samples machined with the tool inclined by 1° (by 50–70%) compared to the results with perpendicular tool positioning, which were in the range of 2.00 to 5.30 μm for all tested conditions. Similar relations were found for *Sq*, *Ssk*, and *Sku*.
- The texture aspect ratios (*Str*) for the machined surface did not present significant changes (the differences for the same cutting conditions were about 15–20%).
- Roughness parameters are not sufficient for comparison of the complex topography of the surface after machining because of their inhomogeneity and sensitivity to the measured place, which affect skewness, kurtosis, etc.
- Study of the surface parameters revealed that no crucial variable of surface topography was linked to the fatigue results of the studied material.

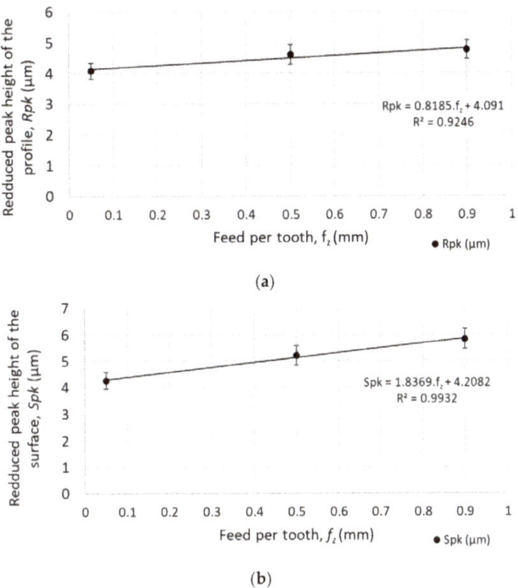

Figure 14. Mean height of peaks above the core material for different feed speeds (at cutting speed v_c = 200 m.min^{-1}): (**a**) *Rpk* - mean height of peaks above the core material for a profile; (**b**) *Spk* - mean height of peaks above the core material for a selected area.

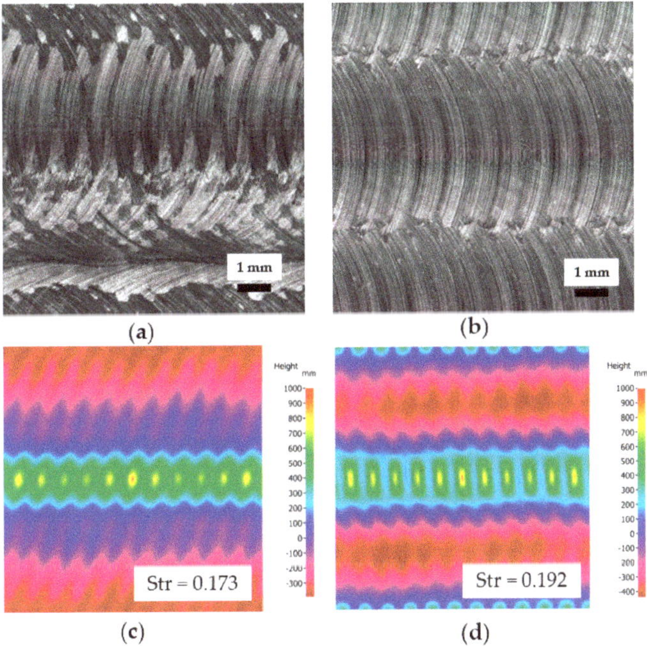

Figure 15. An example of the surface topography of the machined samples (f_z = 0.9 mm, v_c = 200 m.min^{-1}, a_p = 0.2 mm, a_e = 6.0 mm, dry cutting): (**a**) Without a tool inclination; (**b**) Tool inclination of 1°; (**c**,**d**) Corresponding autocorrelations and texture aspect ratio values of the surface structures.

3.3. *Force Loading Analysis of High Feed Face Milling and Induced Severe Plastic Deformation*

The parameters of primary shear deformation and its rates were very high, as can be seen in Figure 16. Primary shear deformation decreases with increase of the cutting speed as well as with increase of the feed speed (feed per tooth).

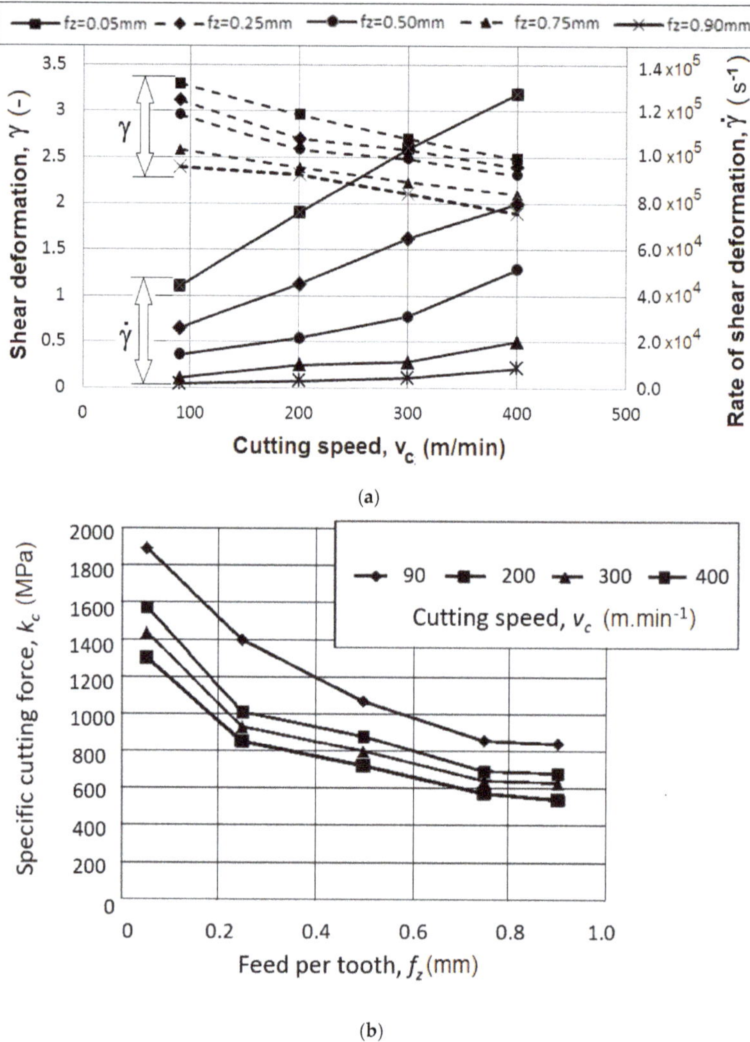

Figure 16. (**a**) The calculated mean values of the shear deformation and the rate of shear deformation in the primary zone; (**b**) Specific cutting forces for the given cutting conditions.

An explanation of this phenomenon is not easy and other works dealing with dislocations are ongoing. Meanwhile, we tentatively propose that there exists a small region over which there is a sudden proliferation of high angle boundaries in the microstructure of the material as it is deformed into the chip [29].

3.4. Fatigue Testing and Frature Surface Analysis

The effect of the milling parameters on the fatigue life was striking even if the surface topography after machining was not the key factor affecting the fatigue crack nucleation.

The important decrease of the fatigue life of specimens machined with higher feed rates while keeping the same cutting speed (an increase from feed per tooth $f_z = 0.05$ mm to $f_z = 0.90$ mm) can be seen in Figure 17. This decrease of the fatigue life may be caused by the severe plastic deformation achieved in the smallest chip cross sections and machined at the highest cutting speeds (v_c = 200 m.min^{-1}).

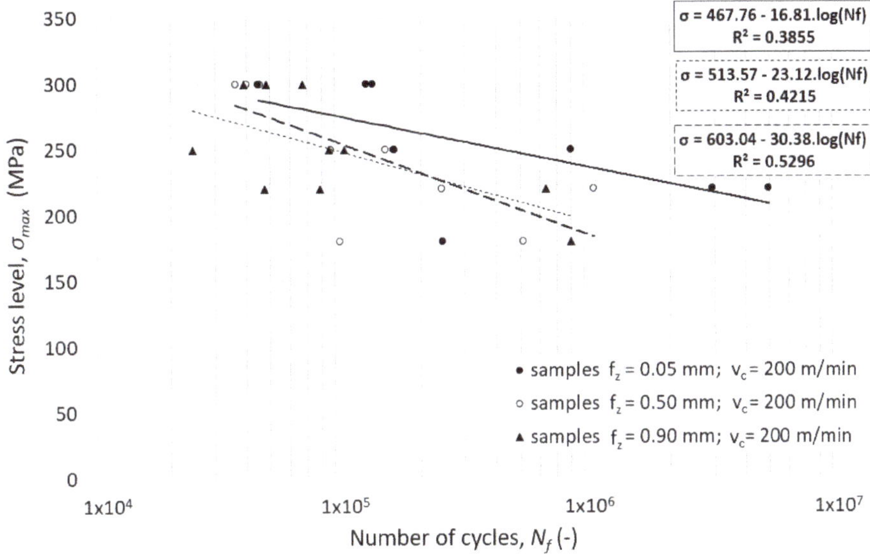

Figure 17. The effect of the feed speed (feed per tooth) on the fatigue life of alloy 7475-T7351—flat un-notched specimen (v_c = 200 m.min^{-1}).

The effect of the higher cutting speed on the fatigue life of the specimen is evident in the case of combination with lower feed speed, as presented in Figure 18. Use of higher feed speed increases the fatigue life for both low and high cycle fatigue modes. However, this effect is suppressed by combination with a high feed cutting strategy.

Slight inclination of the cutting tool (only 1°) resulted in a 29–64% reduction of total cycles (for specimens machined with the combination of the cutting speed v_c = 90 m.min^{-1} and different feed speeds, as indicated in Figure 19).

Regardless of the cutting conditions, the fatigue cracks were always initiated in the large intermetallic particles which occurred in different morphologies (as large particles, elongated particles, or clusters of intermetallic particles). Fatigue cracks were mostly initiated in the intermetallic particles located in the vicinity of the machined surface regardless of the stress level. An example of the fracture surface is presented in Figure 20.

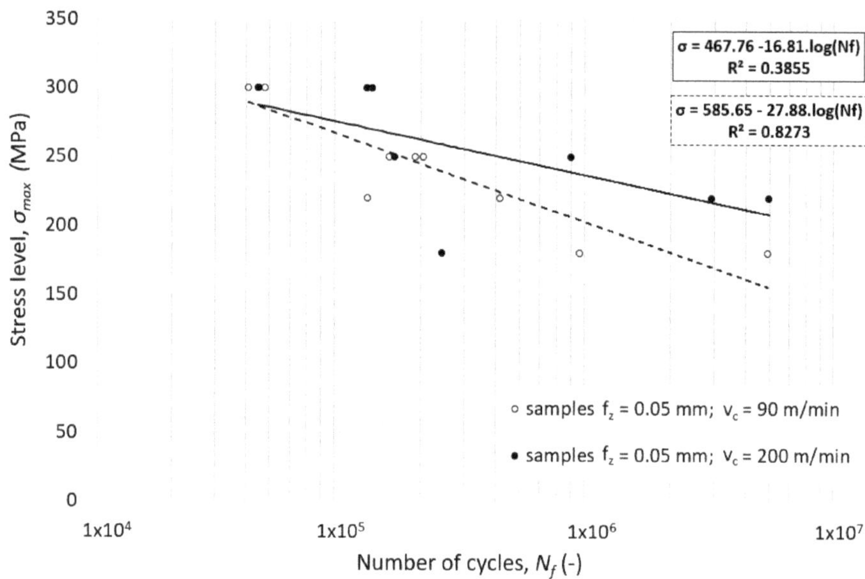

Figure 18. The effect of the cutting speed combined low feed cutting on the fatigue life of alloy 7475-T7351—flat un-notched specimen.

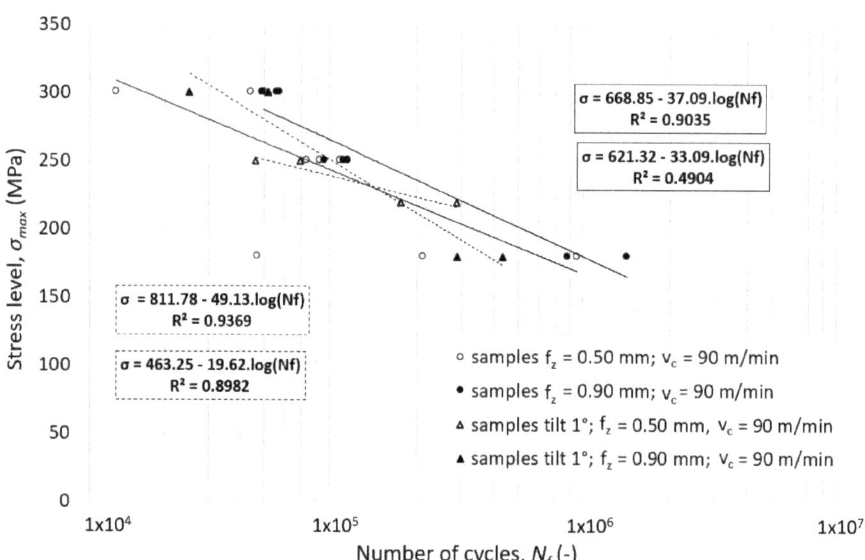

Figure 19. The effect of the tool inclination on the fatigue life of alloy 7475-T7351—flat un-notched specimen.

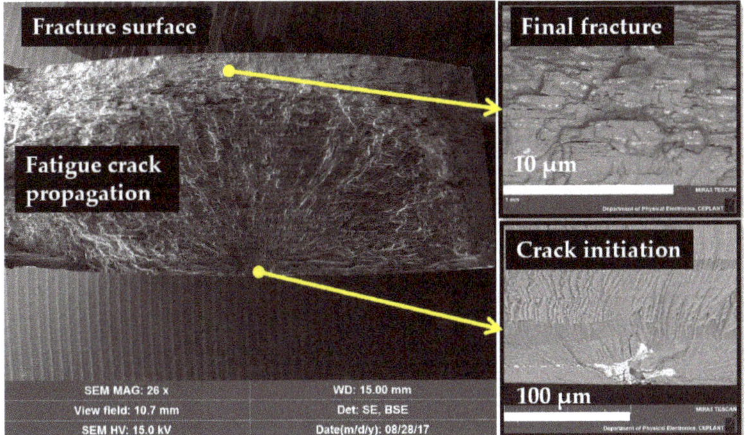

Figure 20. Fracture surface for a specimen machined with the following parameters: f_z = 0.90 mm and v_c = 90m.min^{-1}, stress level 250 MPa.

3.5. *In Situ Testing*

3.5.1. In Situ Tensile Testing

Crack propagation from large intermetallic particles to the material matrix was observed even before reaching the tensile strength limit (484 MPa), as presented in Figure 21. Large intermetallic particles were the main source of the local stress concentration regardless of the severe plastic deformation caused by the machining process. Cracks propagated at the angle of 45° to the direction of the maximum shear stresses.

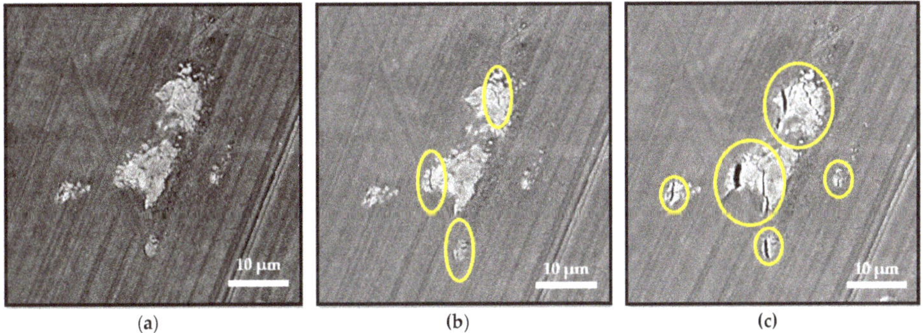

Figure 21. Observation of particle cracking: (**a**) Intermetallic particle before tensile loading; (**b**) Intermetallic particle at yield strength (415–419 MPa); (**c**) Intermetallic particle at tensile strength limit (484 MPa). The loading axis was horizontal.

The evolution of the engineering strain distribution under tensile loading is shown in Figure 22. The average engineering strain at the area of the intermetallic particle at yield strength was in the range of 0.80% to 0.90%, and upon reaching the maximal engineering strain (1.15–1.20%), local crack initiation was observed.

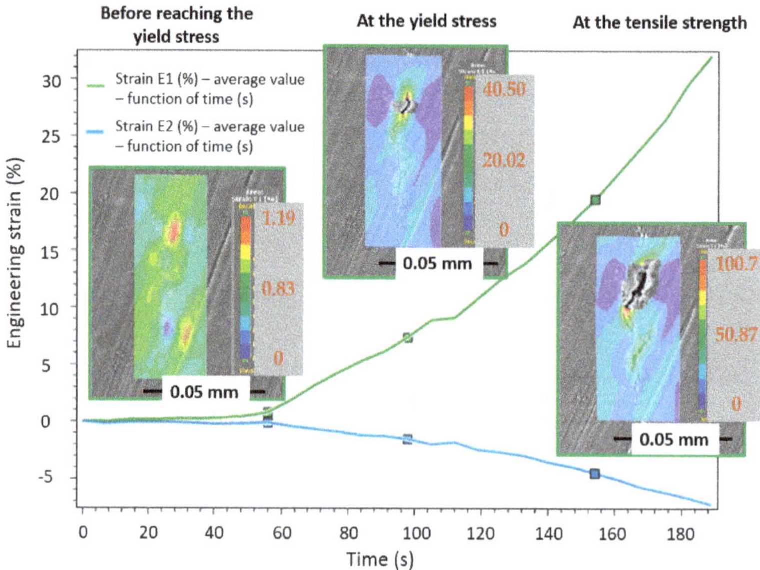

Figure 22. DIC analysis of the engineering strain distribution in the intermetallic particle during tensile loading: at the beginning of the tensile loading, at the yield strength, and at the tensile strength limit.

3.5.2. In Situ Cyclic Testing

During in situ cyclic testing, initial local fatigue cracks were observed in some intermetallic particles before the fatigue life (1000 cycles) was reached, regardless of severe plastic deformation induced by the milling process. Local fatigue cracks were initiated in the core of the intermetallic particles, and with the rising number of cycles, the fatigue cracks propagated locally to the boundary of the intermetallic particles and the material matrix. Fatigue testing was interrupted at the fatigue level of 6000 cycles. Up to this number of fatigue cycles, the short fatigue cracks remained inside the intermetallic particles and did not propagate further. DIC analysis confirmed that once the local engineering strain reached values of 1.15–1.20% (at 1000 cycles, in this case), the local fatigue cracks were initiated, as demonstrated in Figure 23.

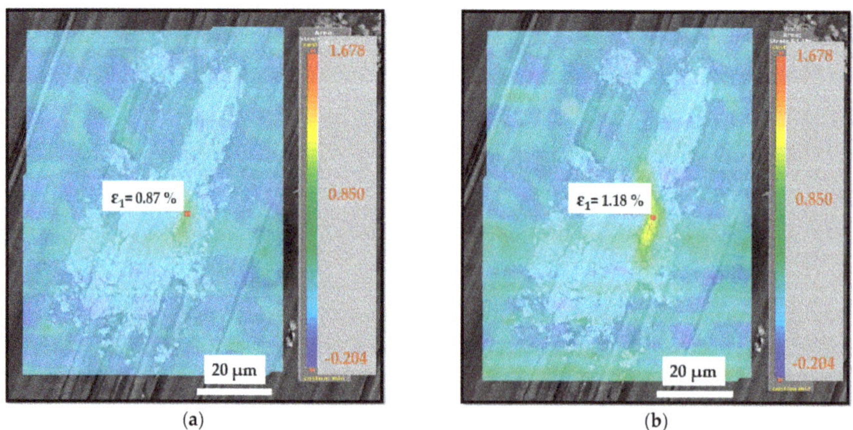

Figure 23. Engineering strain distribution (**a**) at 100 cycles; (**b**) at 500 cycles.

4. Discussion

The observations from the experimental machining, surface analyses, and fatigue testing confirm similar results as those of Ojolo et al. [17] and Novovic et al. [18].

Surface topography analysis confirmed that the roughness parameters increase with the increase of the feed speed (feed per tooth). The increase of the cutting speed caused a decrease of the surface roughness parameters. This result partially confirms the observation of Ojolo et al. [17].

The increase of the feed speed increases surface topography parameters such as average height of the selected area (*Sa*), root-mean-square height of the selected area (*Sq*), or maximum valley depth of the selected area (*Sv*). The roughness parameters were found to be similar for both strategies (perpendicular and inclined by 1°); however, a greater increase of roughness parameters was observed while using higher feed speeds for samples machined with the tool inclined by 1°. The average height of the selected area (*Sa*) showed higher values for samples machined with the tool inclined by 1°. This parameter was not adequate, considering that maximum valley depth of the selected area (*Sv*) was higher for samples machined by a tool positioned perpendicularly to the machined surface.

The trends of specific cutting force and shear deformation confirm a reduction of plastic deformation with increasing cutting speed, but more intensive deformation with reduction of feed per tooth. In other words, the intensity of plastic deformation is higher for shallow cuts and higher cutting speeds.

The highest fatigue resistance was observed at samples machined with the highest cutting speed (v_c =200 m.min^{-1}) and lowest feed per tooth (f_z = 0.05 mm). A decrease of the fatigue life of specimens machined with higher feed rates while keeping the same cutting speed was observed (increase from feed per tooth f_z = 0.05 mm to f_z = 0.90 mm). This decrease of the fatigue life may be caused by the severe plastic deformation achieved in the smallest chip cross sections and machined at the highest cutting speeds (v_c = 200 m.min^{-1}).

Slight inclination of the cutting tool (1°) resulted in reduction of the total cycles for specimens machined with cutting speed v_c = 90 m.min^{-1} and different feed speeds. This reduction may be caused by plastic deformation caused by teeth not engaged in the cut, as in the case of face milling perpendicular to the machined surface. This plastic deformation may impose beneficial compressive residual stresses into the machined surface and thus increase the fatigue life.

Therefore, the cutting conditions can affect the material removal rate, but a more serious impact can be seen in terms of the surface quality and the resistance to mechanical loading. The effect of inclusions is very serious, and materials used for dynamic loading should be carefully analyzed not only in view the surface integrity, but also considering the occurrence of the phases, which confirms the results of Piska et al. [20]. The effect of material hardening and thermal softening when cutting should be studied further in terms of the density and arrangement of dislocations, stacking fault energy, and other atomic hardening or softening mechanisms.

5. Conclusions

The application of very advanced laboratory facilities yielded the following results:

- The 7475-T7351 aluminum material was suitable for dynamic mechanical loading with good machinability when milling with the special monolithic cutter SECO JHF 980 Special.
- The quality of surface parameters and fatigue resistance improved when higher cutting speeds (200 m/min) and low feeds per tooth (0.05 mm) were used, and extreme shear deformation (γ = 2.5) and deformation rates (1.2×10^5 s^{-1}) were achieved.
- The measured values of *Sa* correlated with the *Ra* parameters in the trends according to the cutting conditions and proved to be featureless multiplications of the *Ra* values for all tested conditions. Nevertheless, no significant variable of the surface topography linked to the fatigue results was found.

- The fatigue resistance of the samples machined with the standard perpendicular position of the tool to the machined surface (i.e. without any spindle inclination) was greater than the results for samples machined with the inclined tool. Therefore, surfaces with more complex topography seems to be beneficial; however, new studies with a material with minimal intermetallic inclusions are needed.
- The crucial and decisive factor for crack nucleation can be seen in the coarse intermetallic inclusions (Al_7Cu_2Fe; Al-Cr-Fe-Cu-Si) in sizes from 2 μm up to 20 μm, suppressing the effect of the surface parameters after machining.
- Further study of the dislocation mechanism responsible for deformation hardening and softening are required. Research relating surface layer depths and grain size are ongoing.

Author Contributions: Conceptualization, methodology, writing, investigations in machining, original draft preparation, project administration, funding acquisition: P.O. and M.P. (Miroslav Piska), material analyses M.P. (Martin Petrenec), investigations in the in situ REM measurements: J.D., fatigue investigations: J.H., P.S.

Funding: This research was funded by FME BUT Brno, Czech Republic as the "Research of perspective production technologies", FSI-S-19-6014 and by the project CEITEC 2020, LQ1601, Ministry of Education, Youth and Sports of the Czech Republic.

Acknowledgments: This research work was supported by the FME BUT Brno, Czech Republic as the "Research of perspective production technologies", FSI-S-19-6014 and by the project CEITEC 2020, LQ1601, Ministry of Education, Youth and Sports of the Czech Republic.

Conflicts of Interest: The authors declare no conflict of interest.

References

1. Dietrich Altenpohl, G. *Aluminium: Technology, Application, and Environment, a Profile of a Modern Metal*, 6th ed.; Wiley-TMS: Washington, DC, USA; Weinheim, Germany, 1999; p. 319.
2. Verma, B.B.; Atkinson, J.D. Study of fatigue behaviour of 7475 aluminium alloy. *Bull. Mater. Sci.* **2001**, *24*, 231–236. [CrossRef]
3. Static and Dynamic Fracture Properties for Aluminum 7475 T7351: Final Report. Ohio. 1975. Available online: http://www.dtic.mil/dtic/tr/fulltext/u2/a014353.pdf (accessed on 1 October 2019).
4. Alloy 7475 Plate and Sheet. Available online: https://www.spacematdb.com/spacemat/manudatasheets/alloy7475techplatesheet (accessed on 1 October 2019).
5. Kunčická, L.; Terry, C.L.; Casey, F.D.; Kocich, R.; Pohludka, M. Synthesis of an Al/Al_2O_3 composite by severe plastic deformation. *Mater. Sci. Eng. A* **2015**, *646*, 234–241. [CrossRef]
6. Naizabekov, A.B.; Andreyachshenko, V.A.; Kocich, R. Study of deformation behavior, structure and mechanical properties of the AlSiMnFe alloy during ECAP-PBP. *Micron* **2013**, *44*, 210–217. [CrossRef] [PubMed]
7. Kunčická, L.; Kocich, R.; Ryukhtin, V.; Cullen, J.C.T.; Lavery, N.P. Study of structure of naturally aged aluminium after twist channel angular pressing. *Mater. Charact.* **2019**, *152*, 94–100. [CrossRef]
8. Kocich, R.; Kunčická, L.; Král, P.; Macháčková, A. Sub-structure and mechanical properties of twist channel angular pressed aluminium. *Mater. Charact.* **2016**, *119*, 75–83. [CrossRef]
9. Kunčická, L.; Kocich, R. Deformation behaviour of Cu-Al clad composites produced by rotary swaging. In 5th global conference on polymer and composite materials (PCM 2018). In *IOP Conference Series-Materials Science and Engineering*; Curran Associates, Inc.: Red Hook, NY, USA, 2018; Volume 369.
10. DeGarmo, E.; Black, J.; Kohser, R. *Materials and Processes in Manufacturing*, 8th ed.; Prentice Hall: Upper Saddle River, NJ, USA, 1997.
11. Humár, A.; Technologie, I. Technologie Obrábění-1.Část-Studijní Opory pro Magisterskou Formu Studia. 2003, p. 138. Available online: http://ust.fme.vutbr.cz/obrabeni/opory-save/TI_TO-1cast.pdf (accessed on 1 October 2019).
12. Beňo, J. *Teória Rezania Kovov*; Strojnická Fakulta TU Košice: Košice, Slovakia, 1999; ISBN 80-7099-429-0.
13. Driensky, D.; Fúrik, P.; Lehmanová, T. *Strojní Obrábění I. 1*; SNTL: Prague, Czech Republic, 1986; p. 424.

14. Kunčická, L.; Kocich, R.; Dvořák, K.; Macháčková, A. Rotary swaged laminated Cu-Al composites: Effect of structure on residual stress and mechanical and electric properties. *Mater. Sci. Eng. A* **2019**, *742*, 743–750. [CrossRef]
15. Kunčická, L.; Kocich, R.; Strunz, P.; Macháčková, A. Texture and residual stress within rotary swaged Cu/Al clad composites. *Mater. Lett.* **2018**, *230*, 88–91. [CrossRef]
16. Broek, D. *Elementary Engineering Fracture Mechanics*; Springer: Dordrecht, The Netherlands, 1986; ISBN 978-940-0943-339.
17. Ojolo, S.J.; Orisaleye, I.J.; Obiajulu, N. Machining Variables Influence on the Fatigue Life of End-Milled Aluminium Alloy. *Int. J. Mater. Sci. Appl.* **2014**, *3*, 391–938.
18. Novovic, D.; Dewes, D.K.; Aspinwal, W. The Effect of Machined Topography and Integrity on Fatigue Life. *Int. J. Mach. Tools Manuf.* **2004**, *44*, 125–134. [CrossRef]
19. Koster, W. Effect of Residual Stress on Fatigue of Structural Alloys. In Proceedings of the Third International Conference, ASM International, Indianapolis, IN, USA, 15–17 May 1991.
20. Piska, M.; Ohnistova, P.; Hornikova, J.; Hervoches, C. A study of progressive milling technology on surface topography and fatigue properties of the high strength aluminum alloy 7475-T7351. In Proceedings of the 17th International Conference on New Trends in Fatigue and Fracture, Cancún, Mexico, 25–27 October 2017; Springer International Publishing AG: New York City, NY, USA; pp. 7–19, ISBN 978-3-319-70364-0.
21. Chemin, A.; Marques, D.; Bisanha, L.; Motheo, A.J.; Filho, W.W.B.; Ruchert, C.O.F. Influence of Al7Cu2Fe intermetallic particles on the localized corrosion of high strength aluminum alloys. *Mater. Des.* **2014**, *53*, 118–123. [CrossRef]
22. Davis, J.R. *Aluminum and Aluminum Alloys*; Illustrated Edition; ASM International: Ohio, MI, USA, 1993; pp. 351–416.
23. Priya, P.; Johnson, D.R.; Krane, M.J.M. Precipitation during cooling of 7XXX aluminum alloys. *Comput. Mater. Sci.* **2017**, *139*, 273–284. [CrossRef]
24. The Magazine about Alicona-Metrology. *Alicona Focus Variation*, 8th ed. Available online: https://irp-cdn.multiscreensite.com/ebc29cc1/files/uploaded/Alicona_FOCUSvariation_magazine_2018_EN.pdf (accessed on 20 October 2019). (In English).
25. *ISO 25178-606:2015, Geometrical product specification (GPS) - Surface texture: Areal Part 606: Nominal characteristics of non-contact (focus variation) instruments*; ISO copyright office: Geneva, Switzerland, 15 June 2015.
26. *ASTM E466–15, Standard Practice for Conducting Force Controlled Constant Amplitude Axial Fatigue Tests of Metallic Materials*; ASTM International: West Conshohocken, PA, USA, 1 May 2015.
27. *DIN EN 6072, Aerospace series - Metallic materials - Test methods - Constant amplitude fatigue testing*; German Institute for Standardisation: Berlin, Germany, 1 January 2011.
28. 3D Functional Parameters. Available online: https://www.michmet.com/3d_s_functional_parameters.htm (accessed on 1 October 2019).
29. Shankar, M.R.; Chandrasekar, S.; Compton, W.D.; King, A.H. Characteristics of aluminum 6061-T6 deformed to large plastic strains by machining. *Mater. Sci. Eng.* **2005**, *410–411*, 364–368. [CrossRef]

MDPI
St. Alban-Anlage 66
4052 Basel
Switzerland
Tel. +41 61 683 77 34
Fax +41 61 302 89 18
www.mdpi.com

Materials Editorial Office
E-mail: materials@mdpi.com
www.mdpi.com/journal/materials

www.ingramcontent.com/pod-product-compliance
Lightning Source LLC
LaVergne TN
LVHW070124170726
843515LV00007B/1745

* 9 7 8 3 0 3 6 5 0 0 7 6 8 *